CONSERVATION OF
BIOLOGICAL RESOURCES

This book is dedicated to Ione and Thomas,
as representatives of the next generation

Conservation of Biological Resources

E.J. Milner-Gulland

Ecosystems Analysis and Management Group,
Department of Biological Sciences,
University of Warwick, Coventry CV4 7AL, UK

and

Ruth Mace

Department of Anthropology, University College London,
Gower Street, London WC1E 6BT, UK

*With case studies
contributed by other authors*

**Blackwell
Science**

© 1998 by
Blackwell Science Ltd
Editorial Offices:
Osney Mead, Oxford OX2 0EL
25 John Street, London WC1N 2BL
23 Ainslie Place, Edinburgh EH3 6AJ
350 Main Street, Malden
 MA 02148 5018, USA
54 University Street, Carlton
 Victoria 3053, Australia
10, rue Casimir Delavigne
 75006 Paris, France

Other Editorial Offices:
Blackwell Wissenschafts-Verlag GmbH
Kurfürstendamm 57
10707 Berlin, Germany

Blackwell Science KK
MG Kodenmacho Building
7–10 Kodenmacho Nihombashi
Chuo-ku, Tokyo 104, Japan

The right of the Authors to be
identified as the Authors of this Work
has been asserted in accordance
with the Copyright, Designs and
Patents Act 1988.

First published 1998

Set by Semantic Graphics, Singapore
Printed and bound in Great Britain
at the University Press, Cambridge

The Blackwell Science logo is a
trade mark of Blackwell Science Ltd,
registered at the United Kingdom
Trade Marks Registry

DISTRIBUTORS

 Marston Book Services Ltd
 PO Box 269
 Abingdon, Oxon OX14 4YN
 (*Orders*: Tel: 01235 465500
        Fax: 01235 465555)

USA
 Blackwell Science, Inc.
 Commerce Place
 350 Main Street
 Malden, MA 02148 5018
 (*Orders*: Tel: 800 759 6102
         781 388 8250
     Fax: 781 388 8255)

Canada
 Login Brothers Book Company
 324 Saulteaux Crescent
 Winnipeg, Manitoba R3J 3T2
 (*Orders*: Tel: 204 224-4068)

Australia
 Blackwell Science Pty Ltd
 54 University Street
 Carlton, Victoria 3053
 (*Orders*: Tel: 3 9347 0300
     Fax: 3 9347 5001)

A catalogue record for this title
is available from the British Library

ISBN 0-86542-738-0

Library of Congress
Cataloging-in-publication Data

Milner-Gulland, E.J.
   Conservation of biological resources /
 E.J. Milner-Gulland, Ruth Mace; with case
 studies contributed by other authors.
        p.   cm.
   Includes bibliographical references
   (p.   ) and index.
   ISBN 0-86542-738-0
   1. Biological diversity conservation.
 2. Sustainable development.
 I. Mace, Ruth. II. Title.
 QH75.M5235 1998
 333.95′16—dc21   97-31751
                                CIP

For further information on
Blackwell Science, visit our website:
www.blackwell-science.com

Contents

Part 4: Making Conservation Work

List of Contributors

Leonid M. Baskin *Institute of Ecology and Evolution, 33 Leninsky Prospekt, Moscow 117071, Russia.*

Thomas M. Butynski *Zoo Atlanta, Africa Biodiversity Conservation Program, National Museums of Kenya, PO Box 24434, Nairobi, Kenya.*

Sophie des Clers *Renewable Resources Assessment Group, Imperial College of Science, Technology and Medicine, 8 Princes Gardens, London SW7 1NA, UK.*

Stephen R. Edwards *IUCN Sustainable Use Initiative, IUCN, 1400 16th Street NW, Washington, DC 20036, USA.*

Joel Freehling *Department of Anthropology, State University of New York, USA.*

R.E. Gullison *Renewable Resources Assessment Group, Imperial College of Science, Technology and Medicine, 8 Princes Gardens, London SW7 1NA, UK.*

Anne Gunn *Wildlife and Fisheries Division, Department of Resources, Wildlife and Economic Development, Government of the Northwest Territories, Yellowknife, Canada.*

Julie P. Hawkins *Environment Department, University of York, York YO1 5DD, UK.*

Jan Kalina *Zoo Atlanta, Africa Biodiversity Conservation Program, PO Box 24434, Nairobi, Kenya.*

Kathy MacKinnon *Environment Department, The World Bank, 1818 H. St NW, Washington, DC 20433, USA.*

Stuart A. Marks *Safari Club International, 4800 West Gates Pass Road, Tucson, AZ 85745, USA.*

M. Norton-Griffiths *Centre for Social and Economic Research on the Global Environment, Department of Economics, University College London, 6 Gower Street, London WC1E 6BT, UK.*

Andrew R.G. Price *Ecosystems Analysis and Management Group, Department of Biological Sciences, University of Warwick, Coventry CV4 7AL, UK.*

Callum M. Roberts *Environment Department, University of York, York YO1 5DD, UK.*

Vivienne Solis Rivera *IUCN Regional Office for MesoAmerica ORMA, P.O. Box 1161-2150, Moravia, Costa Rica*

Preface

Conservationists are a diverse group. They include academics and practitioners, idealists and theoreticians. Academic ecologists have tended to focus on issues of theoretical interest. These often have limited practical application to reducing the main threat to wild species—the direct or indirect impact of humans making use of their environment. At the other end of the spectrum, many conservation practitioners have become advocates of a particular conservation philosophy based on little or no theoretical analysis, or even empirical evidence that their approach works. 'Sustainable use' is one area in which theory and practice have not been well integrated. Those in favour of conservation based on consumptive use and those advocating conservation through preservation have entered into many polarised, heated but often sterile arguments. The battle to list all elephants on Appendix 1 of CITES, thereby banning the ivory trade, was one case where this debate came to wide public attention—a debate that is still current. We felt the need for an examination of the conservation of biological resources, recognising that use is inevitable (extensive, exploitative and sometimes devastating use) and trying to address how such use might be contained within sustainable limits.

Thus, the aim of this book is to present a broad and inter-disciplinary view of the issues surrounding the conservation and use of wild living resources. It is aimed at anyone interested in the human dimension of conservation, but especially at Masters students or final-year undergraduates taking courses in conservation, environmental management, ecological economics or related subjects; conservation professionals (including managers, policy-makers and researchers); and non-specialist readers who have followed the debate about conservation and sustainable use. It could be used for a one or two term course in conservation and resource use, or for a graduate seminar series.

The book is divided into four parts: an introduction to the issues, the theoretical background, case studies, and a discussion of future developments which draws on the case studies and the theory. We assume no prior knowledge of the field, although students with a basic understanding of calculus and population ecology will find the theoretical material easier. The theoretical material is best presented in class, with the case studies used as the basis for discussion in an accompanying seminar series. One intention of this book is to demonstrate how important it is that conservationists have an understanding of both the ecological and the social issues affecting resource use, and how they interact. Therefore, we recommend that courses cover all the theoretical material discussed in the book, rather than selecting sections.

At the front of the book there are lists of the mathematical notation, acronyms and conventions used in the book, and at the back there is a glossary of terms. The words in italic in the text are highlighted in order to emphasise particular points, not necessarily because they appear in the glossary.

Acknowledgements

The authors would like to express their grateful appreciation to Jim Cannon, Georgina Mace and Marc Mangel for reviewing the manuscript. We thank the Environmental Investigation Agency, Jacquie McGlade, Robin Milner-Gulland, Sarah Robinson and Katriona Shea for their advice. The Department of Biological Sciences, Warwick University, the Department of Anthropology, University College London, and the Royal Society provided us with financial and administrative support. Photos were kindly provided by Jo Abbot, Karl Ammann, Lynn Clayton, the Environmental Investigation Agency, W. Giesen, Kim Hill, Loida Pretiz, Michael Schmitt and Ken Wilson, as well as by the case study authors. We thank Alison Milner-Gulland for the beautiful cover illustration. We are especially grateful to Martin Williams and Mark Pagel for all their help and support while we were writing the book. Particular thanks go to Ian Sherman at Blackwell Science, for all his encouragement and help.

The case study authors would like to thank the following people. *Butynski & Kalina:* We are indebted to Samson Werikhe, Kelly Stewart, Katie Frohardt, Liz Williamson, Liz Macfie, Wolfgang von Richter, Bernd Steinhauer-Burkart, Netzin Gerald-Steklis and Juichi Yamagiwa, as well as to the Uganda Wildlife Authority, Office Rwandais du Tourisme et des Parcs Nationaux, and Institut Congolais pour la Conservation de la Nature, for much help and unpublished data. We thank Esteban Sarmiento, Liz Williamson, Jim Sanderson, Wolfgang von Richter, Liz MacFie, Tom Struhsaker, Debra Forthman, Richard Wrangham, Truman Young, Jeff Hall, Bernd Steinhauer-Burkart, Angela Meder, Juichi Yamagiwa, Netzin Gerald-Steklis, Dieter Steklis, John Oates, Esmond Bradley-Martin, Jim Else, Dietrich Schaaf, James Fuller and Ursula Karlowski for their valuable comments and suggestions on the manuscript. The generous support of Zoo Atlanta, the National Museums of Kenya, and the IUCN Eastern Africa Regional Office is gratefully acknowledged. *des Clers:* Research on the Falklands Islands Fisheries at Imperial College is funded through a multi-annual programme by the Falkland Islands Government, Fisheries Department, Stanley. Encouragement and advice from John Barton and all at FIFD are gratefully acknowledged. *Norton-Griffiths:* Many colleagues in Kenya helped me with data and ideas for this work, including E. Barrow, D. Berger, L. Emmerton, J. Grunblatt and B. Heath.

Price, Roberts & Hawkins: Andrew Price wishes to thank staff at several Maldivian institutions for their assistance: the Ministry of Tourism, in particular Ismail Firaq; the Ministry of Planning, Human Resources and the Environment; the Marine Research Section of Ministry of Fisheries and Agriculture, especially Charles Anderson, Susan Clark and Hassan Maniku. Valuable discussions were also held with Bill Allison. Financial support from Maxwell Stamp plc, Nethconsult/Transtec, the European Commission and the Asian Development Bank is acknowledged with thanks. Studies in Saba by Callum Roberts and Julie Hawkins could not have been done without the help of many people and organisations. We are very grateful to staff of the Saba Marine Park and Saba Conservation Foundation for their long-standing support, in particular Susan White, Tom van't Hof, Kenny Buchan and Percy ten Holt. Financial support has been provided by the Saba Marine Park, WWF (Netherlands), University of Puerto Rico Sea Grant Program, British Ecological Society and Overseas Development Administration (UK).

List of Symbols

Notation

A	A payment made at some time in the future
α	Power constant
β	Over-compensation parameter
c	Unit cost of input
δ	Discount rate
d_i	Decision i
E	Effort expended by harvesters
E_t	Effort expended at time t
E_∞	Open-access equilibrium effort
E_Π	Static profit-maximising equilibrium effort
Eu	Expected utility
ε_{ij}	Cross-price elasticity of demand
ε_I	Income elasticity of demand
ε_P	Own price elasticity of demand
F	Population growth rate
$\mathfrak{I}$	Fine for illegal harvesting
H	Number of individuals removed by harvesting (harvest rate)
$H*$	Optimal harvest rate
H_t	Harvest rate at time t
I	Income
K	Carrying capacity
λ	Arithmetic rate of population increase
N	Population size
N_E	Minimum viable population size under depensation
$N*$	Optimal population size
N_0	Initial population size
N_∞	Open-access equilibrium population size
N_t	Population size at time t
P_i	Price of good i
P_E	Equilibrium price
$P*$	Optimal price of a transferable permit
p	Unit price of output
p_i	Probability of outcome i

Π	Profit
Q_i	Quantity of good i demanded
$Q*$	Optimal quantity of resource use
Q_E	Equilibrium quantity
Q_P	Quantity of resource use that is optimal for a private user
Q_0	Quantity of resource use that is optimal for society
q	Catchability coefficient
r	Intrinsic rate of population increase
R	Maximum point of stock–recruitment relationship, as a proportion of K
$s*$	Optimal subsidy
t	Time
$t*$	Optimal Pigovian tax
τ	Time lag
T	Generation time
θ	Probability of receiving a fine
u	Utility
x_i	Payoff from a decision i
y_i	Decision-maker's state
ω	State of nature

Mathematical Conventions

Δx	A small change in the value of x
$\dfrac{\mathrm{d}x}{\mathrm{d}y}$	The differential of x with respect to y
$\displaystyle\int_0^\infty [\ldots]\,\mathrm{d}t$	Integration over time, from time 0 to time ∞
$x \propto y$	x is proportional to y
∞	Infinity
$\displaystyle\sum_a^b$	Sum from a to b
x'	The differential of (i.e. result of differentiating) x
x_y	The partial differential of x with respect to y
$x(y)$	x is a function of y
$x = f(y)$	x is a function of y
$|x|$	The magnitude of x (e.g. $|-2| = |+2| = 2$)
$\approx$	Approximately equal to
$\ln(x)$	Natural log of x
e^x	Exponential of x
x_t	The value of x at time t

Acronyms

ADMADE	Administrative Management Design
CDN$	Canadian dollars
CFCs	Chlorofluorocarbons
CITES	Convention on International Trade in Endangered Species
CPUE	Catch per unit effort
DRC	Democratic Republic of Congo (formerly Zaire)
EIA	Environmental Investigation Agency
EU	European Union
FAO	Food and Agriculture Organisation of the United Nations
FSC	Forestry Stewardship Council
GATT	General Agreement on Tariffs and Trade
GDP	Gross domestic product
GNP	Gross national product
GNWT	Government of the North-West Territories
GOK	Government of Kenya
IGCP	International Gorilla Conservation Programme
IIED	International Institute for Environment and Development
IMF	International Monetary Fund
ITQ	Individual Transferable Quota
IUCN	World Conservation Union
IWC	International Whaling Commission
KWS	Kenya Wildlife Service
MEC	Marginal external costs of production
MNPB	Marginal net private benefit
MPC	Marginal private costs of production for the producer
MSC	Marginal social costs of production
MSY	Maximum sustainable yield
NGO	Non-Governmental Organisation
NWT	North-West Territories of Canada
PA	Protected area
PRA	Participatory rural appraisal
PV	Present value
RMP	Revised management procedure
TAC	Total allowable catch
TCM	Traditional Chinese medicine
UN	United Nations
UNEP	United Nations Environment Programme
WCMC	World Conservation Monitoring Centre
WTMU	Wildlife Trade Monitoring Unit at WCMC
WTO	World Trade Organisation
WWF	World Wide Fund for Nature

Part 1
Introduction to the Conservation
of Biological Resources

Burning forest, South-East Asia. (Photo by Kathy MacKinnon.)

Introduction

Conservation is about people. Most wild species are found in areas where people either live, or venture in to harvest resources. It is the impact of human societies on natural ecosystems that has created the field of conservation biology. This is a field that spans many disciplines; some of which can be addressed without directly considering human behaviour, but in most practical situations, it is the activities of individuals, communities, or commercial ventures that are likely to pose the most direct threats to biodiversity, and thus demand the majority of conservationists' attention. Our aim in this book is to investigate how these human actions are influencing biological resources, which we take to mean both wild species that are harvested directly or used in some non-consumptive way, and natural ecosystems that are directly threatened by the uses people wish to make of them. We go on to investigate what can be done to limit use effectively, so that the resources can be conserved.

Historical background

Humans have always made use of other species. Over the last 10 000 years, our food has come increasingly from cultivated plants and animals; this has resulted in a rapid expansion in the human population. However, wood for fuel and building, grass for grazing livestock, wild-caught fish and meat and gathered fruits and seeds have remained an important part of our subsistence base. Especially in the developing world, wild species still provide essential goods and services that are not available from any other source.

Throughout evolutionary history, humans were the most likely cause of the extinction of many species, even though we lived as hunters and gatherers at much lower population densities than today. However, whilst extinctions have always occurred, the rate of loss of wild species has accelerated dramatically in the last two centuries; this is largely due to various major technological advances over that period. In the 19th century, extensive seafaring and colonisation of islands caused very large numbers of island-adapted, endemic species to become extinct (UNEP 1995). They could not compete with introduced species, many were not adapted to flight from predators, and were easy to harvest to extinction. In the 20th century, extinctions on islands continued at a slightly slower rate (due to the fact that most island endemics were already lost), but extinctions on continents started to increase dramatically (WCMC 1992). This process, which continues, is due to several, related phenomena. Habitats are being turned from those that are relatively favourable to biodiversity (such as wetlands) to those that are not (such as agricultural monocultures and urban developments). Species are being threatened directly by harvesting, which is becoming more and more technologically advanced. Even when species are not driven to extinction, the size of populations and the number of viable sub-populations are being dramatically reduced by these processes. Tropical forests are being clear-felled for hardwood, with the loss of most of the associated species. In the mid-1980s humid tropical forests were being lost at an estimated rate of 10 million hectares per year, and the situation for dry tropical forests is just as serious (UNEP 1995). Fish are being extracted from the sea in such large numbers that several species once thought common, such as cod, are now classified as threatened by IUCN – the World Conservation Union. For more exotic species, the trade in wildlife products for traditional medicines, pets or ornamentation has proved a serious threat to many of the populations that remain. CITES, the Convention on Trade in Endangered Species, which came into force in 1976, has been steadily upgrading species after species to levels where trade is more and more restricted, indicating greater and greater threat of extinction from human use.

Human population growth is at the root of the threat to natural resources. Population growth inevitably leads to an increased demand for resources, and an indefinite increase in that demand is not compatible with conserving our resource base. The last century has seen a particularly dramatic rise in human population. But the relationship between population and consumption is not always a simple one. Technological advances lie behind both human population growth and the impact each community has on the environment. At present, the populations which consume the most are those from wealthy countries where family sizes tend to be small; where population growth is rapid, families are typically living frugal lives. An increase in standard of living is generally associated with a decline in population growth rate, although the precise relationship between economic development and fertility decline is complex and not fully understood. This demographic transition to smaller family sizes started two centuries ago in Europe and North America, much later in Asia and South America and very recently in Africa. Although population growth is still rapid, the rate of growth now appears to be slowing in every country in the world. This offers the prospect, if distant, of an eventual stabilisation of the human population. However, our impact on wild living resources is not showing the same trend; there is no evidence that our rate of exploitation of the natural world is decelerating. As economies grow and harvesting becomes more technologically efficient, consumption is fast outstripping the limits that would sustain biodiversity at, or even near, present levels.

Biodiversity is the variability of nature at all levels, from genes, through individuals and species to ecosystems, including biological processes as well as entities. Biological conservation means maintaining biodiversity, which involves the prevention of species extinction, but it is much broader than this. Conservation is distinct from preservation because it involves recognising the dynamic nature of biological systems, and allowing them to change and evolve. When species are considered as resources, then conservation frequently has a slightly different meaning; the emphasis is usually on sustaining the resource so that profitable exploitation can continue.

Defending resources from exploitation by other groups of people has been one of the underlying motivations for tribalism and nationalism, and associated conflicts, throughout history. Within communities, rights to exploit certain areas or to hunt particular prey were sometimes limited to certain lineages or age-grades within society. Particularly prized resources (such as favoured game species) were sometimes reserved by leaders and élites, who did not permit less powerful members of society to harvest them. Some of the first nature reserves were sacred sites, others were hunting grounds that powerful people set up as protected areas in which they could practise their sport. Many of the more traditional systems of resource conservation have not survived exposure to modern economic pressures.

Fisheries were one of the first arenas where the conservation of wild species became an important issue in the eyes of national governments: sustainable use was the priority. Reproducing organisms, like fish, are renewable resources. It has long been appreciated that some level of harvesting could threaten the species being harvested, but that there must be a lower level of harvesting that should be sustainable; fish taken from the sea will be replaced by the offspring of those left behind to reproduce. Understanding population dynamics, including the impact of harvesting, is the basis of the science of population biology. The greater and more reliable the harvest, the better for the businesses and communities living off fish, which constitute an important section of the economy of many nations. To find the optimal balance between the yield from harvesting and sustainability became the main aim of fisheries science.

Despite the expenditure of considerable scientific and political effort, the recent history of fisheries has not been a happy one. More than 70% of the world's commercially important marine fish stocks are described by the Food and Agriculture Organisation of the UN as either fully fished, over-exploited, depleted, or slowly recovering (FAO 1995a). The collapse of numerous fisheries in the second half of the 20th century led to some doubt as to whether the models of sustainable harvesting produced by fisheries scientists were of any value. Unforeseen factors, such as ecological fluctuations or a large proportion of 'by-catch' (non-target species that are thrown back into the sea but never counted in the official measure of the size of a catch) were seriously distorting estimates of the impact of fishing effort on ecosystems. The collapse in 1992 of Canada's Grand Banks cod fishery, off the coast of Newfoundland, with the loss of at least 40 000 jobs, was a recent disaster where scientists were attacked for failing to predict the crash. But fisheries scientists argue that whilst mistakes have certainly been made, they learn from them, and it is the politicians who are failing to implement policies that would be sustainable (Rosenberg *et al.* 1993). The focus of scientific effort now places less emphasis on trying to understand the population biology, and more on developing methods of assessing stocks that can be applied even when important biological parameters are unknown, and also on methods for controlling catches that are easier to enforce. Economics and politics are being included with population biology and ecology in a more multidisciplinary approach to understanding fisheries. Unfortunately politicians have always tended to favour the short-term interests of the fishermen over conservation. As fishing communities suffer financial hardship due to declining stocks, subsidies are frequently introduced to support them. A recent estimate valued worldwide government subsidies to the fishing industry at US$54 billion per year, whilst the total worldwide catch is worth just US$70 billion per year (Kempf *et al.* 1996). These subsidised fleets are thus exploiting fisheries at a rate even faster than that caused by market forces alone.

The quantitative approach of fisheries managers does not extend widely into other areas of conservation. Forestry is a notably exception, where tools similar to those used to devise sustainable harvests in fisheries have been developed. This is largely due to the fact that both forestry and fisheries are major industries. Where resources are considered of less commercial importance, conservation initiatives have usually had a different emphasis. For most of the terrestrial ecosystems that governments want to conserve, protected areas have proved an important conservation tool. Reserves have frequently been established as havens from all consumptive use, and are often maintained largely for recreational or aesthetic reasons. This approach has had some significant conservation successes, and fisheries managers are beginning to consider protected areas as another option for conserving marine resources (Roberts 1997); marine reserves have not to date been widely employed. Nevertheless, those interested in conserving terrestrial wildlife are becoming increasingly uneasy with the idea of protected areas as a conservation tool. Problems arise when the motivation for protecting an area comes from outside the local community, with whom the reserves are deeply unpopular. This happens when the local population does not have much to gain from the wider national or international benefits derived from protecting the resource; nor are they particularly interested in the values upon which the reserve may have been founded. They have pressing short-term needs that could be alleviated if they could use the products of the reserve for their own livelihoods. Frequently, the reserves are in areas which they traditionally exploited and from which they have only recently been excluded. If local people feel that they gain nothing from conserving a resource and that they have a right to exploit it, the costs (financial and political) of enforcing its protection may be beyond the reach of governments. Therefore, for both ethical and practical reasons, areas fully protected from human exploitation are facing an uncertain future.

In the international arena, related ethical and political arguments have arisen. Poor countries have argued that it is unreasonable for wealthy countries to expect them to forego environmentally damaging developments, when those wealthy countries have in fact based their own economic success on environmentally damaging industries.

Sustainable development

Since the 1980s, the great majority of conservation initiatives have attempted to address the issue of sustainable use, within the framework of sustaintable development. IUCN has been an important influence in pushing sustainable use up the conservation agenda, and Table 1 summarises some of their major initiatives over the last two decades. There is a recognition in all these initiatives that the main justification for conserving natural resources is their economic importance, both to present and future generations of people. Whilst other beliefs about the importance of conservation, based on aesthetics, ethics, religion or science, are strongly upheld by many, it is clear that many, if not most, of the wild resources that are directly threatened by human exploitation are economic goods. Economic forces play an enormous role both in their exploitation and in the success or failure of efforts to conserve them. Conservationists have had to consider more than ever before how the apparently conflicting goals of economic development and conservation can both be met, and whether the former could be used to help the latter.

The term 'sustainable development' became a buzz-word in the 1980s. By referring to two concepts that are universally popular, it became a phrase that few could disagree with – perhaps we could have our cake and eat it? The enthusiasm that was generated persists today, for good reason, but some of the initiatives that were spawned were naive. The use of species came to be seen as a conservation tool in itself in some spheres. Creating new markets for wild products, or seeking new ways of increasing their market value, became the explicit aim of many conservation projects.

Table 1 IUCN Sustainable Use Initiatives. (Adapted from J. Robinson & S. Edwards, pers. comm.)

Year	Source	Level of understanding	Present status
1980	World Conservation Strategy (IUCN, UNEP, WWF)	Recognised most human use was not sustainable, but argued that natural populations produce a 'surplus' that could be harvested sustainably. Sustainable use was a goal	Simplistic biological model (assumed populations were 'capital' and harvest was 'interest')
1990	Conservation of wildlife through wise use as a renewable natural resource (IUCN)	Assumed global guidelines could be drawn up to discriminate between uses that were sustainable and those that were not	Not correct. Ignores local complexities and multi-dimensional nature of the problem
1992	Caring for the Earth: a Strategy for Sustainable Living (IUCN, UNEP, WWF)	Acknowledges that harvesting reduces population density, but argues that populations can be managed for sustainable use. Social equity also put forward as a goal	Simplistic ecological model (e.g. insufficient recognition of environmental uncertainty)
1996	Factors enhancing sustainability: a report of the activities of the IUCN Sustainable Use Initiative (IUCN)	Acknowledges that most use is to the detriment of biodiversity in general. Argues that sustainable use is not determinant; there is a multitude of configurations of biological, social and economic conditions influencing sustainability	IUCN's current hypothesis. Still vague on how economic forces influence the sustainability of use

The precise mechanism by which this new market value would improve the conservation of the resource, rather than threaten it, was not necessarily thought through. The later IUCN initiatives became more cautious in their tone, the latest acknowledging that use is likely to be ecologically harmful.

Unfortunately, 'sustainable development' is such a non-specific term that it can be (and is) used to describe almost any economic activity based on exploiting renewable resources. To a rural development agency, a sustainable project is normally taken to mean a project that will not fall apart after the explicit actions of the development agency have ended; thus the main emphasis is usually on social sustainability. Rural development projects can frequently describe themselves as integrating conservation and development with few real changes to the conventional development agenda. For example, income-generating schemes set up on the boundaries of nature reserves are frequently described as conservation efforts, but in fact can prove conservation hazards as new immigrants are attracted to the area by the new job opportunities, putting the natural resources under additional pressure. To a commercial company, the term 'sustainable' may refer only to the timescale over which they generally consider their profit, which may mean little more than 'economically viable'. In both these cases, the timescales considered are much shorter than that over which an ecologist might consider something to be 'ecologically sustainable'. Sustainability has an ecological, economic and social dimension, and, beyond that, defies a simple definition (Gatto 1995). Ecological systems are dynamic and it is difficult to predict their future course even without exploitation by humans. Nothing remains unchanged indefinitely, whether it is used or not. Thus, the acceptable timescale over which something is considered sustainable depends in part on the objectives of whoever is defining the term.

After an international summit meeting in Rio de Janeiro in 1992, the Convention on Biological Diversity was launched in 1993, which all governments were invited to ratify. This convention is founded on three objectives: conservation, sustainable use and the equitable sharing of benefits from biodiversity. It is argued that they are mutually reinforcing objectives, and cannot be considered in isolation. It is one of our aims in this book to explore how mutually reinforcing these objectives currently are, and where they are in conflict.

The book covers the ecological, economic and social dimensions of the conservation and use of wild living resources. It is divided into two main sections. The first section covers the theoretical issues involved in the use of renewable resources. This involves drawing theory from the fields of population biology, ecology and economics, combined with insights from politics and anthropology. The underlying models are presented so that the dynamics of the behaviour of both harvested populations and their harvesters can be better understood. We present the simplest, heuristic models from both population biology and bio-economics, and then consider more complex biological and economic models which take into account the numerous interactions that occur and influence the likelihood of sustainable use. The use of these models to devise conservation measures in real ecosystems generally requires a series of modifications that are specific to the system concerned; we describe a range of considerations that may be crucial to understanding how a particular system is likely to behave. Technical and social problems that can arise are presented, ranging from the interpretation of imperfect data to the political challenges that arise in community-based conservation initiatives. Parts of the material in this section can be found in either ecology or economics textbooks; other material is scattered across diverse sources, including a range of academic journals and the 'grey literature' of applied conservation publications. To address the practical demands of conservation, conservationists have been forced to venture across the boundaries of many of the conventional academic disciplines, which we bring together in this section of the book.

The second half of the book presents case studies from contributors, which describe a range of situations in which wild species are being exploited. All the contributors have been professionally involved in conservation either as practitioners, consultants or campaigners. They work for organisations as diverse as IUCN, UNEP and the World Bank, Safari Club International and Zoo Atlanta. Some of these case studies illustrate applications of concepts from ecology and economics, others illustrate social problems that arise. Another important part of their value derives from the fact that they illustrate the range of perspectives that can be found amongst conservationists. But, whilst each group of conservationists has distinctly different approaches, their ultimate aim is usually similar. It is clear that those in different spheres are emphasising different issues, and that there may be lessons that can be learnt by one group from another. We bring together in one book both the theory and practice of the many different constituents of the conservation movement. Our hope is that this breadth of knowledge can be applied as widely as possible to the urgent objective of conserving wild species for future generations.

Part 2
Theoretical Background

Mahogany logs at sawmill, Bolivia. (Photo by Ted Gullison.)

Chapter 1: The Ecological and Economic Theory of Sustainable Harvesting

1.1 Introduction

Models are tools for understanding the mechanisms behind observed phenomena. The modeller forms a hypothesis about what the key factors driving the system are, and uses a model to explore the consequences of this hypothesis. The predictions of the model are compared to data; if they fit the data, this suggests that the model is correctly specified. Good models are robust and give clear predictions. Robustness means that small changes in parameter values do not cause large changes in model predictions, which is important because there is always uncertainty surrounding parameter estimates. Clear predictions are also important – if several possible underlying mechanisms give similar results, then the modelling process has not advanced our understanding of the system.

Effective conservation of a harvested resource requires an understanding of how that resource responds to exploitation. Although it might be possible to treat each case as unique and find a sustainable harvesting strategy by trial and error, this would be both inefficient and dangerous. A body of theory has been built up over the last few hundred years that allows natural resource managers to predict the effects of harvesting in a general way. The majority of the early research concerned the impact of harvesting on the population dynamics of a single species, chiefly in fisheries and forestry. Faustmann developed a formula for the optimal rotation length for forestry plantations in 1849, and Gordon (1954) developed the first incarnation of the modern fisheries model which is now the foundation of resource harvesting theory. Clark (1976, 1990) was a pioneer in consolidating and extending this early theory. The early models are still the basis

13

of much of harvesting theory today. Economic and sociological research has allowed us to model the incentives for harvesters to hunt at particular levels and formulate policies for regulating resource use.

It is therefore possible to underpin our discussions of conservation and sustainable use with the theoretical and practical knowledge accumulated over many decades of more or less successful management of renewable resource stocks (Rosenberg *et al.* 1993). In theory, we should be able to predict the levels at which unregulated industries will harvest renewable resources, and what the effects of various management regimes will be on harvesters' incentives. Of course, the degree to which reality and theoretically derived predictions match is another matter, particularly in more complex systems. There is still work to be done on the effects of harvesting several different species at once, and on the more diffuse ecological effects of hunting on other organisms in the ecosystem. The situation is even less clear with the effects of non-consumptive use (use that doesn't involve harvesting), and analysis has not really progressed beyond a case-by-case treatment. However, the foundations laid by the bio-economic modelling of simpler situations are still relevant to these ecologically more complex cases. Those who ignore the current body of theory on the sustainability of resource use are doomed to continue to act on a case-by-case level, and much of the discussion of sustainable resource management is indeed missing out on the broader understanding that science can provide.

In this chapter, a simple model of the population dynamics of a single species is presented. The model is then modified to include harvesting either a fixed number of individuals or a fixed proportion of the population. Different harvesting strategies and harvesting levels produce equilibrium population sizes which differ in stability. The usefulness of the concept of maximum sustainable yield is discussed, and it is demonstrated that it has flaws both from biological and economic perspectives. We summarise some of the harvesting strategies that managers commonly employ, and highlight their advantages and disadvantages. The discussion then moves from assuming that a manager has control over harvesting levels to a consideration of harvesting by individuals in a market economy. The structure of the market affects the harvesting level and thus the equilibrium population size of the harvested stock. In particular, the time horizon over which harvesters plan is a major determinant of the equilibrium stock size.

1.2 A simple model of population growth

The key assumption behind all sustainable harvesting models is that populations of organisms grow and replace themselves – that is, they are *renewable resources*. Otherwise, it would be impossible to harvest them sustainably, and they would be treated theoretically in a similar way to reserves of fossil fuels. Some species, such as blue whales or mahogany, may grow so slowly that, to all intents and purposes, they should be treated as non-renewable resources by harvesters, but this is unusual.

Another assumption of renewable resource harvesting is that populations of organisms do not continue to grow indefinitely; they reach an equilibrium population size, which occurs when the number of individuals matches the resources available to the population. The population then remains stable at that size. This seems self-evident (because the world is not several kilometres deep in the fastest growing bacteria) – populations do not grow indefinitely. The equilibrium size of the population is its *carrying capacity*.

Given that populations increase to a limit, a function that describes the way in which they grow is needed. The function will be bounded, because populations do not grow at the two extremes: when there are no individuals there to reproduce, and when there is an equilibrium number of individuals, so all the resources are in use. One of the most convenient assumptions to make about the growth rate between these two limits is that population growth is sigmoidal (Fig. 1.1a). This is the assumption made in the *logistic equation* of population growth. We shall explore the logistic equation in some depth, although it is not the perfect representa-

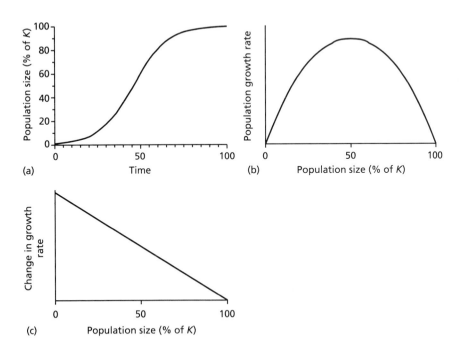

Fig. 1.1 Curves of the logistic equation. (a) Sigmoidal population growth, as described by the logistic equation. (b) The parabolic relationship between the population growth rate and the population size. (c) The linear relationship between the change in population growth rate and the population size.

tion of population growth. There is evidence that some populations do grow in a logistic fashion towards a stable equilibrium: Fig. 1.2(a) is a commonly cited example of a population of yeast growing logistically in the laboratory (although the fact that the data were recorded in 1927 gives a hint that perfect logistic growth is not that easy to find!). Other data come from less controlled conditions, but still seem to show a fair fit to the logistic curve (Fig. 1.2b). The equation describing logistic growth is:

$$N_t = \frac{K}{1 + \left(\dfrac{K - N_0}{N_0}\right)e^{-rt}} \tag{1.1}$$

The parameter values are:
N_t = The population size at time t.
K = The carrying capacity of the population.
N_0 = The population size at time zero.
r = The intrinsic rate of population increase (the rate at which the population grows when it is very small).

In the case of the yeast population in Fig. 1.2(a), K is estimated as 665, r as 0.531, and the starting population size as 10. The equation for population growth is thus:

$$N_t = \frac{665}{1 + \left(\dfrac{665 - 10}{10}\right)e^{-0.531t}}$$

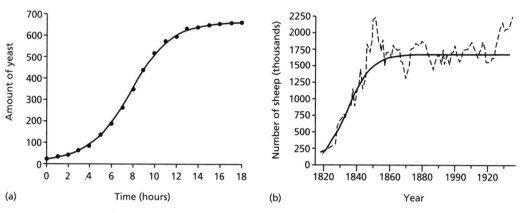

(a)   Time (hours)

(b)   Year

Fig. 1.2 Examples of fitting population growth data to the logistic equation (from Renshaw 1991). (a) A laboratory population of yeast. (b) Sheep in Tasmania. Note that economic factors will have an important influence on the population size in this case, particularly after the initial growth phase. The fitted equation is:

$$N_t = \frac{1670}{1 + e^{240.8 - 0.131t}}.$$

From this, the population size at any point can be calculated. For example, 10 hours after the growth starts, the population size is predicted to be 502.

Differentiating this equation with respect to time is useful for developing the analysis. This is a way of finding the slope of a curve at a particular point (Bostock & Chandler 1981); so differentiating Equation 1.1 gives an expression for how the rate of population increase changes as t increases. At first the population grows fast, but the growth rate slows gradually as time goes on, until equilibrium is reached with a growth rate of zero (Fig. 1.1a). Figure 1.1(b) shows how the population growth rate (in numbers of individuals) varies with population size. The equation for Fig. 1.1(b) is the differential of Equation 1.1:

$$\frac{dN}{dt} = rN\left(1 - \frac{N}{K}\right) \tag{1.2}$$

dN/dt means the differential of N with respect to t, or more simply, the change in N with a small change in t. Equation 1.2 is the usual way in which logistic growth is represented mathematically. Differentiating Equation 1.2 with respect to population size gives an expression for how the rate of change in the number of individuals changes with population size:

$$\frac{dF}{dN} = r - \frac{2r}{K}N \tag{1.3}$$

where F is the right hand side of Equation 1.2. Equation 1.3 shows why the logistic equation is so convenient mathematically; the rate of change in the population growth rate declines linearly as population size increases (Fig. 1.1c). Another way to express 'the rate of change in the population growth rate' is as the 'per-capita growth rate'.

Equation 1.2 has some important features. At very low population sizes, the value of N/K is small, so the population growth rate is approximately equal to rN, meaning that the population is growing almost exponentially at a rate r (in exponential growth, the population size is multiplied by the same constant in each time-period, so if r was 2 it would double each year). Despite this, the population growth rate is low (low values on the y-axis of Fig. 1.1b). This is usually because, although each individual in the population is reproducing at a high rate, there are few reproducing individuals present. When the population is very large (set $N = K - 1$ in Equation 1.2), then the value of N/K approaches 1, so the part of the equation inside the brackets approaches zero. The population growth rate is again very low, either because each individual is hardly reproducing at all, or mortality rates are high. Between these two extremes there is a balance between enough individuals being present and these having a high enough per-capita growth rate. The population growth rate is at a maximum when $N = K/2$.

1.2.1 The uses of the logistic equation

Part of the appeal of the logistic equation is that it describes population growth using very few parameters: time from the start of the growth (t); population size (N); carrying capacity (K); and the intrinsic rate of increase (r), but the measurement of these parameters is not necessarily simple. The logistic equation is a continuous time model, but data are collected at intervals. These intervals should be relevant to the population – probably years for large mammals, but days or weeks for a laboratory yeast population. Using the wrong time units obscures the underlying functional form of the population's growth. Population size is the number of individuals in the population. The definition of population size can also be problematic; for example, should the population be counted at its minimum size each year, or at its peak of abundance? Like time, population size is not expressed in whole numbers in the logistic model but varies continuously, although this is only a problem if the population being modelled is very small.

The two key parameters in the model are carrying capacity and the intrinsic rate of population increase. Carrying capacity is a concept that is simple in theory but slippery in application. Population sizes generally fluctuate over time, and K is then taken as the average size of the population at equilibrium (the horizontal part of the solid line in Fig. 1.2b). It is unlikely that it will have an exact value for many populations, and is more realistically thought of as the bounds within which a population fluctuates. Even if populations fluctuate greatly in size, they will have some upper size limit beyond which they cannot increase further. However, in some cases, carrying capacity of a population cannot be measured, and the parameter K is not meaningful. These include:
- *Density-independent population growth*. A population grows at the same rate indefinitely, but a large proportion of individuals are wiped out by occasional catastrophes.
- *Environmental stochasticity*. If populations are found in a very variable environment, they may never reach a stable carrying capacity, despite there being an interaction between density and resource availability. The variation in population size will mask any tendency to stabilise at an equilibrium.
- *Complex interactions between density and resource availability*. In some systems there may not be a straightforward two-way interaction between available resources and population size. In arid rangelands rainfall is very variable and is the main determinant of ecosystem dynamics. The density of grazers is determined by vegetation availability, but may have little reciprocal effect on the vegetation until a sudden transition to an overgrazed state occurs (Mace 1991).
- *Meta-population structure*. Some species (especially of insects) have a

high rate of colonisation of new sites as well as a high extinction rate within sites. When a site is colonised, the population grows and produces propagules that may colonise other suitable sites. When the population at a particular site has exhausted the available resources, it dies out. Carrying capacity is often not meaningful at the population level. There may be an equilibrium meta-population size, determined by the number of suitable sites for colonisation, but this is hard to measure.

• *Unstable internal dynamics.* A population never reaches an equilibrium size because of the instability inherent in its population dynamics. It is the population regulation itself that is causing the population to fluctuate. Some such systems are described as chaotic.

The fact that carrying capacity is not reached does not mean that the population size is unregulated by resource availability. In the cases listed above, only when the population is growing at the same rate regardless of population size is it completely unregulated. Carrying capacity is most likely to be a meaningful concept for species with low intrinsic rates of population increase, which tend to have more stable dynamics, as discussed in section 2.1.4. In these species, the variance in population size is normally significantly lower than the mean population size. The mean is therefore an adequate approximation to the equilibrium population size. Carrying capacity is frequently expressed simply as a number, or population size may be expressed as a percentage of carrying capacity. However, it is important to bear in mind its precise meaning, which is the equilibrium, unexploited, density of a particular population in a particular area. Carrying capacity is not intrinsic to a particular species, although the likely range of carrying capacities is. For example, the carrying capacity of elephants in a national park in the savannah areas of East Africa might be $2\,km^{-2}$, while the carrying capacity of the same species of elephant might be less than $0.5\,km^{-2}$ in a Central African forest – the habits of the species dictate that carrying capacity is never high, but it varies fourfold with habitat type (Burrill & Douglas-Hamilton 1987).

Whilst K is problematic in its interpretation and varies between populations of the same species, at least its biological meaning is relatively easy to comprehend. This cannot be said for the intrinsic rate of population increase, r; yet r is often cited as the key parameter that needs to be known if a population is to be harvested sustainably. The intrinsic rate of population increase is best comprehended as the rate at which the population grows (the difference between births and deaths) when the population is very small and growing exponentially, as in Equation 1.2 when $N = 1$. Mortality from factors other than resource constraints (such as predators or climate) is ignored, although in reality these factors might be very important. In practice, primary data on the value of r are rarely obtainable. In particular, it will be hard to obtain a value for r for populations already stable at intermediate or high population sizes. Usually r has to be inferred, either by fitting a population model to data with all the other relevant parameter values known, or by using fecundity data from zoo populations together with assumptions about natural mortality rates (Caughley 1977; Robinson & Redford 1991).

Despite these drawbacks, the logistic equation is used as the basis for virtually all theoretical work on population dynamics. This is because:

• Most populations appear to grow in a way that is not dissimilar to logistic growth. However, any function that describes rapid population growth to an equilibrium will look similar to the logistic equation at first glance. More careful analysis may bring out significant deviations from logistic growth.

• It is very convenient for analytical solutions, and is one of the simplest functional forms that can be assumed for population growth. It has a linear form and requires only two parameters to be estimated. This means that the logistic equation can easily be built upon to produce simple analytical models for population dynamics – models that have been shown to be relatively good descriptors of real-world population dynamics.

• It is used for historical reasons; the pioneering research into population dynamics used it. Science works by building on previously established foundations.

These reasons interact to make the logistic equation a powerful and widely used tool in theoretical population dynamics. Other functional forms for population growth have been used (Getz & Haight 1989), but we will use the logistic equation for the sake of simplicity and in order to make use of previous theoretical results. Most of the results we present are relatively easily generalised to these other functional forms.

Whatever functional form is used, it will generally involve negative feedback; that is, the growth rate of the population in the next time-period depends on its current size. This negative feedback is called *density dependence*. Density dependence is a population-level amalgamation of processes occurring between individuals, when there is competition for scarce resources among members of a species. These scarce resources are usually assumed to be food, but another common one is space (e.g. rock space for barnacles to settle on or nesting sites for seabirds). Increases in competition for these scarce resources leads to increases in the mortality or decreases in the fecundity of individuals. The functional form of density dependence depends on these individual-level processes. Not all limiting resources produce a logistic density-dependent curve; the logistic equation is really only appropriate for limiting resources with a rate of renewal independent of the population size (such as when food supplies are renewed daily in laboratory populations). In general, food comes in the form of another species, such as grass, which has a rate of renewal dependent on the population size of the species eating it. If this dependence is strong, then predator–prey models are needed instead of a simple logistic equation.

1.3 Maximum sustainable yield and its limitations

The simplest way to model harvesting is to modify the logistic equation so that a certain number of individuals is continuously removed:

$$\frac{dN}{dt} = rN\left(1 - \frac{N}{K}\right) - H \tag{1.4}$$

If the number of individuals removed (H) is constant, then the population is at equilibrium when the number of individuals removed is equal to the population growth rate (Fig. 1.3a). If a proportion of the population is removed, then $H = aN$, and the equilibrium situation is as shown in Fig. 1.3(b). The equilibrium population size under a particular harvesting rate can be found as follows: from Equation 1.4, the population is stable when it is not growing, i.e. when $\frac{dN}{dt} = 0$. This happens when the population growth rate is the same as the harvest rate:

$$rN\left(1 - \frac{N}{K}\right) = H$$

In Fig 1.3a, the hunting rate H_1 leads to two possible equilibrium points: a low population size N_a and a high one N_b. However, H_2 leads to only one equilibrium point, N_{MSY}, which is the population size that produces the maximum growth rate (half the carrying capacity with logistic growth). The offtake level H_2 is called the *maximum sustainable yield* (MSY), because

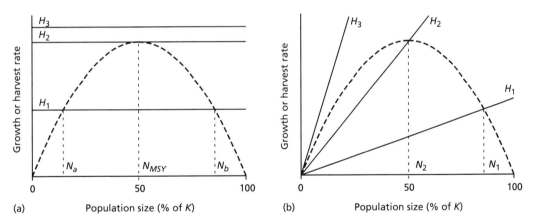

(a)   Population size (% of K)   (b)   Population size (% of K)

Fig. 1.3 The equilibrium population sizes under various hunting rates. The solid lines represent the number of individuals removed from the population at particular values of H. The dashed line is the population growth rate at a given value of N. The equilibrium population sizes occur when the number of individuals removed by hunting is equal to the population growth rate (where the growth curve and the hunting rate curve intersect). Three hunting rates, H_1 to H_3, are shown for each hunting strategy. (a) Hunting removes a constant number of individuals. (b) Hunting removes a constant proportion of the population.

it is the largest yield that can be taken from a population at equilibrium – if H was higher than H_2 then offtake would exceed the population's capacity to replace itself at any population size (H_3). In Fig. 1.3(b), there is only one equilibrium population size for a given slope of H. If H is too steep to cross the growth rate curve (H_3) there is no equilibrium population size, because the harvest rate is higher than the population growth rate at all values of N. This rate of harvesting is not sustainable.

Analysis of the stability of these equilibria in the face of environmental fluctuations gives an insight into their usefulness as targets for a harvester. Inspecting the equilibrium point producing population size N_b in Fig. 1.3(a), assume a perturbation slightly increases the population size, perhaps due to an especially good breeding year. This leads to the hunting rate slightly exceeding the population growth rate, because density dependence makes the population growth rate decline (thus the line for H_1 is above the dashed population growth rate line). This means that more individuals are being removed than are being added to the population, and the population size moves back to N_b. Similarly, if the population size was reduced below N_b, density dependence would lead to the growth rate increasing, and the population would return to equilibrium. So the equilibrium point H_1, N_b is stable, because the population returns to it after a perturbation. Figure 1.4 shows the outcome of this analysis for each equilibrium point. It can be seen that, for a constant H with two equilibria, the higher population size equilibrium point is stable and the lower unstable (because any perturbation leads to the population size moving away from the equilibrium). The MSY equilibrium is semi-stable – a small increase in population size is compensated for, a small decrease leads to

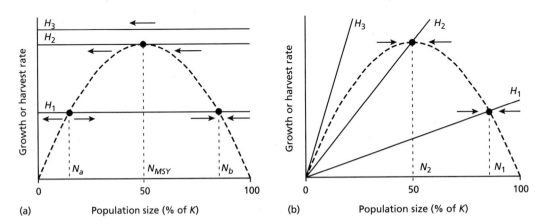

Fig. 1.4 Stability analysis for the equilibrium population sizes shown in Fig. 1.3. (a) Hunting removes a constant number of individuals. (b) Hunting removes a constant proportion of the population.

extinction if H is not decreased. Harvesting at MSY is dangerous because it is on a knife-edge – any small population decline leads to positive feedback, with the population declining rapidly to extinction if the number harvested stays the same. If H varies linearly with N (Fig. 1.4b), then all the equilibrium points are stable. The MSY harvesting rate produces the same kind of equilibrium as all the other sustainable rates. IUCN recently suggested that sustainable use projects should harvest at a maximum rate of $r/2$ (Prescott-Allen & Prescott-Allen 1996), which is equivalent to MSY (H_2 in Fig. 1.4b). This is not a safe suggestion, because harvesting at MSY is dangerous, although taking a constant proportion of the population that is equivalent to MSY is safer than taking a constant number of individuals at MSY.

Even this very simple analysis leads to some useful observations. The first two are quite general; the other two apply specifically to the logistic model.

• Starting to harvest a previously unharvested population will always lead to a decrease in the population size. It is impossible for a harvested population to remain at carrying capacity. The harvested population will either stabilise at a new lower equilibrium size or, if the harvest rate is too high, decline to zero. The model does not specify how long it will take for the population to stabilise at its new equilibrium – this depends on the values of r and H. Thus, a decreasing population size does not in itself indicate that a population is being unsustainably harvested, rather that it is not at equilibrium. This point is ignored in most of the literature on sustainable use.

• The reason why populations can be sustainably harvested is that they exhibit a density-dependent response. This means that at any population size below K, the population is producing a *surplus yield* that is available for harvest without reducing population size. Density dependence is the regulatory process that allows the population to return to equilibrium after a perturbation. The logistic equation assumes that density dependence takes the form of smooth negative feedback. If this is not the case, small changes in harvesting rate can lead to sudden large changes in population size, which could be dangerous for the population. The form of density dependence assumed is a key component of model predictions about the effects of harvesting on population size.

• If harvesting involves removing a constant number of individuals (Fig. 1.4a), then harvesting at a level greater than the MSY leads to a rapid population decline to extinction. Harvesting below the MSY level leads to a stable equilibrium population if the starting population is above the unstable equilibrium population size. The further below the MSY level the harvest is, the safer it is, because the two equilibrium population sizes are further apart. This reasoning lies behind a common suggestion that harvesting should not exceed half the MSY level. If a population is over-exploited, the harvest rate needs to be reduced below the current population growth rate to allow the population to recover. The fastest recovery occurs when $H = 0$.

• If harvesting rate is proportional to population size, a single stable equilibrium population size results (so long as the harvesting rate is low enough to cross the population growth curve at some point). A slight increase in harvesting rate leads to a slight decrease in population size – there is a smooth and gradual relationship between the two, with no rapid changes. Thus, in the logistic model, a strategy which involves harvesting a certain percentage of the population each year is safer than one which involves harvesting a fixed number of individuals each year, so long as the harvesting rate is not too high (the proportion taken must be less than r). There is no need for wide safety margins to guard against unstable equilibria if the relevant parameter values are known.

1.3.1 The determinants of harvesting level

So far, the assumption has been that the harvesting rate is simply set at a particular level. In practice, the level of harvesting may not be under the perfect control of a resource manager. A more useful model would there-fore include the economic parameters which determine harvesting level, so that the effect of harvesting on population size could be predicted in situations when resource managers do not intervene to set the level of H.

The simplest such model starts by assuming that harvesting rate is proportional to population size (Fig. 1.4b), but that harvesting rate is also determined by harvesting effort:

$$H = qEN \tag{1.5}$$

Here, qE has replaced a, the percentage of the population killed, that was used in Fig. 1.3(b). E, the *harvesting effort*, is treated as a variable. Harvesting effort can be measured in various ways, such as the number of days spent hunting, the number of fishing boats working in the area, or the number of each type of gun or snare used. In a given harvesting situation, data on effort need to be collected in appropriate units. The constant q is the *catchability coefficient*, which is a measure of the ease of catching an individual of a particular species. For example, if a single snare is set in a previously unhunted area, q is the probability that an individual of a particular species is caught in it. Alternatively, if a single fishing boat goes out for a day, q is the proportion of the total fish population that it catches. The units of q depend on the measure of effort chosen. Equation 1.5 says that if either effort expended by harvesters or population size increase, then the number of individuals killed will increase.

By substituting H/qE for N into Equation 1.4, and assuming that the population is at equilibrium ($dN/dt = 0$), we obtain an expression for yield

(H) as a function of effort:

$$H = qKE\left(1 - \frac{qE}{r}\right) \tag{1.6}$$

Comparing Equation 1.6 and Equation 1.2 highlights their similarities – both represent parabolas. Thus, as effort increases, yield increases to a maximum and then declines to zero. At $E = 0$, the population size is at its maximum, K. As effort increases, population size decreases and more and more effort is needed to catch the remaining individuals until, at maximum effort, the population size reaches zero. The yield–effort curve is effectively a mirror image of the population growth rate–population size curve (Fig. 1.1b). This is a very useful result because yield and effort are parameters that can easily be measured and come directly from harvesting data. Population size is, by contrast, very hard to measure directly, fish and forest-living mammals being obvious examples of cases where individuals are difficult to count. If this model were to hold, simply measuring yield and effort should be enough to discover the underlying stock–production relationship, without the difficulty and expense of population monitoring.

One much-used concept in the literature on resource use is *catch per unit effort* (CPUE), the ratio of yield to the effort expended. Trends in CPUE give an indication of trends in stock size, because Equation 1.5 can be rearranged as:

$$\frac{H}{E} \propto N.$$

Therefore if the yield from a population is declining while the harvesting effort remains unchanged, the population size can be assumed to be declining as well. This relationship could be a very useful indirect measure of population trends. A non-declining CPUE has been promoted as a good indicator of sustainable harvesting. However, it is important to remember the assumption that has been made to get to this useful relationship – that yield is directly proportional to effort and population size. In the early stages of harvesting, a declining CPUE does not necessarily mean harvesting is unsustainable, because the population may just be declining to an equilibrium size. Conversely, a stable CPUE could be due to undetected increases in technological efficiency, giving a false picture of stability as the population declines. The effects of aggregative behaviour on harvesting costs can also lead to misleading trends in CPUE. If a population aggregates into large groups even at low population sizes, or if its location is always predictable, then the costs of harvesting are not related linearly to population size. The CPUE does not decline as population size declines, because small populations are not much harder to harvest than larger ones. The population declines dramatically without any reduction in harvesting

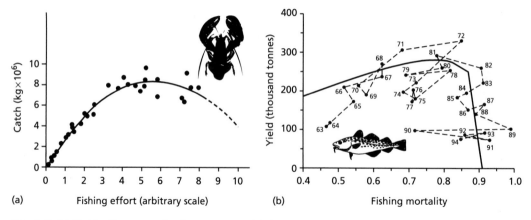

Fig. 1.5 (a) Yield–effort curve for the western rock lobster population (From Begon *et al.* 1996b). (b) Yield–fishing mortality curve for the North Sea cod population. Numbers on the data points are the year of catch. (From Cook *et al.* 1997.)

effort, and without warning signs, because catches continue to appear healthy. These conditions are common in fisheries.

Figure 1.5(a) suggests that, for the western rock lobster fishery, the parabolic relationship between yield and effort holds quite well at relatively low effort levels. There are few data above the MSY level, but those recorded seem to show an increase in the variance around the yield–effort relationship as effort levels increase. These high levels of effort are often the levels that are of conservation interest but there are rarely data available to plot the yield–effort curve. This example shows how important it is not to extrapolate theoretical relationships beyond the regions with supporting data. Data are available from the North Sea cod fishery, which has been severely over-harvested (Cook *et al.* 1997; Fig. 1.5b). Here the theoretical relationship between yield and fishing mortality rate seems to hold well, despite the high level of variance around the curve. Fishing mortality rate is related to effort; in the simple logistic model it is qE. In the case of the cod, the dynamics of the population are such that the theoretical relationship predicts a sharp collapse after MSY is reached.

1.3.2 A comparison of management strategies for sustainable harvesting

The theory presented thus far allows us to specify a few simple guidelines for managers of harvested populations. These will be refined as other considerations are introduced, particularly social and economic ones (section 4.3) However, the simple surplus yield models suggest that the following harvesting strategies have different advantages and disadvantages. The strategy which is the most appropriate in a particular situation depends on the manager's priorities.

Harvest a constant number

Figure 1.3(a) shows the strategy of harvesting a constant number of individuals. The strategy might be implemented by a manager setting a fixed quota each year. The advantages are:
- Harvesting a constant number guarantees the same yield each year. The users of the resource know how much they will be able to produce each year.
- The strategy is very simple to administer. So long as the yield is set at an appropriate level to start with, no data need to be collected to calculate the yield each year.

The disadvantages are:
- The strategy is based on the assumption that stock size is deterministic, with no natural variation due to factors like climate. There are assumed to be no measurement errors in the calculation of the model parameters. If either of these assumptions are untrue and the population is being harvested at or near the MSY level, then the strategy could lead to rapid population decline, reinforced by positive feedback. Similarly, if the population is at an unstable equilibrium, any decline in population size will accelerate unless the level of harvest is reduced.
- In order to avoid population collapse, a large margin of error is needed. Harvests should not exceed about 75% of the MSY level, depending on the population variability and other characteristics of the species concerned. Harvesting well below the MSY leads to a significantly lower yield than would be possible under other management strategies.

Harvest with a constant effort

Harvesting with a constant effort is another commonly used management option. It might involve a fixed length of hunting season, a specified type of hunting equipment, or a specified number of hunters. Des Clers (see Chapter 8) discusses a fishery which is managed by controlling the number of fishing licences issued, but where managers retain the right to close the fishing season early if necessary. Equation 1.5 shows that harvesting with a fixed effort leads to an offtake that is proportional to population size.

The advantages are:
- Harvesting with a constant effort is easier to control than harvesting a constant number of prey. Effort levels, such as the number of boats fishing in an area, are usually easier to monitor than offtake levels (the amount of fish caught by each boat).

• The strategy is safer and more efficient than harvesting a constant number, and allows higher yields because large margins of error are not needed. Fine-tuning the effort level doesn't destabilise the system, allowing experimentation and corrections.
• It can be administered without monitoring population size or determining the relationship between population size and the population growth rate. Yield should automatically track population size, with the catch per unit effort decreasing as population size declines.

The disadvantages are:
• There is an incentive for the hunters to use technological innovation to circumvent the regulations. If the number of days hunting is restricted, the hunters will try to increase yields by using more efficient technology. This undermines the policy by leading to constant pressure to increase effort, and is economically inefficient because the cost of the innovation is an unnecessary expenditure at the group level, though unavoidable for each individual.
• Yields vary from year to year depending on the population size.
• There is a rather indirect relationship between policy and biology. Lack of biological transparency can make hunting restrictions less palatable to hunters.
• Any major violations of the assumption of logistic growth could lead to unexpected population collapse.

Harvest a constant proportion of the population

Harvesting a constant proportion of the population is equivalent to fixing a hunting mortality rate. For example, in the Northwest Territories of Canada, 3–5% of the caribou and muskox populations can be killed each year (see Chapter 13). Theoretically, this strategy should have the same effect as harvesting with a constant effort, as Equation 1.5 shows.

The advantages are:
• There is a clearer relationship between harvesting a constant proportion of the population and the biology of the system than for constant effort, so it is a more transparent policy.
• Harvesters do not make constant attempts to circumvent the regulations if they are free to use any technology they please. This makes it economically more efficient than constant effort.

The disadvantages are:
• The manager needs to know the population size, in order to set the number that can be killed that year. This means that pre-harvest monitoring is required to determine population size. Pre-harvest censuses are an expense for managers. However, they might not be regarded as such a disadvantage by conservationists, as an annual census provides regular assurance that the population is healthy.

- As with constant effort, yields vary from year to year.
- As with constant effort, any major violations of the assumption of logistic growth could lead to unexpected population collapse.

Constant escapement

Constant escapement strategies involve leaving a fixed number of breeding individuals at the end of each hunting season. The harvesters can remove the surplus over and above this number each year.

The advantages are:

- Constant escapement is the safest option biologically. It may be particularly useful for annual species, or species with very variable population dynamics. *Loligo* squid are an annual species, and so are managed by the Falklands Islands government using a constant escapement strategy (see Chapter 8). If, in one year, all breeding individuals are accidentally removed before they reproduce, the population has been exterminated. A constant escapement policy should prevent this happening. In longer-lived species, immature individuals provide a buffer against over-harvesting, so a constant escapement policy may be less necessary.
- The strategy does not rely on the assumption of logistic growth.

The disadvantages are:

- Constant monitoring is needed throughout the hunting season, so that the season can be closed as soon as the population reaches the escapement level. In the Falkland Islands, *Loligo* stock size is assessed weekly from mid-season onwards (see Chapter 8).
- Yields and season lengths are very variable.
- A high level of expertise is required in order to analyse the data.

Which of the above options is chosen for a particular population depends on the manager's priorities. Some common considerations are:

1 Stability of yield.
2 Conservation of the harvested stock.
3 Size of yield.
4 Ease of enforcement.
5 Profitability of the enterprise.

If stability of yield was the main priority, constant numbers might be chosen. Conservation of the harvested stock suggests the selection of a constant proportion or constant escapement, size of yield a constant proportion, ease of enforcement a constant effort, and profitability a constant proportion. However, a manager is likely to have more than one objective, which are unlikely to be mutually compatible. In particular, the stability and the size of the yield often conflict. Rather than trying to maximise both, trade-offs need to be made, and a compromise strategy formulated.

1.4 Profitability and the market

In section 1.3.1, the level of harvesting effort was introduced as an important determinant of the yield which is obtained from harvesting a population. Section 1.3.2 set out some possible strategies by which a resource manager could regulate harvesting. Next, we explore what determines the level of effort that harvesters wish to expend. So far, effort has been treated as an externally determined parameter, but in the real world, the sustainability of renewable resource use is not determined arbitrarily, but by the interplay between the biology of the exploited species and market forces. A bio-economic model is needed to capture these two sides of the coin. Analysing the economics of harvesting leads to insights into the likely equilibrium size of harvested populations under a range of market conditions. This range includes the situation when a sole manager has full control of the harvesting of the resource, as well as when harvesting is an uncontrolled free-for-all.

The level of harvesting can be predicted by adding a term for the profits made from harvesting to the population model. The profits made from harvesting are calculated as the revenues from harvesting, which depend on the yield, minus the costs of harvesting, which depend on the effort expended. Costs are defined rather differently in economics than in ordinary life, using the concept of *opportunity costs*. The opportunity cost of an action is the cost of not doing whatever you could have done instead of what you decided to do. For example, Norton-Griffiths (see Chapter 11) discusses the preservation of Kenya's wildlife in terms of the opportunity cost of not developing the land it occupies for agriculture. The costs of harvesting include not only the direct outlays involved, but also the cost of the harvester's time. The opportunity cost of working as a harvester is the wage the harvester would have received in the best-paying alternative employment. If the harvester's revenues are not covering this wage, then it would be better to leave harvesting and move into the best-paying alternative employment instead. Thus, 'profits' actually means the extra money that the harvester is earning over and above that which could be earned in other feasible jobs. This definition of costs is important when one considers that individuals in economically undeveloped areas may not have much alternative employment other than hunting or fishing; thus their opportunity costs are very low. Individuals in areas with more developed economies have more employment opportunities and so higher opportunity costs. Other things being equal, over-exploitation is more likely when opportunity costs are low, because low costs lead to an equilibrium point at lower stock sizes and higher harvester numbers than would be the case in an area where alternative employment is available. In Fig. 1.3(b), high opportunity costs of harvesting might produce the line H_1, low opportunity costs the line H_2.

The integration of a hunting community into the wider economy is likely to have two important effects on costs, which work in opposite directions – increasing opportunity costs because the opportunities for wage-earning employment improve, and decreasing actual costs of hunting due to the introduction of new technology. Frequently the latter outweighs the former. Many fisheries follow the pattern of gradual cost reduction as their equipment improves, and this contributes to the over-exploitation of a once lightly exploited resource. Des Clers (see Chapter 8) discusses this problem with reference to large commercial fishing vessels which are continually increasing their catching power. Freehling and Marks (see Chapter 10) describe the opposite effect. In the Luangwa Valley, Zambia, in the first half of this century, the use of muzzle-loading guns became widespread, because the guns were a symbol of prestige. However, the adoption of this new technology reduced the hunting success of local hunters, compared to using traditional weapons like snares and poisoned arrows.

A simple assumption about profits is that:

$$\Pi = pqEN - cE \tag{1.7}$$

where Π is the profit from harvesting, p is the price per unit of offtake (usually per harvested individual) and c is the cost per unit of effort expended, for example the cost of a day's hunting. The yield from hunting, H, is qEN (Equation 1.5). Equation 1.7 implies that the total costs of hunting increase as population size decreases, because more effort is required to harvest low-density populations. Equations 1.4 and 1.7 can be combined to produce a complete bio-economic model of the equilibrium state of the harvested system. A bio-economic equilibrium involves two separate equilibria, biological and economic. The biological equilibrium is that described in section 1.3.1 – the number harvested equals the population growth rate, so the population size is constant over time. The economic equilibrium occurs when the amount of effort expended is constant over time, because the individual harvesters cannot improve their profits by increasing or decreasing their effort. One common assumption is that at equilibrium, profits are zero. This is not as strange an assumption as it appears, due to opportunity costs – zero profits mean that harvesters are earning the same as they would earn in the best-paying alternative profession they could go into. At equilibrium, the population growth rate is zero because the population is stable. So we can write:

Profits are zero: $pqEN - cE = 0$

Population growth is zero: $rN\left(1 - \dfrac{N}{K}\right) - qEN = 0$

Rearrange and substitute: $\Rightarrow N_\infty = \dfrac{c}{pq}, E_\infty = \dfrac{r}{q}\left(1 - \dfrac{c}{pqK}\right)$ \hfill (1.8)

where N_∞ is the population size at equilibrium, and E_∞ is the equilibrium level of effort. The solution E_∞, is shown in Fig. 1.6, along with the solution obtained under the assumption that instead of being zero, profits

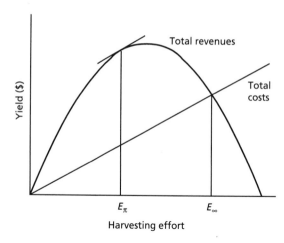

Fig. 1.6 Equilibria for the bio-economic model. Revenues are related parabolically to effort as in Equation 1.6, while costs are linearly related to effort. Profit is the area between the two curves (Revenues–Costs). The equilibria shown are the profit-dissipating equilibrium E_∞ and the profit-maximising level E_Π.

are maximised. This is denoted by E_Π and occurs when the distance between the revenue curve and the cost curve (revenues minus costs) is maximised. An important observation is that the profit-maximising solution occurs at a lower effort level (and so a higher population size) than the zero-profit solution. It also occurs at a population size above the MSY level. Harvesting at the MSY level was the standard management recommendation until recently. We have seen that it is an unsafe strategy biologically, due to the semi-stable nature of the equilibrium, and now realise that it is not likely to be the most profitable strategy either. The inclusion of the costs of harvesting, which increase as stock size decreases, means that maintaining a larger population size is a better strategy for the profit-maximising harvester than harvesting at MSY. Harvesting at MSY is only the profit-maximising strategy if costs are zero or unrelated to stock size. Otherwise, MSY will not be popular either with conservationists or with commercial harvesters, however convenient it might be mathematically.

The zero-profit equilibrium E_∞ is sustainable, in as much as it is a stable equilibrium if the environment is deterministic and the stock–production relationship is smooth. However, harvesting at the zero-profit level is unsatisfactory both for biological and economic reasons. The resource is biologically over-harvested if the equilibrium point is on the descending portion of the revenues curve, which happens if harvesting costs are sufficiently low. The population might be in danger of sudden collapse from stochastic events, and in any case the productivity of the resource will be low. Also, it would be more economically efficient, both for the harvesters and for society, if the profits from harvesting were maximised, rather than being dissipated. When benefits from the resource are being lost, the resource is considered economically over-harvested.

The key determinant of whether harvesters maximise profits or dissipate them is the *market structure* of the industry. One extreme of market structure is when a sole owner has full control of the resource and can set the harvesting level – this is called a *monopoly*. In this case, the producer can charge any price, knowing that the amount of the good that consumers will buy depends on the price set. The owner could be a private individual, a community or the state; for example, the state was a monopoly exporter of game products in the Soviet Union (see Chapter 14). A likely aim of a sole owner is to maximise profits, and thus to harvest at the profit-maximising equilibrium E_Π. We are making important assumptions here, particularly that prices are fixed at a constant value and that there is no time dimension – both will be discussed later. The other extreme is that the resource is *open-access*; anyone can harvest it. This leads to the other extreme of harvesting at E_∞, because as long as there are profits to be made, new harvesters enter the industry. If losses are being made, some harvesters leave the industry. The equilibrium number of harvesters, and so the equilibrium effort level, is when profits are exactly zero. Open-access harvesting is also known as harvesting under conditions of *perfect competition*. The price in an open-access market is generally lower than the price in an equivalent monopoly market. After the break-up of the Soviet Union, a competitive market in exports of game products has arisen there, causing concern for the populations of several game species (see Chapter 14).

Quite specific conditions must be fulfilled if an industry is to operate at the extremes of either monopoly or perfect competition (Begg *et al.* 1984). Perfect competition, in particular, is rarely found in reality. In order for it to exist, individual buyers and sellers must have no influence at all on the market price for the good. Then an individual can just assume that the price of the good is independent of how much of it they personally buy or sell. This happens when there are a large number of individual buyers and sellers, each selling an identical product. The classical example of perfect competition is a large fruit and vegetable market with a lot of individual stall-holders selling the same type of produce. If one stall had higher prices than the others, they would not sell anything at all. Any lowering of prices would be impossible because a lower price would not cover the sellers' costs. Many cases of small-scale harvesting probably fit this scenario fairly well, but global markets in products such as mahogany, discussed by Gullison (see Chapter 6), are also largely competitive. Monopoly markets are found wherever there are barriers to entry into resource harvesting, which may be caused by the type of resource, or by regulation and legislation. For example, the area where the *Loligo* squid population is found for most of the year is owned by the Falkland Islands government, which can charge as much as it likes for licences to fish there (see Chapter 8). Most markets are somewhere between these two extremes of monopoly and open access, although they usually resemble one or the other more closely. In particular, a competitive market can develop even when there are only two producers present, if they are engaged in a price war which drives profits down to zero. One important intermediate type of market structure is *oligopoly*, where there are only a few producers (Box 1.1). In an oligopoly, the price of the good is determined both by an individual's actions and by the actions of all the other producers in the market. Oligopoly leads to a constant tension between collusion and competition. Collusion allows the producers to put their prices up and share monopoly profits, but competion occurs for the largest possible share of those monopoly profits. Oligopoly is sometimes found in resource harvesting, for example when several countries are harvesting the same stock of fish. Whether oligopoly leads to a competitive market with zero profits, or to a monopoly-type market with firms sharing the profits, depends on the situation.

1.5 Harvesting over time

The bio-economic models described so far give a static result rather than a dynamic equilibrium. In other words, there is no mention of time. In particular, no account is taken of how long it might take to reach equilibrium. Time is important, because resource owners trade off present and future revenues. For example, if a manager took control of a population that had been harvested down to the open-access equilibrium, the first task might be to reduce effort levels from the profit-dissipating level E_∞ to the profit-maximising level E_Π. This would lead to an immediate drop in profits, followed by a gradual increase to the new higher level as the stock recovered. The question for a manager is then how gradual would the increase in profits be, and would it be worth waiting for it? The concept of *time preference* becomes important here – profit tomorrow is assumed to be worth less to an individual than profit today. This concept is discussed further in section 3.2.3. Interest rates are a product of time preference. Banks give savers interest on their money to encourage people to save with them. So £100 invested today at 5% interest will pay out £128 in 5 years time. This extra £28 is the incentive to the saver to defer spending the £100 for 5 years; this incentive is needed because the value of an amount of money is less the longer you have to wait to spend it. The usual assumption is that the value of money declines with time according to a negative exponential (Fig. 1.7):

$$PV = Ae^{-\delta t} \tag{1.9}$$

where PV is the present value of some cost or revenue A incurred at a future time t. δ is the *discount rate*, and is equivalent to the interest rate (the rate is called 'interest' when value is increasing over time, 'discount' when value is declining over time).

The harvesting model can be generalised to include time by assuming that both stock size and harvest rate can vary over time. To keep the analysis simple, unit costs and prices are still assumed to be constant. The profits made in each time-step must be *integrated* to find the total profits

Box 1.1 An oligopoly – the ivory cartel of the Southern African states

Fig. B1.1 Ivory store room, Singapore Sunchfong Factory. (Photo by EIA.)

An example of an oligopoly is the cartel in ivory that has been proposed for the ivory trading countries of Southern Africa (Swanson 1989). The cartel was proposed after the international ivory trade was banned in 1989, with the idea of promoting conservation by reducing the number of elephants killed for the ivory trade, and paying for conservation by ploughing the money from the trade back into elephant management. The idea's proponents reasoned that a cartel would allow the governments of member countries to control the trade in ivory, reduce ivory hunting to a sustainable level, and share the monopoly profits from selling ivory legally at a high price. The cartel would have monopoly rights to sell to consumer governments, so as to maximise the ivory price and prevent undercutting by illegal traders. International legislation under the Convention on International Trade in Endangered Species (CITES) could be used to restrict the trade and ensure a sustainable offtake. The cartel would be economically beneficial to its members, but not to other ivory producing countries or to the majority of consumers. Its success would hinge on various factors:
- The political acceptability of the idea to non-members of the cartel.
- The potential profits to be made, which depend on how high a price can be charged for ivory.
- The cartel's ability to resist competition from cheaper illegal ivory.
- The cartel's ability to resist break-up caused by internal divisions (the usual fate of cartels).
- The idea's acceptance by the international community as a valid conservation method.

continued

Box 1.1 *Continued*

In 1997, three Southern African states (Namibia, Botswana, Zimbabwe) gained permission from CITES to sell their ivory to one buyer (Japan). Ivory trading will not start until March 1999, and remains banned between all other CITES countries. These producers are now able to form a cartel, although the cartel's power to influence the ivory price will be severely limited by the fact that there is only one buyer, who has the power to refuse to buy any ivory if the price is too high. Whether or not this ivory trade proves a threat to elephant populations will depend on the extent to which the countries concerned can prevent illegally acquired ivory seeping into the trade, either from within their countries or from elsewhere.

---

that are made over time. Integration is the reverse of differentiation, and can be thought of conceptually as finding the area under a curve (Bostock & Chandler 1981). It is summation in continuous time: in a continuous flow of profits, the distance from the x-axis to the profit curve at any point represents the profits made at that point (Fig. 1.8). To find the total profits made over a period of time, calculate the area under the curve (integrate) between these two points. To find the total profit made in perpetuity, integrate from $t = 0$ (the present) to $t = \infty$. This is written as:

$$PV = \int_{t=0}^{t=\infty} [e^{-\delta t} \Pi_t] dt \tag{1.10}$$

Putting Equation 1.10 into words: the present value of all the harvests from the resource in perpetuity is the integral of the profit made in each future time period, multiplied by the discount factor (so that the further into the future the profits are, the less they are worth).

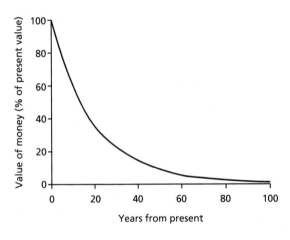

Fig. 1.7 The exponential decline in the value of an amount of money A depending on how far into the future it is received, at a discount rate of 5%.

Fig. 1.8 An illustration of the principle of integration. The curve shows profits over time. At any particular point of time (say time *a*), the profit received is the distance of the curve from the *x*-axis. The profits over the period *a–b* are thus represented by the shaded area. Integration of the profit curve from one point to another finds total profits by calculating the area under the curve.

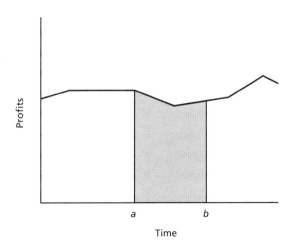

Profit, Π, depends on the stock size *N* and the harvesting effort *E* at time *t*, so that Π_t can be replaced in Equation 1.10 by Equation 1.7, but with *N* and *E* at a particular time shown as N_t and E_t. The bio-economic model is then generalised so that profits in perpetuity are maximised, given that stock size and harvest rates are always greater than or equal to zero:

$$\max PV = \int_{t=0}^{t=\infty} [e^{-\delta t}(pqE_t N_t - cE_t)]dt \tag{1.11}$$

given that: $N_t \geqslant 0$, $H_t \geqslant 0$

With some minor rearrangement, this becomes a problem in *optimal control theory*. The equation is solved to find which harvest rate H_t maximises *PV* given these constraints (Burghes & Graham 1986; Clark 1990). In the case of the simple harvesting model presented here, the solution gives an implicit value for the equilibrium steady-state population size *N**:

$$F'(N^*) - \frac{c'(N^*)F(N^*)}{p - c(N^*)} = \delta \tag{1.12}$$

The superscript ′ denotes that a function has been differentiated with respect to *N*, and the parentheses in *F(N)* and *c(N)* that the population growth rate *F* and harvesting costs *c* are functions of *N*. Equation 1.12 is hard to interpret in this form, but it can be rearranged as:

$$\int_0^\infty [e^{-\delta t}\Pi_t'(N^*)]dt = p - c(N^*) \tag{1.13}$$

Equation 1.13 means that equilibrium is reached when the amount of money gained from harvesting one extra individual (the right-hand side) is just offset by the future losses due to that individual not being available to produce offspring for future harvest (the left-hand side). In economic language, at equilibrium the marginal immediate gain from harvesting equals the present value of the marginal future loss. The concept of the amount gained (or lost) from increasing or decreasing production by a tiny amount is important in economics. The expression for changes in profits from tiny changes in production is *marginal* gains or losses. Finding a marginal gain or loss is equivalent to differentiating. This generalisation to a dynamic equilibrium has the same logic as the static profit-maximising equilibrium – harvest until the costs of taking one more individual (caused by the reduction of stock size making future harvesting more expensive) are equivalent to the revenues received for that individual.

Further analysis of Equation 1.12 shows that another way to describe market structure is to incorporate time preference (Clark 1990). Under perfect competition, producers place a value of zero on future revenues, which is equivalent to having an infinite discount rate (Box 1.2). If the resource is open to anyone who wishes to exploit it, then the harvesters look only to short-term gains and have no interest in increasing future productivity, because individuals that one harvester leaves to replenish future stocks will simply be harvested by another. Thus, the effort level is E_∞, the subscript denoting the infinite discount rate. Monopolists are able to profit-maximise over time, because they are sure that the increased future yields will be there for them and will not be taken by someone else. Thus, a monopoly harvester should preserve a larger stock size than an open-access harvesting system. How much larger the stock size is depends on the monopolist's discount rate. A person with a zero discount rate would harvest at E_Π. E_Π is thus the most conservative effort level, which is only adopted by harvesters who value the future equally with the present. This may seem reasonable at first glance, at least from a conservationist's point of view, but in fact it is an extreme assumption. For example, a payment of £100 now in order to receive £101 in 20 years time would be worth making, if the payer had a zero discount rate. With a discount rate between zero and infinity (the usual case for a monopoly harvester), the optimal stock size depends on the discount rate – the higher the discount rate, the lower the optimal stock size. An important point raised by Clark (1973) is that it can be in the long-term interests of the sole owner of a harvested population to hunt it to extinction. This only happens under rather special conditions, the main one being that the growth rate of the population is always below the prevailing interest rate in the economy. This could be a problem for very slow-growing resources that potentially will never yield a competitive return on an owner's investment. The rapid depletion of the population is then an economically rational act. Gullison (see Chapter 6) demonstrates that the rate of increase in a mahogany tree's value is so much less than the prevailing interest rate in the countries where it is found that a sustainable harvest is unlikely to be economically worth-

Box 1.2 The Ache – an example of a very high
discount rate

Fig. B1.2 An Ache woman
and child. (Photo by Kim
Hill.)

The Ache are an indigenous group living in Paraguay, who used to live as
foragers and were then settled on reservations in the 1970s, where they
barely participated in the cash economy. In late 1988, they received legal
title to their reservations. Rather than encouraging them to conserve their
resources, this led them into a spree of selling off all the valuable timber
on their land. By mid-1989 they had made large amounts of cash from
this and started to run their own store. People had new clothes and
houses, and spent large sums on food, gambling and soft drinks. By mid-
1990, the boom had ended, and the Ache had returned to their previous
lifestyles. The store had closed down, and no one made any major cash
purchases. In 1992, an assessment of household assets showed that the
Ache's net worth was lower than at the previous survey in 1979. They had
not accumulated any property since 1979 and continued to be among the
poorest people in Paraguay. The 1988 boom might never have happened
(Hill & Hurtado 1996).

continued

Box 1.2 *Continued*

This sequence of events suggests that the Ache placed a near-zero value on future revenues, and thus that their discount rate was very high. They were using their resources for short-term gain only. It is often assumed that giving people ownership of their resources automatically leads them to value the future productivity of the resources highly, and thus to conserve them. Clearly this is not always the case.

---

while. This result holds irrespective of whether the market is open-access or monopoly, and probably applies to many other tree species as well.

1.6 Supply and demand

People will only produce a particular good in response to a demand for that good. This is as true for the hunter who uses time and effort as inputs in order to supply dinner for the family as it is for a firm manufacturing sausages. But unlike ordinary economic goods, the size of a harvested population is affected (through harvesting effort) by the amount of the good that harvesters wish to supply. Two functions need to be estimated in order to estimate the amount of a good supplied to a market. One is how the costs of supply vary with the amount produced. The other is how the quantity demanded varies with the price of the good. Together, these functions determine the equilibrium price of the good. The shapes of these functions have implications for how the market responds to regulations that aim to reduce the supply of the good, such as cartels. Thus, the shapes of a harvested good's supply and demand functions have important implications for policies that aim to promote its conservation and sustainable use.

The *production function* defines the relationship between the amount of the good produced and the costs involved in producing it. In the simple bio-economic model, the production function was assumed to be linear. The total cost of producing the good increased linearly as the amount of effort put into harvesting it increased (Equation 1.7). This relationship was simplistic in that it assumed *constant marginal costs* – that the cost per unit of effort was constant however much effort was put in. This is unlikely to be true – for example, there are often economies of scale. As the scale of a harvesting operation increases, it is possible to introduce cost-saving technologies like larger boats or vehicles. Equation 1.7 also assumed no *fixed costs*. This is unrealistic because some costs, such as the maintenance of harvesting equipment, or the salaries of managers, need to be paid regardless of the amount of harvesting that goes on. The model also assumed a *constant price* throughout. This is a fair short-term assumption for open-access harvesters supplying only a small quantity of a good to a large market. However, if a significant proportion of the market is being described, or if prices change through time, then the assumption of fixed prices is not tenable, irrespective of whether the harvesting is done in an open-access system or by a monopolist.

The market model of supply and demand describes the relationship between costs and prices and the quantity of the good supplied and demanded. It can be used to derive the equilibrium price and quantity of the good supplied. It is a static equilibrium analysis, like the previous simple bio-economic model, meaning that the long-term equilibrium is being described, not the dynamics of getting there. The basic model is shown in Fig. 1.9. Demand is assumed to slope downwards with the quantity consumed. If a good is very hard to obtain, consumers are prepared to pay a very high price per unit of it; if the market is swamped with a good, all the consumers have enough and are not prepared to pay much for the good. Conversely, the supply curve usually slopes upwards because the higher the price, the more of the good the supplier wishes to sell. The equilibrium is reached in this way: if the price is too high, the suppliers make more of the good than is wanted. In order to clear excess stock, they cut their prices, which has the dual effect of increasing the amount of the good that consumers demand and decreasing the amount that suppliers wish to sell. Similarly, if the prices are too low, there is excess demand that is unsatisfied. Shops are always running out of the good, so the suppliers realise that the price could be increased. This price rise both cuts the amount of the good the consumers want and increases the amount suppliers want to supply, and equilibrium is reached.

The total revenues obtained by the seller are given by the unit price paid for the good multiplied by the quantity of the good demanded (the area $P_E Q_E$). *Consumer surplus* is the amount of extra satisfaction that consumers receive from the goods that they have purchased, over and above the amount they have paid for. This arises because sellers can only charge one

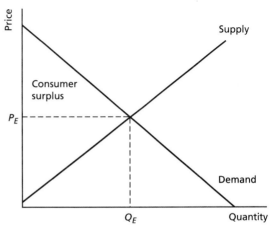

Fig. 1.9 The market model of supply and demand. The supply and demand curves intersect at the market equilibrium. The equilibrium price, P_E, and quantity demanded, Q_E, are shown.

price, however much different individuals would be prepared to pay for the good. The total economic value of a good at market equilibrium is the area under the demand curve from $Q = 0$ to $Q = Q_E$. This is divided up into the revenues that cover the costs of production (the area under the supply curve) and the economic gain from the market. The economic gain is divided into the profits to the manufacturer (the area between the supply curve and the line P_E) and the consumer surplus (the area between the line P_E and the demand curve). The market equilibrium is called the 'socially optimal' level for the market, because it is the point at which the economic gain from the market (the area between the supply and the demand curves) is maximised. In a competitive market, the equilibrium is reached automatically through individuals pursuing their own best interests. In a non-competitive market, the equilibrium is not reached automatically. A sole supplier of a good has the power to set the price of the good. If the sole supplier is a private monopolist, the price is set at whichever level maximises the firm's private profits. This is not usually at the socially optimal level where the supply and demand curves cross, but at a higher price, leaving demand unsatisfied.

The demand for a good is not just determined by its price, despite the emphasis on price in the market model. Price is highlighted in this model because it both determines the quantity supplied and is determined by the quantity demanded, while the other variables are externally determined in the wider economy. In the model, changes in these external variables shift the supply and demand curves to the left or right, thus affecting the equilibrium price and quantity. The major external variables affecting demand are the price of related goods, consumer income and consumer tastes. Related goods can be substitutes or complements – for example, the prices of rhino horn, water buffalo horn and saiga antelope horn on the traditional Chinese medicine market are probably linked because saiga antelope horn and water buffalo horn are being promoted as possible substitutes for rhino horn. However, saiga antelope horn is generally used in traditional medicines as a complement to rhino horn. Consumer income can be a major influence on demand for a good – in fact it seems that in Japan, consumer income was the only significant determinant of demand for ivory over the period 1950–85, while the price of ivory had no discernible effect (Milner-Gulland 1993). This information is important not only for those planning ivory cartels but also in predicting possible future increases in demand for ivory as other Asian countries become richer. Changes in consumer tastes can also be important, as was demonstrated by the virtual shutting down of the markets for furs in the UK and ivory in the EU and USA. This was largely achieved by animal rights campaigners, who managed to change social perceptions enough for the majority of people to regard buying these items as unacceptable.

Like the demand curve, the supply curve is also affected by variables other than price. These variables include technology, input costs and regulations. Improvements in technology or reduction in input costs shift the supply curve to the right because they make it possible to supply more of a good at a given price. This can be crucial to the sustainability of harvesting because the model predicts that technological improvements such as better, more modern weapons will increase the equilibrium quantity harvested. Finally, regulations affect the supply curve, preventing producers from adopting otherwise cheaper technologies, and thus move the supply curve to the left, leading to higher equilibrium prices and lower quantities supplied. Regulations requiring hunters to use traditional weapons would have this effect. MacKinnon (see Chapter 5) discusses programmes in Irian Jaya, Indonesia, where local hunters have collaborated with a conservation organisation to develop regulations that limit hunting to traditional weapons. This allows the locals to continue to hunt in their usual way, but makes it illegal for outsiders to come into the area and hunt game using more efficient technology.

1.6.1 Elasticity

Elasticity is a parameter which describes the shapes of supply and demand curves, and their responsiveness to changes in key variables. For example, the price elasticity of demand measures the responsiveness of demand to changes in price. It is defined as the percentage change in the quantity of a good demanded with a 1% change in price: a 1% fall in price that leads to a 2% increase in the quantity demanded has an elasticity of −2. Elasticity is related to the slope of a curve, but unlike the slope, it is dimensionless. This allows elasticities to be compared between goods with different units, and between different-sized markets. Elasticity can be calculated for both supply and demand curves. Norton-Griffiths (see Chapter 11) shows that Kenyan pastoralists increase their supply of livestock to the market by 0.7% for every 1% increase in meat prices. However, we shall use the price elasticity of demand as our example, because it is the most commonly used of the elasticities. If the demand for a good is 'elastic' with respect to price, this means that the quantity demanded is strongly affected by price; the magnitude of the elasticity is greater than one. If demand for a good is inelastic, the quantity demanded is not greatly affected by price, and the magnitude of the elasticity is less than one (Fig. 1.10).

Although slope and elasticity are related, a constant slope does not lead to a constant elasticity. This is because if demand is linear, and so has a constant slope, a change in price of 1 unit leads to the same change in quantity demanded, whatever the original price and quantity were. Because the change in quantity demanded with a change in price is not dependent on the quantity demanded, the percentage change is not constant. For example, if a change in price of 1 unit always leads to a change in quantity

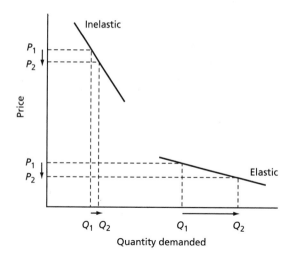

Fig. 1.10 Inelastic and elastic demand curves. If demand is inelastic, a small change in price leads to a small change in quantity demanded, while if demand is elastic, a small change in price leads to a large change in quantity demanded.

of 5 units, the line has a constant slope, but the percentage change is 50% if the quantity is changing from 10 to 15, and 5% if it is changing from 100 to 105. Elasticity is a measure of percentage change, not actual change, so elasticity is not constant when demand is linear. If the demand curve is linear, the demand is elastic at low quantities demanded (the same change in price leading to large percentage changes in quantity), and inelastic at high quantities demanded.

Elasticity is most conveniently calculated using the formula below, for small ΔP:

$$\varepsilon_P \approx -\frac{|\Delta Q|/Q}{\Delta P/P} \tag{1.14}$$

where Δ means a change in a variable, $|\ |$ means the magnitude of a variable (regardless of sign), P is the price, Q is the quantity demanded, and ε_P is the own price elasticity of demand. Similar equations can be derived for any other elasticity. For example, the cross-price elasticity of demand for a good i with respect to changes in the price of good j is:

$$\varepsilon_{ij} \approx \frac{|\Delta Q_i|/Q_i}{\Delta P_j/P_j}$$

while the income (I) elasticity of demand is:

$$\varepsilon_I \approx \frac{|\Delta Q|/Q}{\Delta I/I}$$

The cross-price elasticity of demand measures how strongly the demand for one good is affected by changes in the price of another good. If the goods are substitutes, such as beef and antelope meat, an increase in the price of one good should increase the quantity of the other good demanded, and ε_{ij} is positive. If the goods are complements, like fish hooks and fishing line, an increase in the price of one good should decrease the quantity of the other good demanded, and ε_{ij} is negative. The income elasticity of demand measures how a change in income affects the amount of a good demanded. The income elasticity of most goods is greater than zero, so the amount demanded increases as people's income increases. These are called 'normal goods'. 'Luxury goods' have $\varepsilon_I > 1$, so an increase in income leads to a more than proportional increase in the amount of the good demanded. Ivory and foreign travel are examples of luxury goods. If a good is particularly associated with poverty, like inferior cuts of meat, then an increase in income leads to a decrease in the amount of the good demanded ($\varepsilon_I < 0$). These are called 'inferior goods'.

The own price elasticity of demand, ε_P, is largely determined by how easy it is to substitute another good for the good in question. If there are a lot of similar goods, demand is elastic, because a small price increase causes consumers to switch to the substitute goods. Tastes and social customs are also important. For example, if a product is regarded as frivolous, then consumers have elastic demand for it and buy much less of it if the price rises. If the product is seen as a necessity, people continue to demand it despite price rises – it has inelastic demand. Long-run demand tends to be more elastic than short-run demand, because in the long run, consumers can change their behaviour and substitutes can be found, while in the short run, continuing to use a good despite price rises might be unavoidable.

This discussion of elasticity is particularly relevant to tourism. Consider the effects of charging entrance fees for national parks. Tourism is generally a luxury good, with demand that is elastic to price. Raising prices for a whole tour package is likely to put customers off. However, within the package, the effect of raising prices for entrance to a particular park depends on the availability of substitute experiences in the vicinity. If there are few alternatives available, demand for visits to the park may be inelastic, and raising entrance fees may increase the revenues from the park. However, in the long run, and especially if an international clientele is targeted, competition will be strong and demand will become increasingly elastic. The consumers will find substitute goods – perhaps tour operators will rearrange their itinerary to visit a cheaper park. In order to retain the park's market niche in the long term, one strategy is to find a specialist niche, without any obvious substitutes. If the park offers a unique experience, or is in a particularly convenient location, consumers will still demand it despite higher prices. This is the strategy that was used for the mountain gorilla projects in Central Africa, and which brought in substantial revenues to the Rwandan, Ugandan and Zairean governments (see Chapter 12). However, the highly elastic nature of demand for tourism also means that it is a volatile market, subject to unpredictable influences. Price *et al.* (see Chapter 9) show how tourist revenues in the Caribbean islands were hit by a global slump in tourism following the Gulf War in 1991, and again by hurricane damage in 1995; Butynski & Kalina (see Chapter 12) describe the desperate state of the mountain gorilla tourism projects caused by the wars in Rwanda and Zaire.

1.6.2 Supply and demand for renewable resources

The theory of supply and demand in fisheries was discussed by Clark (1990), for the simple bio-economic model presented in section 1.3. Introducing supply and demand into the model allows the assumption of a constant price to be relaxed. The supply curve expresses the quantity supplied in terms of the price: $Q = f(P)$. The quantity of a harvested resource that is supplied is simply the yield H. The supply curve in Equation 1.15 shows how much yield is supplied from an open-access resource at equilibrium at a given price:

Profits are zero: $pqEN - cE = 0$

Population growth is zero: $rN\left(1 - \dfrac{N}{K}\right) = H$

Rearrange and substitute: $\Rightarrow H = \dfrac{rc}{pq}\left(1 - \dfrac{c}{pqK}\right)$ (1.15)

Note that this method of constructing a supply curve is not standard, but it is valid for this model because the costs of harvesting are assumed to be linearly related to yield. Clark (1990) discusses the effect of non-linear costs further. Figure 1.11 shows the supply curve, which has the unusual property that it is backward-bending at high prices.

The effect of the backward-bending supply curve on the stability of the equilibrium population size and yield depends on the elasticity of the demand curve. For the elastic demand curve D_1, the equilibrium is reached at point A (Fig. 1.12a). If consumer incomes increased, or tastes changed, causing the demand curve to shift up to D_2, the increase in price would cause the effort to increase, and so the yield to decrease. However, the equilibrium would still be stable. If the demand curve is inelastic (Fig. 1.12b), there is the potential for serious instability in the market. A relatively small shift in the demand curve, from D_1 to D_3, could change a sustainably exploited resource being harvested below the MSY level (point A), to one that was severely over-harvested (point E). Demand curve D_2 has three possible equilibrium points – the stable equilibria B and D, and the unstable equilibrium C. C is unstable because a small shift in effort that moves the harvesters down the supply curve leads to supply exceeding demand, and so to a reduction in effort until point B is reached. A small shift up the supply curve leads to demand exceeding supply and a corresponding increase in effort to point D. Figure 1.12(b) is of concern because it suggests that, under inelastic demand for a harvested good, a

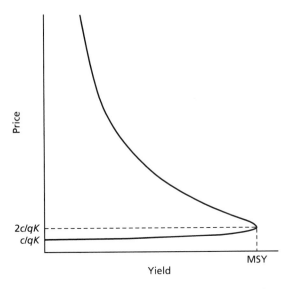

Fig. 1.11 The supply curve for a harvested renewable resource, under open-access harvesting. The curve crosses the y-axis when the quantity supplied is zero (the population is at carrying capacity). When price is so low that $p = \dfrac{c}{qK}$, the quantity inside the bracket in Equation 1.15 is zero, and it is not economically worthwhile to harvest. As price increases, so does the quantity supplied, until MSY is reached. At this point the yield is maximised: $\dfrac{\mathrm{d}H}{\mathrm{d}p} = 0$, so by differentiation, $p = \dfrac{2c}{qK}$. As p increases to infinity the right-hand side of Equation 1.15 gets smaller and smaller. The curve bends backwards as price increases further because higher prices lead to increased harvesting effort (Equation 1.8). As effort increases past E_{MSY}, the population declines, and so does the yield (Fig. 1.5).

small shift in demand from D_1 to D_3 would lead to a rapid expansion of effort as the harvesters moved up the supply curve from point A to point E. Yield would increase temporarily and then decline to the new equilibrium at a higher price, low population size and low sustainable yield. This description of the dynamic behaviour of the market accords well with observations of fishery collapses following increases in demand (Glantz 1986).

So far, the discussion has focused on the open-access supply curve. A supply curve can be generated in a similar way for a sole owner (Clark 1990). The shape of the supply curve depends on the discount rate (Fig. 1.13). If $\delta = \infty$ then the same curve is generated as in the open-access case because future yields are not considered by the harvester. If $\delta = 0$, then the supply curve is an ordinary upwards-sloping curve, because however high the price, it is not economically optimal to harvest more than MSY each year if future yields are just as valuable as present yields. An upwards-sloping curve does not generate any instabilities. In between $\delta = 0$ and

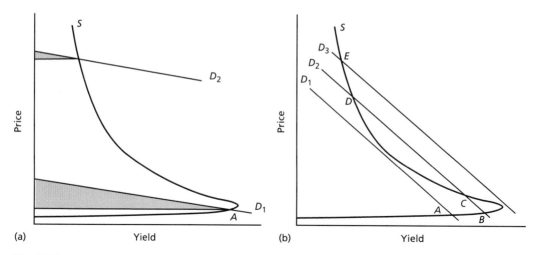

(a)   Yield   (b)   Yield

Fig. 1.12 (a) Supply and demand for an open-access resource, under elastic demand. The shaded areas are the consumer surpluses generated by each demand curve. (b) Supply and demand for an open-access resource, under inelastic demand. Three demand curves, D_1–D_3, are shown, and their equilibria A–E.

$\delta = \infty$, the degree of backwards bend depends on the value of δ, but is less steep than in the open-access case. Therefore Fig. 1.13 shows that, in general, sole owners are less likely to be affected by market instability than markets based on open-access resources. The backward-bending supply curve is potentially important because it shows how unstable markets can be for harvested populations, but, as yet, it has been rather overlooked in practical resource management.

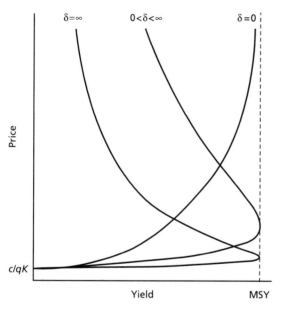

Fig. 1.13 The supply curve for a sole owner harvesting a renewable resource. The shape of the supply curve depends on the discount rate: at a discount rate of ∞, the open-access supply curve is generated; at a discount rate of zero, the supply curve approaches MSY asymptotically. In between, the degree of backwards bend increases as the discount rate increases.

1.7 Summary

• The key assumption behind sustainable harvesting models is that populations grow and reproduce themselves. Population growth is density dependent, so lowering population size by harvesting causes a compensatory increase in population growth rate. These extra individuals can be harvested sustainably.

• The most commonly used function for modelling population growth is the logistic equation. The logistic equation has drawbacks, particularly the difficulty of estimating its two parameters, the carrying capacity and the intrinsic rate of population increase, but it is still the basis for most population models.

• Starting to harvest a previously unexploited population always leads to a reduction in population size, followed either by stabilisation at a lower equilibrium population size, or by extinction. The logistic equation assumes smooth negative feedback between the population size and the population growth rate. Harvesting a constant number of individuals above MSY leads to rapid extinction. Harvesting at MSY is dangerous because MSY is a semi-stable equilibrium. The further below MSY the harvest level is set, the lower the yield and the safer the population is. Harvesting at a rate proportional to population size is safer than harvesting a constant number of individuals because all equilibria are stable.

• The yield from harvesting is usually modelled as being proportional both to the population size and to the effort expended by harvesters. The catch per unit effort can be taken as a measure of the harvested stock's population size, but it is often misleading.

• Harvesting strategies available to managers include harvesting a constant number of individuals, harvesting with a constant effort, harvesting a constant proportion of the population, and harvesting the population down to a threshold population size. Each strategy has advantages and disadvantages. The best strategy in a particular case depends on the manager's objectives. However, objectives often conflict and compromises are required.

• The profitability of harvesting must be considered if the level of harvesting is to be correctly predicted. Monopoly harvesters generally aim to maximise profits, while if harvesting is open access, profits are dissipated. Profit-maximisation generally leads to a higher population size than profit-dissipation, because costs increase as the harvested population declines. Other types of market structure include oligopolies, in which a tension exists between competition and co-operation to obtain monopoly profits.

• Monopoly harvesters maximise profits over time, while open-access harvesters maximise short-term gains. Monopolists trade off current and future revenues, discounting future revenues, so that revenues are worth less the further in the future they are received. The higher the monopolist's discount rate, the lower is the equilibrium population size of the harvested stock, because the present harvest is more valuable than the increased future harvests that could be obtained if current harvest rates were reduced. Open-access harvesting leads to the same population size as a monopolist harvesting at a discount rate of infinity.

• The amount of a good that a harvester wishes to supply depends on consumer demand for the good, because revenues depend on the price received for the good. The demand function describes the price that consumers are prepared to pay at each quantity of the good supplied, and generally slopes downwards. The supply function is the cost of supplying the good at each quantity supplied, and it generally slopes upwards. The equilibrium price and quantity supplied is where the two functions cross. Other factors, such as tastes, incomes and technology, determine the positions of the curves. The elasticity of the curves determines how responsive supply and demand are to changes in price and quantity of the good.

• Renewable resources are unusual because they have backwards-bending supply curves at high prices. Combined with inelastic demand, this can lead to instability in the harvest rate, and possible population collapses, particularly if there is open-access harvesting. The lower the discount rate of a monopoly harvester, the less steeply the curve bends back, and the less instability there is in the market.

Chapter 2: Harvesting and Ecological Realities

The aim of developing a harvesting model is to capture the essential aspects of the biology and economics of a system, such that the behaviour of the system over time can be understood. The sustainability of use can then be correctly assessed. The basic bio-economic model developed in Chapter 1 has given many qualitative insights into the dynamics of harvested populations, but it is inadequate for making quantitative predictions in most real-life situations. In this chapter, we discuss the main modifications that are used to obtain a more accurate picture of the sustainability of resource use, and under which circumstances each of these modifications is needed. There is a trade-off between the simplicity of a model and its realism. A realistic model, that includes all the key factors acting on the system, may give useful results that have relevance to the real world. On the other hand, models need to be simple, because:

- A simple model is more tractable; if a model can be solved analytically, the dynamics of the system can be better described and understood.
- In a simple model, the reasons for results are transparent, so the model is fulfilling its purpose of providing a testable mechanistic description of the process under study.
- The more parameters there are which need to be estimated, the more error there is in the model, and the less trustworthy the results.

The model that was used in Chapter 1 is of the simplest type: a lumped, continuous, deterministic model for a single species. A *lumped* model is one in which the whole population is treated as a single variable N, which effectively assumes that all individuals are identical. *Structured* models, on the other hand, divide the population into classes, commonly by age, or by sex, life-stage, spatial location or size class, depending on which are the most relevant differences between individuals. Models can be in either *continuous* or *discrete* time. In continuous time, populations are followed through time, while in discrete time, population size is calculated at intervals, for example, every year. Continuous time is usually more tractable analytically, using calculus, while discrete time can be easier to use in computer simulations. Populations with a defined breeding season, clear periods of higher mortality and a hunting season are well suited to discrete time modelling. Finally, *deterministic* models are those in which all parameters have given values, while in *stochastic* models, one or more parameters are represented by probability distributions.

2.1 Adding ecological realism to the single species model

2.1.1 Non-linear density dependence

The simple bio-economic model is linear in most parameters. Introducing non-linearities can give insights into the dynamics of real populations. One non-linearity which is especially emphasised in the natural resource management literature (e.g. Clark 1990) is *depensation*, also known as the Allee effect. In the standard logistic model, the proportional growth rate (the differential of the stock–production curve, Equation 1.3) is a decreasing function of population size, so that the smaller a population is, the faster it grows. The simple harvesting model presented in Chapter 1 suggested that taking a constant proportion of the population was a safe and stable harvesting strategy. This model had the form:

$$\frac{dN}{dt} = F(N) - qEN \tag{2.1}$$

where N is population size, $F(N)$ is the population growth rate, q is the catchability coefficient, E is the effort, and qEN is the number of individuals harvested, H. In Chapter 1, it was assumed that $F(N)$ was logistic growth. However, if depensation occurs, $F(N)$ is no longer logistic. Under depensation the population growth rate increases with population size, so that smaller populations grow more slowly than larger ones. Depensation is usually assumed to happen only when the population is small, and produces curves such as Fig. 2.1. Critical depensation (Fig. 2.1b) is particularly serious because once the population has gone below the population size N_E, the growth rate is negative, and the population is doomed to extinction, whether or not harvesting has ceased. N_E is called the minimum viable population size [not to be confused with the genetically based minimum viable population size of Soulé (1987)].

Depensation has serious consequences for the safety of the strategy of harvesting a proportion of the population each year. The effect is to make

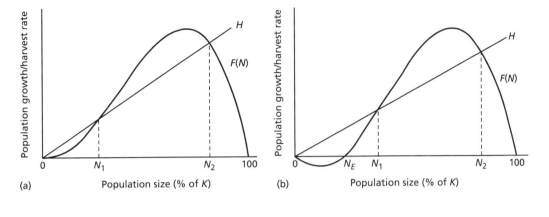

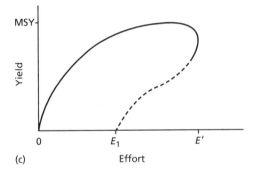

Fig. 2.1 The effect of depensation (the Allee effect) on the parabolic relationship between the population size and the rate of population increase. (a) Non-critical depensation. Harvesting a proportion of the population each year produces two equilibria, N_1 and N_2. (b) Critical depensation. If the population falls below N_E then the population growth rate becomes negative and the population is doomed to extinction. (c) The yield–effort curve for non-critical depensation. If the effort level increases above E', yield declines rapidly and effort must be reduced to E_1 in order to allow the population to recover above the unstable equilibrium level N_1.

the outcome similar to that of harvesting a constant number of individuals with two equilibria; one stable, the other unstable (Fig. 2.1a; section 1.3). The yield–effort curve displays hysteresis – there is a sudden discontinuity in yield as effort is increased, at the point E'. Once effort levels have exceeded this point of discontinuity, it is not enough simply to reduce effort levels below E' to get back to the previous yields (Fig. 2.1c). Yields will only increase if effort levels are reduced further until the population exceeds the unstable equilibrium size N_1. The advantage of harvesting a constant proportion, discussed in section 1.3.2, is that gradual increases or decreases in effort levels have gradual effects on the population size. This gradual change no longer happens if the population exhibits depensation. The curves for critical depensation are even more extreme, so that once the population is below the critical population size N_E, the population declines to zero, regardless of by how much effort is subsequently reduced.

To assess the importance of depensation in the real world, we need to know how common it is, and also the size at which a population is likely to experience it. It is also useful to know what biological mechanisms are behind depensatory population dynamics, so that we can predict which species are likely to suffer from it. Myers *et al.* (1995) carried out statistical analyses of 128 fish stocks for which 15 years or more of data were available, and found only three that showed statistically significant evidence for depensation (two Pacific pink salmon stocks and Icelandic spring-spawning herring). All three had declined to extremely low population sizes, due to climatic conditions, over-fishing, or habitat loss. There were two other stocks for which depensation could not be entirely ruled out, which had also not recovered from very low population sizes. It seems that depensation is only found in commercial fish species at very low population sizes. However, Myers *et al.* (1995) showed that there were 26 stocks that had enough observations at low population sizes to have shown depensation if it was there, and most showed large increases in survival at low population sizes. This is the only large-scale study of the prevalence of depensation that has been carried out to date, and it seems to show that, at least in fish, depensation is not a widespread problem.

Although depensation is probably uncommon, it is worth taking seriously because it could compromise the ability of particular species to recover from low numbers. There have been a few candidates suggested for which depensation is of conservation importance, including the blue whale and fin whale. Both these species failed to increase in numbers as fast as would have been expected under compensatory population growth when hunting ceased (Clark 1990). The underlying biological mechanism suggested was difficulty in finding mates at very small population sizes. The African elephant is another species which has been suggested as a candidate for depensation. Here, depensation is specifically linked to selective hunting for adult males (which have the largest, most valuable trophies). There have been observations of barrenness among females in heavily hunted populations (Poole 1989). Elephants have relatively short breeding periods and, despite their ability to communicate over long ranges, a lack of mating opportunities for females could have led to these observations. Poole (1989) also identifies the social disruption and stress caused by hunting as a possible reason for the lack of reproductive success. Social disruption caused by hunting may not be confined to small populations. These rather complex behavioural effects of hunting, particularly those associated with distorted sex ratios, are better not lumped into the simple population-level factor of depensation, but included more explicitly into a harvesting model. One species for which the circumstantial evidence seems to point to critical depensation playing a role is the passenger pigeon. This species declined to extinction very rapidly from quite large numbers. The original cause of the decline was probably habitat destruction, but it seems likely that a critical population size was required in order for the species to find masting trees reliably. Mast is a patchy and unpredictable food source, and once below the critical size, the population could not find food reliably enough to recover (Bucher 1992). Depensation does not have to lead to extinction, or be found only at small population sizes, however; guillemots show a depensatory relationship between breeding success and density, due to predator swamping (Fig. 2.2).

Non-linearities in the *stock–production relationship* have been explored by Fowler (1981, 1984), who uncovered a relationship between the intrinsic rate of increase per generation and R, the population size at which productivity is greatest divided by the carrying capacity (Fig. 2.3). This relationship is very useful because it predicts the degree of non-linearity that is likely to be found in the dynamics of a particular species. In the simple logistic model, $R = 0.5$; the maximum productivity of the population is at half the carrying capacity. Figure 2.3 shows that this is not a general rule, but that R varies widely between about 0.2 and 0.8. The implications of an R other than 0.5 are shown in Fig. 2.4. If the maximum is below 0.5, then the per-capita growth rate (Fig. 2.4b) shows that the majority of a species' density-dependent response happens at low population sizes. At first, the curve is steep; then at higher densities it is relatively flat (little effect of density on growth rate). Conversely, a species with high R responds relatively little to density at low population sizes, but has a

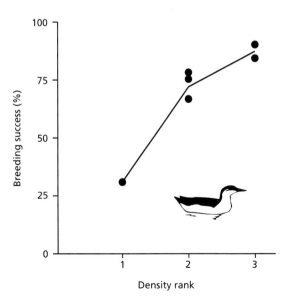

Fig. 2.2 Depensation demonstrated for guillemots. As population density increases, breeding success increases. (From Begon *et al.* 1996b.)

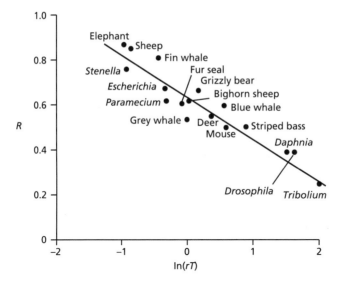

Fig. 2.3 The relationship between R (the population size giving the maximum net productivity, divided by K) and the natural log of the intrinsic rate of increase multiplied by generation time, $\ln(rT)$, for 17 species. A population's generation time is the average age at which reproduction occurs. (From Fowler 1981.)

strong density-dependent response at high population sizes. This concept has been linked to the life-history strategies of different types of species. Those species with high Rs are likely to be K-strategists, limited by their resources, so that crowding leads to a reduction in growth rates only at high population sizes. Species with low Rs are likely to be r-strategists,

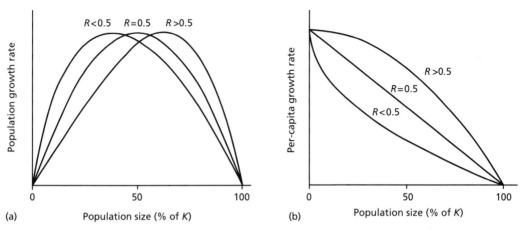

Fig. 2.4 Non-linearity in density dependence as represented by the addition of a power constant α to the logistic equation. (a) The rate of population growth with population size, shown for $R < 0.5$, $= 0.5$ and > 0.5. The linear logistic assumes $R = 0.5$. (b) The per-capita growth rate for the same values of R.

perhaps limited by predators. Large, slow-growing species, such as large mammals, might be expected to have high Rs, while an insect might have a low R.

This non-linearity can be reflected simply in the logistic equation by adding a power constant α. If $\alpha = 1$ then the curve is a simple logistic, with $R = 0.5$. If $\alpha = 0$, there is no density dependence. $0 < \alpha < 1$ gives $R < 0.5$, $\alpha > 1$ gives $R > 0.5$:

$$\frac{dN}{dt} = rN\left(1 - \left[\frac{N}{K}\right]^{\alpha}\right) \tag{2.2}$$

If a harvested species has a value of R that is significantly different from 0.5, then it is important to reflect this in the model. It has implications for the optimal population size under harvesting and for the reaction of a population to harvesting (Box 2.1). The optimal population size for a harvester is related to R. A species with a low R will be harvested to a lower population size than one with a high R. R also has implications for

---

Box 2.1 Harvesting ivory

Fig. B2.1a Elephant in Tsavo National Park, Kenya. (Photo by EIA.)

Basson *et al.* (1991) modelled the optimum elephant population size for a sole owner harvesting ivory. They used a model with logistic population growth, but included non-linear density dependence and a function for tusk growth rate with age. They found that the optimal harvesting strategy to maximise the amount of ivory produced by a population was to leave elephants to die naturally and collect the ivory. This is because there are three non-linearities in the model, with compounded effects. Firstly, the elephant has an R of about 0.8, meaning that the optimal harvested population size is higher than would be expected under linear density dependence. Secondly, tusk growth follows an exponential curve, so that the older the elephant the faster the tusks grow (in terms of biomass). Most species' trophies grow in just the opposite

continued

Box 2.1 *Continued*

way, following a power function in which the growth rate is high at first but then levels off, so that after a certain age, biomass barely increases. Thirdly, the price of ivory per kilogram is not constant, but increases with the size of the tusk, because a carver can do a lot more with one large tusk than with several small ones. These three factors mean that the oldest animals in the population are much more valuable than the youngest ones. Males are much more valuable than females because their tusks are larger (Fig. B2.1b). The elephant is a special case, and it is arguable that the optimal strategy is impracticable because it depends on ivory from natural mortality being easy to find. This would be true in a small national park, but not elsewhere.

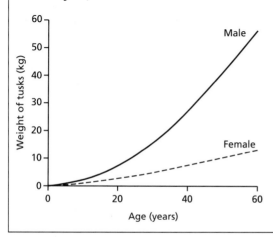

Fig. B2.1b The exponential growth of elephant tusk weight with age. (Redrawn from Pilgram & Western 1986.)

research on density dependence – if most density-dependent change occurs near K, then a relatively heavily harvested population will not exhibit significant density dependence, because it is at a low population size. It would thus be wrong to conclude that density dependence is not an important factor for a population, just because it has not been observed.

2.1.2 Harvesting structured populations

Biological populations are structured, with individuals varying in their age, stage, sex or spatial position. In some species, this structure can have a major effect on the outcome of harvesting (Getz & Haight 1989). *Age structure* has a strong effect on the outcome of harvesting when fecundity and mortality rates vary with age, particularly if there is a time lag between birth and sexual maturity. Harvesting lowers the mean age of a population, because the additional mortality lowers the life expectancy of all individuals. If the lag between birth and sexual maturity is sufficiently long and the hunting mortality sufficiently high, mean age can be lowered to a point at which there is a danger of recruitment failure because all remaining individuals are immature. If older individuals are selected for by hunters (perhaps because they are larger), the reduction in mean age with hunting pressure is even more pronounced. For example, the North Sea cod is a long-lived species which reaches sexual maturity around the age of 4 years. It has been so heavily harvested that some 1 year olds are now harvested, and 2 year olds are fully exploited by the fishery. Only 4% of 1 year olds now survive to age 4 (Cook *et al.* 1997). Rattans are also threatened by over-harvesting, with harvesters cutting stems too young and reducing their ability to re-sprout (see Chapter 5).

Stage-structured populations are structured by life-stage or size. The former include insects, which pass through distinct stages (larva, pupa, adult). Stage rather than age is clearly the key determinant of survivorship and fecundity for these species. Species in which growth is highly plastic and mortality and fecundity depend on size rather than age, such as fish and plants, are correctly described by size-structured models. Thus, forestry models for determining optimal harvesting rates and rotations usually use diameter at breast height as the structuring parameter. This is easy to measure, economically meaningful because value is related to tree volume, and biologically meaningful because fecundity and mortality rates are size-related.

Sex structure is not included in population models as a matter of course, despite often being the most obvious division between individuals in a population. This is because males are not assumed to be limiting on population growth rates, at least in the case of mammals. However, it is often crucial to include sex structure for harvested species in which hunters select by sex, particularly if they select males. This selection might be explicit, for trophies that only adult males bear (such as saiga antelope horns), or implicit for the largest individuals that provide the most meat. As shown in Box 2.1, in the African elephant there is likely to be strong selection for the oldest males, because they have the most valuable tusks; they also have the most meat, which could be important if the local community receives meat as a benefit from wildlife. If a species is polygynous, so that one male mates with many females, then at moderate levels of hunting pressure, targeting hunting on males relieves the pressure on females. Because females provide the recruits to the population, the population growth rate at a given population size is higher if males are targeted than if both sexes are hunted. Therefore, hunting a high propor-tion of males is both less damaging to the population than hunting females, and more profitable. This is true for many hunted ungulate species, although great care needs to be taken that these 'surplus' males really are surplus, and that the proportion of males in the population does not drop so low that it limits fecundity (Ginsberg & Milner-Gulland

1993). If the reproductive system is not polygynous, selection for males might be as damaging to population fecundity as selection for females.

Spatial structure in the harvested population can have major consequences for the population's dynamics, particularly in sedentary species in which interaction with neighbours is much more important to an individual than interaction with distant individuals; plants and corals are obvious examples. Spatial trends in harvesting costs introduce a non-linearity into the relationship between costs and population size and lead to patterns of resource depletion in which the lower cost areas are harvested first (Box 2.2). The optimum population size for a harvester could be higher in a spatially structured population than in a population with constant harvesting costs, so that individuals may survive in remote areas that are very expensive to harvest (Clayton *et al.* 1997). If resources are patchily distributed, it may become worthwhile for harvesters to forage in groups so as to increase their efficiency in searching for prey (Clark & Mangel 1986). Spatial structure in other variables, such as population growth rate, could lead to different harvesting rates being optimal in different sub-populations, and so to different optimum population sizes in each sub-population. Diffusion between areas complicates the picture considerably. If an area is being heavily exploited without diffusion, that area becomes depleted but surrounding areas are unaffected. If, however, individuals diffuse into the depleted area (because the population is well below carrying capacity and so resource levels in the area are high), then the hunting depletes the surrounding areas as well. Baskin (see Chapter 14) discusses how the Lapland Reserve was set up to re-establish the reindeer population in north-west Russia. The population recovered quickly, so harvesting started in the area around the reserve. Six years later, it transpired that almost the whole population had been killed, because most of the reindeer had migrated out of the reserve into the surrounding area, where they were vulnerable to hunting. Woodroffe and Ginsberg (1998) shows that a similar effect has caused carnivores to become extinct much more rapidly in small reserves than in large reserves over recent decades. Those species which ranged furthest were most likely to leave the reserve, and so were at the greatest risk of extermination.

Some species live as *meta-populations*: individuals are distributed among several patches, and the colonisation and extinction dynamics of patches are more significant for the persistence of the species than the dynamics of the populations within the patches. The patch populations reach K at a rate much faster than the migration rate between patches, so patches can be assumed to be either fully occupied or empty (Gilpin & Hanski 1991). Many insects show meta-population dynamics. Badgers seem unlikely candidates for meta-population analysis because their populations reach K rather slowly, but a badger population in a fragmented, agricultural landscape was successfully modelled in this way, because the fragmentation

Box 2.2 Fuelwood gathering in Lake Malawi National Park

Fig. B2.2a Women carrying grass and fuelwood gathered in Lake Malawi National Park. (Photo by Jo Abbot.)

Lake Malawi National Park is a woodland on the southern shores of Lake Malawi. It is a protected area, but villages that existed prior to the establishment of the park have been allowed to remain. Villagers live by farming and fishing, but the expansion of agricultural land in the park is strictly controlled. Firewood can be collected from the woodland, which requires the purchase of an inexpensive permit from the park authorities. The woodland is the villagers' only source of fuel, which is used for essential domestic purposes and for smoking fish to sell.

Aerial photographs indicate a striking increase in sparse woodland, largely at the expense of closed canopy woodland, between 1982 and 1990 (Fig. B2.2b). This has been caused by fuelwood collection. A significant increase has occurred in the population of fishermen in the enclave villages, as fishermen have migrated from the north of the lake where formerly abundant fish stocks have declined due to overfishing. This has led to an increase in the demand for fuelwood from the park; these fishermen pay local women to collect the wood they need for smoking fish.

The spatial pattern of depletion of the woodland resource can be explained by variations in the costs of harvesting according to distance from settlements and the terrain. The pattern is that areas close to settlements, and areas where there are paths in the wood between settlements, are those that have become depleted, all of which correspond to areas of the lowest altitude. Closed canopy woodland on hilltops has, so far, been protected by the additional energy expenditure that women would need to exert to collect wood from those areas. Closed canopy woodland on the nearby islands is also unaffected, due to its inaccessibility (Abbot 1996; Abbot & Mace, in press).

continued

Box 2.2 *Continued*

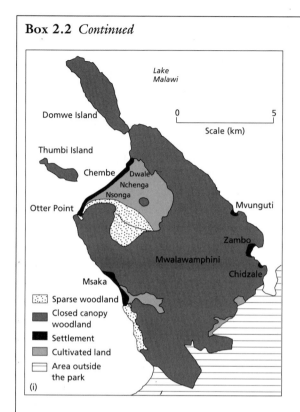

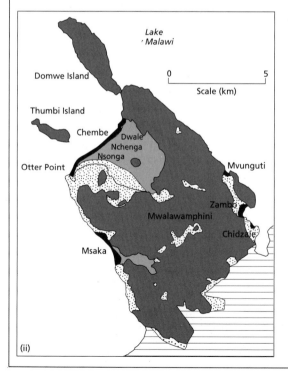

Fig. B2.2b The extent of closed canopy woodland in Lake Malawi National Park in 1982 and 1990, as assessed from aerial photographs. (i) 1982. (ii) 1990.

made patch extinction and recolonisation an important factor in overall population dynamics (Verboom *et al.* 1991). Tuck and Possingham (1994) showed that optimal harvesting of a meta-population involves harvesting the more productive populations less intensively than the less productive populations. This seems counter-intuitive, because in a single population, the more productive the population is, the higher the optimal hunting mortality. The reason is diffusion between populations – if a population is less productive, it is acting as a sink for the more productive population. Taking an extreme example, if a hatchery was producing recruits that went into a river but bred very poorly there, it would be more profitable (given equal costs) to harvest the river very heavily, and leave the hatchery population high so that the river could be re-stocked, than to harvest each population at an equal rate. Tuck and Possingham show that correctly identifying that a stock has a meta-population structure and harvesting it accordingly leads to significantly higher profits than if the stock were incorrectly identified as a single population or two separate populations.

Because individuals are *genetically* different, any preferential harvesting for or against a genetically determined trait is evolutionarily selective for particular genotypes. The genetic effects of harvesting, through selection for individuals with particular characteristics, could well have important long-term consequences for populations that are not immediately apparent. Heavy hunting pressure can cause strong selection on a population, and thus rapid evolution. Gullison (see Chapter 6) presents evidence suggesting that heavy harvesting which concentrates on the largest and best mahogany trees in a forest could lead to genetic erosion of the population. There has been an observed increase in female tusklessness in some populations of African elephants, which may be due to selection, as tuskless elephants have a selective advantage in heavily poached areas (Jachmann *et al.* 1995). Heavily harvested salmon populations may have evolved a smaller mean adult body size as a response to the selective harvesting of larger individuals, while reproductive maturity has been observed at younger ages in harvested cod populations than in unharvested populations (Law 1991). It is difficult to ascertain from observations such as this whether the response to harvesting is ecological or evolutionary. Individuals (particularly males) can reach sexual maturity earlier if there are few adults present, showing phenotypic plasticity rather than genetic change. However, if it is suspected that harvesting could be exerting a selective pressure on a population, then a genetic model might well be needed to examine the long-term consequences of harvesting.

2.1.3 Harvesting in stochastic environments

Stochasticity is random variation. All biological systems involve a certain amount of randomness. Environmental stochasticity is extrinsic to the population, usually involving the effects of climate. Demographic stochasticity is intrinsic to the population, and involves differences between individuals. The latter is only important in very small populations, too small for sustainable use to be feasible. For example, the chance of every reproductively active female in a population producing only male offspring is usually negligible, except in very small populations. A stochastic population model does not merely behave like a deterministic model with some random noise added to it. If the model is a simple, linear one, then it might have 'certainty equivalence'; the mean of the stochastic distribution is the same as the result of the deterministic model. However, if the system is non-linear, as is the case for most population models, certainty equivalence does not occur. For example, the carrying capacity calculated using a deterministic model is higher than the average equilibrium population size in a stochastic model, and a stochastic system has a distribution of population sizes around the equilibrium rather than a single value for K (Getz 1984).

Climatic variation is the most commonly included stochastic variable in harvesting models, because it can cause significant variation in population parameters however large the population size is. Caribou and muskox populations in Canada experience 10-fold fluctuations in population size over a period of several decades; variability in the weather is a major factor in these fluctuations (see Chapter 13). Whether or not the climate is significant for a particular population depends on the relative importance of the variance in the population caused by its internal dynamics (section 2.1.4) and the variance due to environmental noise (May 1973; Beddington 1974). Population characteristics such as a short generation time and discrete breeding seasons make a species more sensitive to stochasticity. The harvesting rate also affects population variability. It is important to consider variation in the population size for conservation and sustainable use, because the more variable the population size, the more variable the sustainable harvest is, and the greater the chance of population extinction. In general, the higher the hunting effort on a stochastic population, the higher the variance in the population size (Box 2.3). This is more pronounced if the population size is less than the MSY level (as seen for the western rock lobster, Fig. 1.5). Engen et al. (1997) suggest that the best harvesting strategies for highly variable populations involve harvesting the population down to a threshold level (a constant escapement strategy, section 1.3.2), rather than using a constant harvesting rate, so that the population is not inadvertently over-harvested.

Another consideration with long-lived species is that harvesting, by lowering the mean age of the population, increases the relative reproductive contribution of the younger age-classes. If a population has stochastic recruitment, this increases the relative variation in recruitment by reducing

Box 2.3 Harvesting the saiga antelope

The saiga antelope is a good example of a species in which harvesting and climatic stochasticity interact. It has a short generation time for an ungulate, with females breeding in the first year of life, together with discrete breeding seasons. The species is long-lived, so there is a strong interaction between harvesting rate and population variance. The stochasticity comes from the climatic extremes of the continental steppe ecosystem where it lives. Frequent, unpredictable droughts and harsh winters cause mass mortality and affect fecundity rates in the following year. The dynamics of the population are therefore significantly influenced by stochastic events. At the same time, the population is heavily hunted for its meat and horns. Previous managers used deterministic models to calculate potential offtake rates in good years, but these rates are clearly over-optimistic for bad years. A sustainable hunting strategy needs to take into account not just the two factors that affect the population dynamics significantly (climate and hunting) but also the interaction between them.

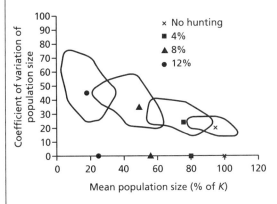

Fig. B2.3 The relationship between population size and coefficient of variation in a model of the saiga antelope population. The contours show the range within which 95% of the results from 50 100-year simulations fall, for each of four hunting mortality rates (none, 4%, 8%, and 12% of the population harvested per year). The overall mean population size and coefficient of variation are shown as a symbol within each contour line, and the deterministic mean population size is given as a symbol on the x-axis for each hunting mortality.

Figure B2.3 shows the results of an age–sex-structured model of the population, with stochasticity in the fecundity and mortality rates, depending on the climatic conditions in a particular year. The mean population size and the coefficient of variation of population size vary between simulations at different hunting mortality rates. This figure shows firstly that the stochastic mean population size is lower than the deterministic mean even when there is no hunting. It also indicates that

continued

Box 2.3 *Continued*

the higher the hunting mortality, the larger the difference between the two means and the higher the variation in the population size, both within and between simulation runs. If a manager wished to harvest the saiga sustainably, then one optimisation criterion might be to minimise the population variance, both to conserve the saiga stock and to give a more stable yield. This could be done by varying harvesting rates according to the most recent climatic conditions, or according to the current population size and age–sex structure of the stock (Milner-Gulland 1994).

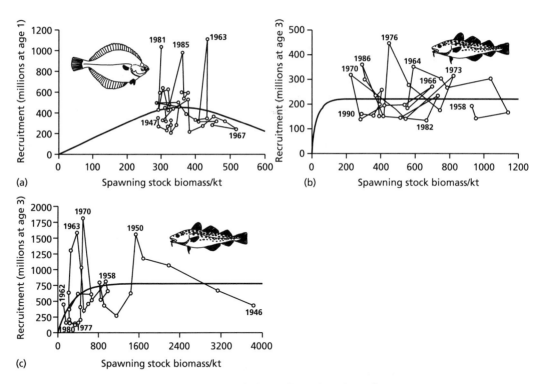

Fig. 2.5 Data on the relationship between stock size and number of recruits over time for three heavily exploited fish species, together with fitted stock–recruitment curves. (a) North Sea plaice. (b) Iceland cod. (c) North-east Arctic cod. (From Shepherd & Cushing 1990.) Stock recruitment curves relate the population of reproductive adults in a year (the spawning stock biomass) to the number of individuals of the youngest age class that can be caught by the fishers the appropriate number of years later (recruits). For example, for North East Arctic cod, the spawning stock biomass in year t is plotted against the number of recruits to the fishery in year $t + 3$. Thus, these are discrete time curves.

the averaging-out effect of having several fertile age-classes. The more fertile age-classes there are, the lower the unexploited variation in recruitment, but the more that variation is increased by harvesting (Reed 1983).

Fish often present particularly severe problems for modellers because of their high population variability. A large part of this may be due to sampling error, as fish are especially hard to count, but some variability is also likely to be due to stochasticity in their population dynamics. Figure 2.5 shows stock–recruitment relationships for three heavily harvested fish populations, together with attempts to fit curves to them. The noise here clearly swamps any deterministic relationships that might exist. In fact, there is still argument about what form, if any, density dependence takes for fish populations (Shepherd & Cushing 1990). The fact that they have survived heavy fishing mortality suggests that there must be some regulatory mechanism at work to increase recruitment rates at low population sizes, but fishing mortality rates may not vary widely enough within one population for the full spectrum of population sizes to be sampled. This means that any stock–recruitment relationship that is derived comes from a rather limited area of the curve as well as having a lot of stochastic variation. Stochasticity is also the driving factor in arid rangeland dynamics. Ellis and Swift (1988) argue that in the arid north of Kenya, highly variable rainfall controls ecosystem dynamics, rather than livestock density. Droughts cause livestock populations to crash. The pastoralists then rebuild their herds as fast as possible, only to be knocked back again in dry years. Livestock numbers hardly ever reach a level that can seriously harm the range. Stable carrying capacity and density dependence are meaningless concepts under these conditions (Mace 1991).

2.1.4 Complex dynamics can result from simple models

The logistic equation assumes perfect compensation and continuous time. That is to say, the population responds instantaneously to changes in its size with a reduction in growth rate (continuous time) and this reduction in growth rate exactly matches the increase in population size (perfect compensation), so that the increase in population size to K is smooth, regardless of the value of r. In reality, populations have a lag in their response to changes in resource levels, which might be very short in species like aphids or long in species like elephants. Species that have the potential to increase very fast (high r) can overshoot their carrying capacity (overcompensation) before the growth rate decreases. In order to include some of these factors, the model can be altered in relatively simple ways. The logistic model can have a time lag added to it:

$$\frac{dN_t}{dt} = rN_t\left(1 - \frac{N_{t-\tau}}{K}\right)$$

(2.3)

Here, the time lag τ is on the density-dependent part of the equation. Alternatively, and usually more realistically, the population can be modelled in discrete time, so that the population size next year (or whatever time-step is appropriate) is a function of the population size this year:

$$N_{t+1} = \frac{\lambda N_t}{(1 + aN_t)^{\beta}} \tag{2.4}$$

This equation produces a similar shaped curve to the logistic equation. However, λ is an arithmetic rate of population increase, relating N_{t+1} to N_t, rather than the geometric rate r, which relates dN/dt to N_t. The relationship between the two is $\lambda = e^r$. a is a constant related to carrying capacity and equals $(\lambda - 1)/K$. β expresses the degree of compensation; if $\beta = 1$, then a discrete-time version of the logistic equation is produced, while if $\beta > 1$, population growth is over-compensatory.

Expressing the logistic equation in discrete time and/or adding time lags has major effects on the outcome of the model. The dynamics of the population have the potential to fluctuate, rather than simply approach K monotonically. Thus, there are two separate mechanisms that could produce the fluctuations observed in natural populations – the external influences of stochastic events such as climate, and the internal dynamics of the population, due to density dependence. May (1975) explored the dynamics of Equation 2.3, and showed that if $\beta = 1$ (perfect compensation), then as in the simple logistic equation, the population increases smoothly to K. If $\beta > 1$, then as β and λ increase, the dynamics become more complex, moving from damped oscillations to stable limit cycles to chaos (Fig. 2.6). As Figs 2.6(d) and (e) demonstrate, although chaotic systems appear unpredictable when plotted as a time series, they are clearly deterministic when plotted in state space. However, it can be very difficult to tell whether the fluctuations observed in a real population are chaotic or stochastic (Sugihara & May 1990). Stochasticity and internal population fluctuations are not necessarily mutually exclusive, making the task of quantifying the importance of each for population dynamics that much harder. They also interact, so that stochasticity doesn't just make oscillations look messy but also changes their frequency and amplitude, and makes extinction more likely at a given set of parameter values (Renshaw 1991; Fig. 2.7).

Although chaotic dynamics are interesting mathematically, they are only likely in species with high population growth rates and strong over-compensation such as the measles virus (Sugihara & May 1990). Chaos is unlikely to be found in vertebrate populations. Many unexploited vertebrate populations tend to have relatively small fluctuations compared to the mean population size. Even large population fluctuations are interspersed with periods of relative stability, and are clearly linked to external factors (Begon et al. 1996a; Fig. 2.8a). Some vertebrates do have clearly

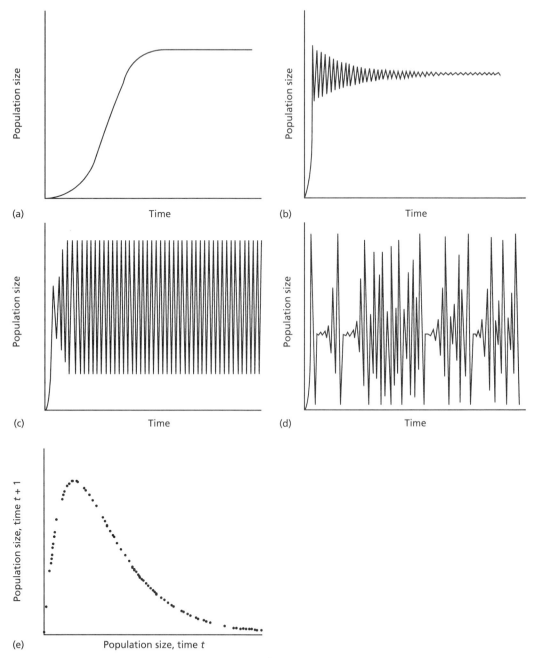

Fig. 2.6 Illustration of the population dynamics obtainable by varying the parameters λ and β in Equation 2.3. As λ and β increase, the cycles get more and more complicated. (a) In monotonic damping, the population increases smoothly to K. (b) Damped oscillations involve progressively smaller overshoots of K with time, until the equilibrium is reached. (c) In stable limit cycles, the oscillations are perpetuated. (d) Chaos. The population fluctuates in a way that seems random but is in fact entirely deterministic, generated solely by the parameter values in Equation 2.3. (e) The same data as (d) but plotted as population size at time t against population size at time $t + 1$, demonstrating that the time series is entirely deterministic.

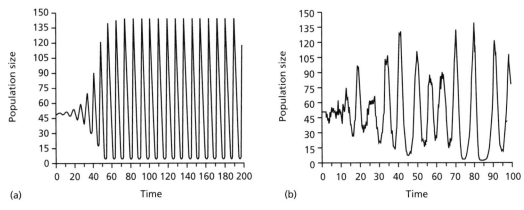

Fig. 2.7 The effect of including stochasticity into a lagged logistic model that would otherwise lead to a stable limit cycle. (a) Without stochasticity. (b) With stochasticity. The lag has had to be reduced from 2 to 1.8 time-steps in order to prevent the population going rapidly extinct. Note the change in frequency and amplitude of the limit cycle. (From Renshaw 1991.)

oscillatory dynamics, for example, some small mammal populations of the tundra, including many lemming populations. The tundra is also the site of the lynx–snowshoe hare population cycles, one of the most famous examples of long-term oscillatory dynamics. Predator–prey pairs often have unstable dynamics, because linking the dynamics of one species strongly to another makes the system inherently unstable. Grouse populations also have oscillatory population dynamics, as may Soay sheep (Grenfell *et al.* 1992; Fig. 2.8b). Thus, although complex population dynamics are unlikely to be the rule among vertebrates, they are not unheard of. Among invertebrates, which have higher rates of population growth, instability in population dynamics is more likely, although over-compensation is not often observed (Renshaw 1991; Fig. 2.8c). Spatial heterogeneity is usually a stabilising influence, which may lead species with the potential for unstable population dynamics to display relatively stable dynamics in nature.

Harvesting generally has a stabilising effect on populations with internally generated complex dynamics, in contrast to its effects on populations with externally generated fluctuations. The removal of a proportion of the population each year damps down over-compensation. Populations with oscillatory dynamics are not usually the best basis for sustainable use, because their dynamics preclude a stable harvest. Yields vary dramatically from year to year. However, grouse show that a profitable business can be based on a species with unstable dynamics, given favourable economic conditions.

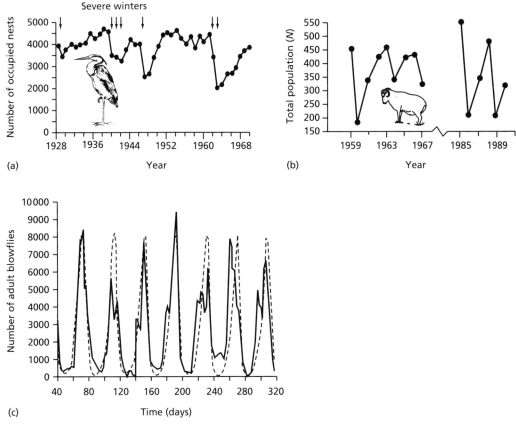

Fig. 2.8 Some examples of natural populations showing various types of population instability. (a) The number of occupied heron nests in England and Wales (a proxy for population size), 1928–70, with the dates of severe winters marked as arrows, showing relative population stability with environmentally induced fluctuations. (From Begon *et al.* 1996a.) (b) The size of the Soay sheep population of St Kilda, which has strongly over-compensatory, lagged dynamics and a high birth rate, leading to population oscillations. (From Grenfell *et al.* 1992.) (c) The population dynamics of a laboratory population of blowflies with a constant daily food supply, showing a good fit to a time-delayed logistic curve with stable limit cycles. (From Renshaw 1991.)

2.2 Interactions between species

It is not ideal to consider a single species in isolation when analysing the sustainability of human use. This is what is usually done, however, and it is testament to the explanatory powers of simple models that they bear any relation at all to the real-world dynamics of exploited species. A single-species model is an adequate simplification for a population in which there is strong intraspecific competition for certain resources (producing density dependence) but where there are only weak interactions with other species, whether they be competitors, prey, predators or pathogens. From the economic side, the model is an adequate simplification if the population is harvested in isolation from other populations of the same or other species. Some exploited populations fulfil these criteria; others do not, and require further analysis. In this section, we give a brief overview of some of the likely effects of human use on systems of two or more species.

2.2.1 Species that interact by being harvested together

The simplest situations to model are those in which the biological interactions between two species are weak or non-existent, but they are harvested together. In this case, they can be treated as two independent populations, but with a combined harvesting effort. This means that the costs of harvesting are also combined, so that the costs relate to the combined population size of the two species. If one species is much more common than another, then the costs of harvesting are not related in any significant way to the size of the rarer population. Thus, the simple bio-economic model considered in section 1.3 can predict the extinction of one of the species at the long-term 'sustainable' open-access equilibrium. In the single-species model, extinction cannot happen at the open-access equilibrium unless harvesting costs are zero or non-linearities are introduced into the model. Writing the model equations out as before, we have:

$$\frac{dN_1}{dt} = r_1 N_1 \left(1 - \frac{N_1}{K_1}\right) - q_1 E N_1$$

$$\frac{dN_2}{dt} = r_2 N_2 \left(1 - \frac{N_2}{K_2}\right) - q_2 E N_2 \tag{2.5}$$

$$\Pi = p_1 q_1 E N_1 + p_2 q_2 E N_2 - cE$$

There are two conditions that must be met for hunting to lead to the extinction of one of the two populations in this model. Firstly, one of the populations (population 1 in this case) must be slower-growing and/or easier to catch than the other:

$$\frac{r_1}{q_1} < \frac{r_2}{q_2} \tag{2.6}$$

Secondly, the cost–price ratio of population 2 must be sufficiently low for hunting to be worthwhile when population 1 has been exterminated:

$$\frac{c}{p_2 q_2} < \bar{N}_2 \tag{2.7}$$

where $\bar{N}_2$ is the size of population 2 at the harvesting effort which exterminates population 1. This result generalises to more than two species, with the relative values of r_i/q_i determining the order of extermination of the species. The carrying capacities K_i do not affect the equilibrium population sizes, but affect the speed of the transition; if, as often happens, the slower-growing species is also found at a lower density, it will be exterminated more quickly. Note that the economic value of the exterminated species is not relevant to the outcome – it could be much more or less valuable than the surviving species. The cost–price ratio of the surviving species is the key economic variable; if it is too high, then harvesting becomes uneconomic before population 1 is exterminated.

This simple model has been used to explain the pressures causing over-harvesting of various species, including blue whales (on the back of the fin whale fishery, Clark 1990) and rhinos. The Luangwa Valley in Zambia saw its rhino population decline rapidly to near-extinction in the early 1980s, coinciding with heavy elephant poaching in the area. Rhino horns were much more valuable than elephant tusks (1.7× as valuable per kill), but modelling showed that the rhino population was small enough for it not to be worthwhile for organised gangs to poach specifically for rhinos. On the other hand, it was highly profitable to hunt the more numerous elephants. Hunting both species together was only slightly more profitable than hunting elephants alone. The poaching gangs captured in the area were large, because hunting specifically for elephants requires carriers for the bulky tusks. As predicted by the model, they were caught with tusks and the occasional rhino horn. In neighbouring Zimbabwe, on the other hand, small gangs went poaching specifically for rhino horn because the rhinos were still numerous. Small gang sizes are preferable for poachers because they are less visible to guards and wildlife, and can be used for rhino hunting because the horns are easier to carry than tusks. Therefore, the economic analysis showed that the rapid decline in rhinos in the Luangwa Valley was not caused by poaching specifically for rhinos, but by opportunistic killing of rhinos by gangs out hunting for elephants. It was the lucrative ivory trade that was driving the rhino decline, not the apparently more lucrative rhino horn trade (Milner-Gulland & Leader-Williams 1992).

If a species is already rare, a low level of opportunistic incidental offtake can have a serious impact on its population size. If the two populations are separated in any way, there is the chance of a refuge from hunting pressure for the more sensitive species. For example, Sulawesi wild pigs and babirusas are two species that are hunted together in Sulawesi (Clayton *et al.* 1997). Because Sulawesi wild pigs are common in both primary and secondary forest, it is possible to dramatically reduce hunting pressure on the endangered babirusa (found only in primary forest) without stopping hunting altogether, simply by hunting only in secondary forest. This could be a way to make hunting more sustainable that might be acceptable to hunters. Fin and blue whale stocks are geographically separated because fin whales tend to feed in more northerly waters than the blue whales, so hunting could be confined to northerly areas (Clark 1990). Other species might be separated in time, perhaps by coming into an area at different times of year, so that a closed season would be similarly helpful.

There are many examples in nature of generalist predators exterminating rare prey species in the presence of more common prey, which act as the predator's main food source (Holt & Lawton 1994). For example, an introduced brown tree snake caused the extinction of three endemic bird species on the island of Guam while preying mainly on lizards (Caughley & Gunn 1996). Human harvester behaviour in these situations is identical to that of other predators.

2.2.2 Harvesting species that interact biologically

Models of interactions between species have concentrated on interactions involving just two or three species. Increasing the number of species leads to complex models that are hard to solve analytically, and may not reveal any more general principles than the simpler models. Competition and predation are two key interactions that have been well studied (Begon *et al.* 1996a), although parasitism is increasingly seen as important; Gunn (see Chapter 13) shows how potentially significant the interactions between caribou and their fly parasites could be, and Price *et al.* (see Chapter 9) cite the drastic effects of disease on the dynamics of Caribbean reefs. Humans themselves can be involved in these interactions – Butynski and Kalina (see Chapter 12) point out the danger that human parasites pose for gorilla groups that are visited by tourists. There is a continuum of interaction strengths, with some species in an ecosystem very weakly linked together, and others entirely dependent on one another There has been very little theoretical work done on the effects of human harvesting on interacting species. It is hard to make generalisations about the likely outcome of interactions between populations, because, depending on the model's assumptions, any type of dynamical behaviour can be generated in the system (White *et al.* 1996). The range of reasonable parameter values that can be found in the natural world is wide enough to ensure that invoking biological realism does not greatly narrow the range of possible outcomes.

Simple models of harvesting competing species show that the effects of competition can be particularly difficult to predict (Clark 1990; Box 2.4). A fishery can collapse due to a shift in competitive dominance before it has even reached MSY. Because the exploited species is dominant, it is the abundant species before the fishery starts, while the competitor is rare, and so is easily overlooked. The apparently healthy fishery can collapse before the existence of a competitor is even noticed. Not every exploited species is going to have this kind of problem (otherwise there would have been many more observed collapses of exploited populations with subsequent takeovers of their niches by another species). The difficulty lies in identifying in advance which exploited species are vulnerable to collapse through competitive exclusion. Fisheries collapses are usually very difficult to assign to a single cause. Sometimes, adverse environmental conditions are suspected to have interacted with heavy fishing pressure to cause a population collapse of the exploited species. An unexploited species may then take over as the dominant population in the area, and stop the exploited species recovering. At other times, competition between two species may have interacted directly with the fishery to cause a population crash, without the need for an environmental perturbation. In the first case, the second species becomes dominant after the fact, because an ecological niche has been left vacant. In the second case, dominance actively switches. The two situations look identical – one population is fished heavily, collapses, and a second species becomes dominant – but the biological processes behind them are different. In both cases, however, the newly dominant species cannot be dislodged by the previously dominant species until a further perturbation happens in the system. In diverse tropical forests, many tree species are thought to be competing for a similar niche. If one species is logged selectively, it may be replaced by any of the other species in the forest, and therefore there is no reason to assume that the harvested species

---

Box 2.4 The collapse of the Pacific sardine and the Peruvian anchoveta fisheries

Along the west coast of the Americas, two groups of fish compete: sardines and anchovies. Although the species vary from place to place, one or other group is usually dominant in any particular area. Seabed core samples suggest that, over hundreds of years, dominance has shifted between the two groups of species, probably due to shifts in environmental conditions. Both sardines and anchovies have been subject to heavy fishing pressure in places where they have been dominant, and both have suffered dramatic collapses and been replaced by the other, with disastrous effects on the fisheries: the Peruvian anchoveta fishery (off the coast of Peru) made up 15% of the entire world's commercial marine yield by biomass in the 1960s (including mammals and crustaceans as well as fish). Its collapse in 1973 was serious enough to have an effect on world food prices. The Pacific sardine fishery off the coast of California collapsed in the late 1940s, but to less global effect.

There has been much controversy about the causes of these collapses. Over-fishing alone is not a good enough explanation for either, as neither seemed to be severely over-exploited at the time of the collapse. In the case of the anchoveta, the most likely contributing factor was the El Niño, a periodic climatic event that causes an influx of warmer water into

continued

Box 2.4 *Continued*

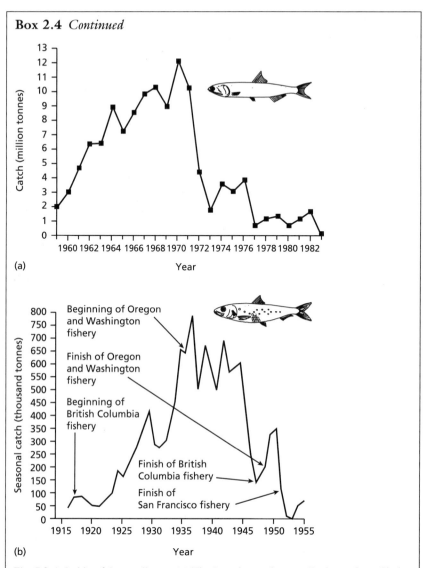

(a)

(b)

Fig. B2.4 Sudden fishery collapses. (a) The Peruvian anchoveta. (Redrawn from Clark 1990.) (b) The sardine fisheries of the North American Pacific coast. (From Glantz 1986.)

their breeding grounds. This leads to very poor recruitment as the anchoveta needs cold water to breed successfully. In the case of the Pacific sardine, one explanation is that the anchovies were inferior competitors able to survive only in small refuge populations. Heavy fishing on the sardines (the superior competitors) led to a non-linear shift in the competitive balance, so that the anchovies were able to competitively exclude the sardines, causing a sudden sardine population crash (Murphy 1977; Glantz 1986; Clark 1990).

will display a positive density-dependent response to harvesting. This means that logging schemes in tropical forests targeted at a single species are very unlikely to be sustainable.

Not all competing species are necessarily similar. For example, there has probably been a shift in dominance between whales and penguins in the Antarctic due to over-harvesting of whales. Baleen whales eat krill (small crustaceans at the bottom of the food chain which are important for many Antarctic species, including penguins). The reduction in whale numbers coincided with an increase in penguin numbers, suggesting that penguins have taken advantage of the increase in food resources caused by a reduction in the number of krill eaten by whales. The question now is whether the whales can reclaim their place in the Antarctic ecosystem or whether there has been an irreversible substitution of penguins for whales. In that case, even now when Antarctic whales are completely protected, the whale population can no longer recover to pre-exploitation levels. The issue is complicated further by the industrial krill fishery, so that humans too are competing for the krill resource (Mangel 1994). This example illustrates the point that the results of exploitation depend on whether the hunter is taking a species that is low in the food chain, and is a resource for many other species, or whether the exploited species is a predator or competitor for other species. The competitors of an exploited species benefit from the increased availability of resources as the population size of the exploited species declines, and their populations will probably increase. The prey of an exploited species may or may not increase in population size depending on whether there are other predators present that can increase to compensate for the removal of their competitor. MacKinnon (see Chapter 5) shows how the bearded pig in Sarawak (a valuable meat source) has declined, not due to direct harvesting, but due to over-harvesting of the fruit trees that it relies on for food. The lower down the food chain the exploited species is, the more likely it is that its removal will cause population declines in several other species. Removing more than one species from an ecosystem will have effects that are even more complex and hard to predict. Fisheries modellers are starting to address the effects of harvesting several species from multi-species predator–prey systems, using techniques such as multi-species virtual population analysis, but there are still many uncertainties (Magnusson 1995).

The exploitation of one species can thus have far-reaching effects on the other species in the community. Studying the ecology of the harvested species will probably highlight the species with which it has the strongest interactions. These are the species that should be singled out for analysis, and the species which have only weak interactions with the exploited species can be ignored. Calculations of the food requirements of individuals from particular species can be used to estimate the magnitude of the resource flows from one component of the ecosystem to another. If the basic structure of the ecosystem remains unchanged after harvesting, it is then possible to predict the direction and the magnitude of the effects of harvesting on the other major components of the ecosystem (Box 2.5). The problems come with species like the Pacific sardine (Box 2.4), whose competitors only gain prominence after harvesting has destabilised the equilibrium.

2.2.3 The ecosystem-level effects of harvesting

Harvesting does not just affect non-target organisms through their population-level interactions with the harvested population. It can also affect the physical processes and structure of the ecosystem. Ecosystem-level effects of harvesting can have long-term impacts on all the component organisms of the ecosystem. There are two categories of effect that are particularly important – the effects of harvesting on nutrient flow and on community structure.

Nutrient flow

Minerals are recycled within ecosystems. Nitrogen, in particular, tends to be very efficiently recycled, because it is often a limiting resource for plants and is easily leached away if it is present as nitrate in the soil. On all but the most nutrient-rich soils, the majority of the nutrients present in an ecosystem are tied up in the vegetation, rather than being freely available in the soil. As an organism dies and is decomposed, the nutrients that were bound up within it are released, and most are quickly taken up by the vegetation before they can be leached away. Harvesting can have serious effects on nutrient cycling. These effects are of two major types – direct removal of nutrients and increased leaching. Removing biomass from an area also removes the nutrients that are tied up in that biomass. Carbon loss is only a problem for ecosystem function if harvesting is removing a large proportion of primary productivity, and disrupting energy flows to

---

Box 2.5 Interactions between fishing and seabirds in the Chagos islands

The Chagos islands are isolated coral atolls in the Indian Ocean, rich in seabirds. The seabirds feed on small fish, as do predatory fish such as tuna. In the 19th century, there was heavy harvesting of these small fish species, leading to large reductions in seabird numbers. In the last few years, the Chagos islands have become militarily important, and so fishing has almost ceased, allowing the seabirds to recover. However, it is possible to predict the effects of fishing restarting on the seabird populations, depending on whether the fishing effort targets predatory fish such as tuna

continued

Box 2.5 *Continued*

or the small fish. The effects on the flows of resources through the ecosystem can be approximated from data on food requirements and population counts of seabirds on the islands (Symens 1996).

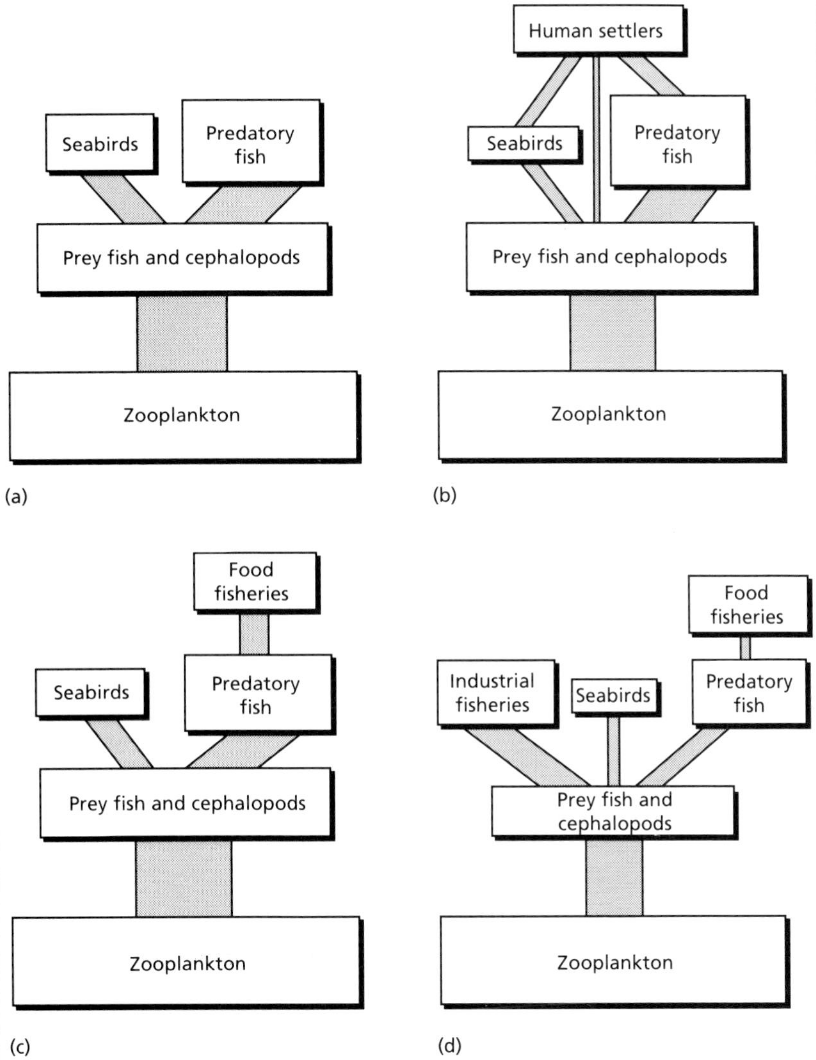

(a)

(b)

(c)

(d)

Fig. B2.5 The flows of energy through the Chagos marine community under various scenarios of human fishing effort. The breadth of the connections between boxes shows the magnitude of the energy flow and the size of the boxes represents the population sizes of the species concerned. The diagrams are indicative, rather than accurate reflections of the magnitude of the flows. (a) Before human settlement. (b) 19th century and first half of 20th century. (c) Present day. (d) Possible future picture, given an increase in industrial fisheries. (From Symens 1996.)

other levels of the food chain. On the other hand, harvests of biomass rich in other nutrients, such as young plant tissues and animal bodies, can represent a constant leakage of nutrients from the ecosystem. Whether this leakage is a problem depends on the relative magnitudes of leakage and recycling. If it is a local subsistence harvest, there will be little concern about nutrient transfer, but large-scale export to other places may be of more concern. Increased leaching is caused by the physical removal of vegetation from an area, leaving exposed ground. Free nutrients in the soil or the topsoil itself can then be washed away (Fig. 2.9). Large-scale deforestation of tropical rainforests has attracted particular attention as this causes erosion and fertility loss from already nutrient-poor soils. In some cases, however, nutrient input to the ecosystem is the problem, rather than nutrient export: Price *et al.* (see Chapter 9) discuss the damage done to coral reefs by nutrient enrichment from human sewage.

Community structure

Human activities can affect community dynamics and structure by disrupting the process of *succession*. Succession is the replacement of one species by another over time in a directional manner. It happens whenever a perturbation creates a new habitat: a fallen tree creating a patch of ground for new tree seedlings to colonise, a dung pat that gradually dries out, a new pond that silts up over time. Grazing by stock, shifting agriculture and fires influence the natural pattern of succession in an area. Human disturbance is not necessarily 'bad' for conservation – long-term grazing has produced grasslands that are now valued for their high plant diversity. Systematic removal of dominant species can promote the diversity of an ecosystem by preventing the exclusion of competitively inferior species. Thus, pioneer species that are shaded as scrub invades can persist in an area that is regularly grazed or cut. Conversely, domestic animals grazing in a woodland can prevent the regeneration of the woodland by eating young seedlings, condemning the woodland to senescence and eventual extermination.

Disturbance is an important force in structuring ecosystems. Natural disturbances such as fire and storms create gaps in the vegetation which can be colonised. Some species are especially adapted to exploit disturbance. For example, pitch pine increases in flammability as it gets older, encouraging fires, which allow it to retain its dominance. Mahogany can only regenerate if large areas are cleared, which means that stands tend to be even-aged, with long periods between disturbance events (see Chapter 6). Human disturbance can also fundamentally alter the species composition and structure of an ecosystem. It is a matter of conservation priorities whether this alteration is considered acceptable. Humans can also have large indirect effects on ecosystem structure by harvesting a *keystone*

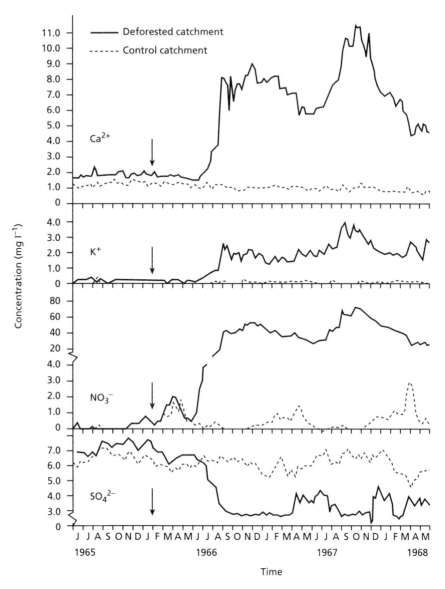

Fig. 2.9 The effects of experimental deforestation on concentrations of mineral salts in the streams flowing from deforested and control catchments at Hubbard Brook Experimental Station. Deforestation took place at the point indicated by the arrow. (From Begon *et al*. 1996a.)

species. Keystone species have a disproportionate effect on their ecosystem, due to their size or their activities. Any changes in their population size have correspondingly large effects on the ecosystem. Beavers are keystone species, because they change water flow substantially by building dams. Elephants are keystone species due to their ability to destroy trees (Box 2.6).

Box 2.6 The keystone role of elephants in savannah habitats

The size of the elephant population in an African national park is a key determinant of the amount of scrub and tree cover found in that park. A high density elephant population promotes open savannah, such as is found in some of the national parks of Kenya. A lower density population produces more wooded parks, such as are found in Zimbabwe. A park manager must decide on the priorities of the park when deciding whether to cull the elephant population. An open park is good for tourists, who can easily view the elephants, and for plains species such as impala. A more wooded park favours scrub-nesting birds and secretive species such as dik-dik.

The role of the elephant in the maintenance of biodiversity was explored by Western (1989a), who analysed plant diversity along a transect from the centre of Amboseli National Park in Kenya to well outside the park. Elephants were being heavily poached outside the park at the time, but were well protected inside. Thus, moving along the transect was the equivalent of moving along a gradient of elephant density, from very high densities in the centre of the park where they were safest from poachers, to virtually no elephants outside the park. As Fig. B2.6 shows, plant diversity peaked at the boundary of the park, an area of moderate elephant density. Western drew the conclusion from this study that an intermediate level of disturbance caused by elephants would best promote biodiversity in the area.

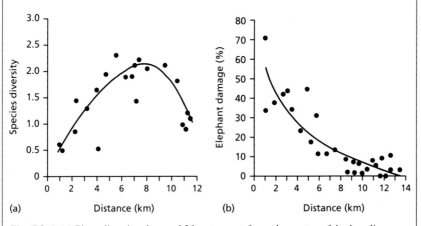

Fig. B2.6 (a) Plant diversity along a 12 km transect from the centre of Amboseli National Park, Kenya. The park boundary is at 6 km. (b) Changes in elephant density along the transect from the centre of the park, measured by elephant damage to acacia trees. (From Western 1989a.)

2.3 Summary

- There is a trade-off between simplicity and realism in population modelling. In general, the simpler the model, the better it is. However, some species require modifications to the simple models in order to represent their ecology correctly.
- Non-linear density dependence includes depensation (inverse density dependence) and skewed growth curves (density dependence expressed at very high or low population sizes). Depensation is probably rare among commercially important species, but is implicated in some cases of populations failing to recover from low levels. Harvesting rates proportional to population size are not safe under depensation. Skewed growth curves change the optimal harvesting strategy and may make density dependence harder to observe.
- Populations may be structured by age, stage, sex, genetics and spatial location. If harvesting selects strongly for particular classes within a population, a structured model is necessary. Selecting by age, stage or sex can affect reproductive performance. Spatial heterogeneity has different effects depending on whether it is the population or the harvesting costs that vary spatially. Harvesting particular genotypes preferentially forces artificial selection.
- Stochasticity has many forms. Climatic variation is most commonly modelled for harvested species, because it can have an impact even in large populations. Harvesting and climatic variability interact; the higher the harvest rate the higher the relative variability in population size, particularly if the population is below the MSY size.
- Simple deterministic population models specified in discrete time or with time-lags may exhibit complex dynamics, if the intrinsic rate of increase or the degree of over-compensation are large. Complex dynamics are found in some harvested species, but chaos is not widespread. Spatial heterogeneity and harvesting are generally stabilising in systems with complex internal dynamics.
- Species are not found in isolation, but multi-species modelling quickly becomes intractable. Simple models of species that interact only by being harvested together show that, if productivities and catchabilities vary greatly between the species, the less productive, more catchable species may be driven extinct while harvesting remains profitable. The effects of harvesting competing species, predators or prey, are difficult to generalise about. There are examples of harvesting possibly contributing to shifts in competitive dominance.
- Ecosystem-level effects of harvesting include long-term nutrient depletion and disturbance. Perturbing an ecosystem by harvesting from it could halt succession. If the harvested species has a major effect on the structure of the ecosystem (such as beavers), harvesting could lead to large changes in ecosystem dynamics.

Chapter 3: Decision-Making by Users of Biological Resources

In this chapter we consider the theoretical basis for analysing human decision-making. In section 3.1 we address those methods that deal with costs and benefits that can be quantified. In section 3.2 we consider some of the other relevant aspects that may not be so easily captured in quantitative terms.

3.1 The theory of decision-making

3.1.1 Classifying decisions

The first step towards formulating effective policies for conserving resources used by people is to understand how people make decisions. The theory of decision-making is helpful as a tool for explaining how people decide which species to exploit, how intensely to harvest them, and what harvesting method to employ.

The conditions under which a decision is made are of key importance for determining both the decision-making process, and how far quantitative analysis can go towards finding the optimal harvesting policy for sustainable use. Two of the main dimensions are time and predictability. A decision can be made in a static or a dynamic framework – either without regard to time, or with time as a variable. There are four divisions of predictability: certainty, risk, uncertainty and chaos.

The framework for decision-making is generally described using the following notation:
- a set D of decisions, d_i
- a set Y of the decision-maker's states, y_i
- a set X of outcomes from the decisions, x_i
- a set Ω of states of nature ω_i

- a set P of the probabilities of the states of nature occurring, p_i, where $p_i \geqslant 0$, $\Sigma p_i = 1$.

The outcome of a decision is a function of the decision-maker's state, which decision is made, and the state of nature: $x = f(y,d,\omega)$. The decision-maker's objective is assumed to be to maximise his or her *utility*, $u(x)$, which is a function of the outcome of the decision. Utility is a vague concept in economics, being the unit of measurement for human happiness. A common proxy for utility is money, so that, if the decision is about the harvesting level of a renewable resource, the objective becomes maximising the monetary yield from the resource. Another measure for utility is *biological fitness* in the Darwinian sense; an organism acts to maximise its genetic representation in future generations. As evolved organisms, our psychology is likely to be broadly adapted to maximise fitness, or at least to prefer actions that would have maximised fitness in our evolutionary past. This measure is being increasingly used in optimal foraging theory and other areas of evolutionary economics (Rogers 1994). The Υ states of the decision-maker are those given factors that affect the decision – for example wealth, location of the household, or family size. The Ω states of nature are the externally generated factors that affect the outcome of a decision – for example whether there is a drought or a harsh winter.

Under *certainty*, the decision-maker knows the outcome of each decision. The state of nature, ω, is known, so it has a probability p of 1. In this case, the decision-maker simply chooses the decision that maximises his or her utility: max $u[f(y,d)]$. This is not necessarily a straightforward procedure, if there are complicated decisions to be made, with many constraints involved. One method that is used is linear programming (Williams 1993).

Under *risk*, there is a set of several possible states of nature, each with a known probability attached to it. The outcome of the decision depends on the state of nature, which is not known until after the decision has been made. In this case, a decision-maker cannot maximise utility, but must instead maximise *expected utility*, $Eu(x)$. A useful tool for representing decisions under risk is a payoff matrix, with each outcome, x_{ij}, listed for a given decision i and state of nature j (Table 3.1a). The outcome, x, is expressed in units of utility (or a proxy such as wealth). The procedure for finding the expected utility for a given decision i is to multiply the payoff, x_{ij}, for each state of nature j, by the probability of that state of nature occurring p_j, and then to add the values of $x_{ij}p_j$ together. The sums for each decision are compared, and the optimal decision is the one that maximises this expected utility.

A simple example is a farmer aiming to maximise the monetary yield obtained per hectare, with a choice of planting three crops (Table 3.1b). If it is a rainy year, potatoes do well; if it is a normal year, wheat does well; if

State of nature ω_j:	ω_1	ω_2	ω_3	
Probability p_j:	p_1	p_2	p_3	
Decision d_i:				Expected utility
d_1				
d_2	x_{21}	x_{22}	x_{23}	$Eu_2 = \Sigma_j x_{2j} p_j$
d_3				
Optimal decision:				$\max_i \Sigma_j x_{ij} p_j$

Table 3.1a Payoff matrix for decision-making under risk.

State of nature:	Rainy	Normal	Dry	
Probability:	0.2	0.6	0.2	
Decision:				Expected utility
Potatoes	100	30	5	39
Wheat	25	70	25	52 (max)
Maize	10	50	90	50

Table 3.1b An example of calculating the optimal decision under risk.

it is a dry year, maize does well. Given that all the fields must be planted with the same crop, which crop maximises the farmer's expected wealth? In this hypothetical example, the crop that maximises the farmer's expected utility is wheat, despite wheat having the lowest maximum yield.

Under *uncertainty*, the decision-maker knows the states of nature, but does not know what the probabilities attached to them are. This is frequently the most realistic scenario for natural resource management. If the probabilities are not known, then the optimal decision cannot be found simply by maximising the expected utility. Two schools of thought exist about how best to tackle uncertainty. *Bayesians*, or subjectivists, contend that there is always some subjective probability available that can be used. These subjective probabilities can be the opinions of experts or of the resource users, as to the probabilities of particular states of nature occurring. The estimates are expressed as prior probabilities, which are used in the optimisation. Every time a decision is made, the prior probabilities are refined into posterior probabilities in the light of the new information that becomes available after the decision. The posterior probabilities are used in the next decision, and so over time there is a learning process, in which the posterior probability distribution assigned to the state of nature sharpens around the true state of nature. Thus, the Bayesian approach to a decision under uncertainty is to assign probability values to the states of nature and then to treat it as a decision under risk. *Classicists*, or objectivists, have a different definition of probability: probability has a mathematical meaning as the limiting frequency of an outcome over time. The more times a coin is tossed, the closer the frequency of heads approaches to 0.5, the limiting frequency. The probability of a coin landing heads is thus 0.5. Under this definition, undefined probabilities exist, and it is not correct to assign subjective values to them. The classicist's approach to uncertainty is to get as far as possible without assigning probabilities, and then to leave a set of admissible decisions from which to choose using a non-mathematical method (Menges 1974). Bayesian methods of decision-making are becoming more widely used in fisheries decision-making, and have the advantage that they are easier for managers to interpret than traditional methods (Clark & Kirkwood 1986). Bayesian methods have an intuitive logic about them, but they do carry the danger that, in some cases, semi-subjective probabilities of unknown quality might be given a status that they do not deserve. The classical method is much more explicit in highlighting the point at which subjectivity comes into the decision, but may ignore relevant data and leave a large set of admissible decisions. The objections to a Bayesian approach are more valid for one-off decisions than for circumstances where the prior probabilities can be modified over time, in line with the evidence.

Under *chaos*, neither the set of states of nature nor their probabilities are known. These are intractable problems, and can only be tackled using a Bayesian approach. Note that this type of chaos is quite different to the deterministic chaos discussed in section 2.1.4.

3.1.2 Individual decision-making

A decision-maker can have various objectives, and the optimal decision is the one that best meets these objectives. A simple objective might be to make money, so the optimal decision is the one that maximises profits. But there are many other objectives that are just as plausible. For example, a resource harvester might wish to maximise sustainable yield over time. Alternatively, a harvester might wish to maximise physical yield (of biomass rather than money). A conservation-minded manager, such as the Falkland Islands government (see Chapter 8), might maximise monetary yield but with a constraint that the population size should not fall below a certain level. A farmer might aim to maximise the household's chances of exceeding the subsistence threshold, so that they have enough to eat until the next harvest, rather than maximising the expected yield from the crops (Mace 1993). Other objectives might be to maximise the chances of output reaching a certain level, minimise the variance in the yield, maximise the rate of throughput... One can imagine many different objectives, involving maxima, minima and constraints. It becomes difficult to make general assessments of what the optimal decision is, given the plethora of possibilities. This is why utility is so convenient for theoretical analysis of decisions. So long as people's objectives can be converted into a utility scale which is internally consistent, then all these different objectives can be collapsed into the single objective of maximising utility. Then mathematical tools like payoff matrices can be used to find the optimal decision. The normal scale of measurement for economic transactions is money, so it is tempting to try to convert utility to money, and then maximise monetary yield, which is easy to measure. Clearly, money is an inappropriate metric for many of the objectives described above; ultimately people may be attempting to maximise their biological fitness. The problems with using money as a scale of measurement for natural resources are discussed in section 3.2.

A person's attitude to risk is an important determinant of the decisions they make. The most analytically convenient assumption to make is that people are *risk-neutral* – they calculate the utilities of the possible outcomes of a risky decision, and weight them by the probability of occurrence to find the optimal decision. For example, a subsistence farmer planting maize in an arid area might expect the crop to fail 4 years in 5, making a loss on seeds and labour of £120, but with a bumper crop making £1000 1 year in 5. Planting the maize is a gamble, with an expected monetary value each year of $(0.8 \times -120) + (0.2 \times 1000) = £104$. The certainty equivalent value (CEV) of a gamble is the amount of money a person is prepared to accept for certain instead of taking the gamble. A *risk-averse* person is prepared to accept a CEV less than the expected monetary value of the gamble, perhaps by taking a job in town that guarantees £50 a year instead of taking the gamble of farming. A risk-neutral person will only accept a job paying at least £104 instead of the gamble, as this is the actual value of the gamble they are being offered, while a *risk-prone* person needs to be offered more than £104 to persuade them not to take the gamble.

The expected monetary value of a risky enterprise is only a valid approximation to people's behaviour if they are risk-neutral. This is likely if the sums of money involved are small compared to total wealth, if the risk is shared by many people, or if there is a long series of similar risks over time. Otherwise, people are generally risk-averse. People may be risk-prone when they are desperately poor, so that taking the risk is the only chance for survival (Box 3.1), or when they are rich, so that the gamble is a very small proportion of their total wealth. The analysis of utility is important in predicting how people's attitudes to natural resources may be affected by their wealth. Poor people who are barely surviving might be more prepared to take risky decisions, such as poaching in a protected area in order to survive. Those who are slightly better off are generally more risk-averse, tending to act conservatively. Usually only relatively well-off people are prepared to experiment, for example with new, perhaps more sustainable, methods of resource use.

Box 3.1 Gambling with goats: an example of the effects of wealth on attitudes to risk

Nomadic herders inhabiting drought-prone areas in northern Kenya keep mixed flocks of sheep, goats and camels. Sheep and goats breed rapidly but are prone to mortality in drought. Camels breed slowly; they are a large unit of resource and are very productive as a source of milk (the staple food) and usually survive drought well. Using a stochastic dynamic programming model, in which the probability of long-term household survival is used as a measure of expected utility, it is possible to show that long-term household survival is maximised if most families keep most of their wealth in the form of camels, and a smaller proportion of their wealth in the form of sheep and goats; but very poor families maximise their survival chances by keeping all their wealth in sheep and goats (Mace & Houston 1989). This is a risk-prone strategy for poor families, that may lead to destitution if a drought were to occur shortly after they fell into poverty, but it has the greatest overall chance of enabling them to build their herds back up to a viable size. If they succeed in this, it is then optimal for them to convert much of their wealth into camels, so that they are better placed to survive future droughts. Comparing the herd composition of families of different wealth from a number of north-east African camel herding groups shows that pastoralist households do this (Mace 1990).

A similar approach can be used to assess when households might switch between different modes of subsistence. In African rangelands, herders, farmers and agropastoralists (people who both farm and keep livestock, at some cost to the productivity of the livestock) can frequently be found together in an area. A stochastic dynamic model, maximising long-term household survival, shows that the optimal mode of subsistence depends not only on the agricultural potential of the land available to a family, but also on their wealth (Fig. B3.1). Returns from livestock increase, and risks

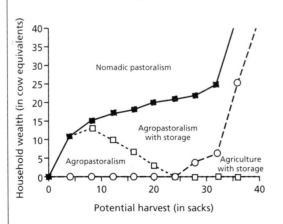

Fig. B3.1 The optimal mode of subsistence for a household, depending on the agricultural potential of land available to them, and their wealth. Wealth can be stored as either livestock or stored grain or a combination of both.

continued

Box 3.1 *Continued*

reduce, in a non-linear fashion with herd size, in a manner similar to an investment account. Therefore, for wealthy families, pure pastoralism is more likely to maximise household survival. For poor families, farming is a safer long-term option. Development agencies are sometimes puzzled when families on agricultural development projects (such as irrigation schemes) in rangeland areas frequently re-invest any profit in livestock, and may leave the scheme altogether if they accumulate a big enough herd, but this is exactly what the model predicts (Mace 1993).

Foraging theory and optimality

One important type of model for individual decision-making is optimal foraging theory. The theory was developed in behavioural ecology, to examine how animals might be expected to behave when foraging to maximise their biological fitness (Krebs & Davies 1991). Fitness is not measured directly, but a currency assumed to approximate to fitness, such as the rate of energy intake whilst hunting, is assumed to be maximised. Most of the models that have been used only consider maximising harvesting rate over the short term. The theory has been successfully used to analyse the behaviour of foraging people. Precise predictions can be made about foraging decisions, such as which types of prey a hunter should hunt and which they should ignore, when a forager should move on from one patch to the next (Kaplan & Hill 1992), what size groups hunters should forage in (Clark & Mangel 1986) and how the arrival of more efficient hunting technology might influence these decisions (Alvard & Kaplan 1991).

These studies have some important messages for conservation. There has been a tendency amongst some anthropologists to create an image of the 'ecologically noble savage' (Redford 1990), whose resource use decisions are assumed to be based on the motive of conserving that resource. Detailed studies of hunting and gathering decisions have used the framework of foraging theory to show that this is not the case (Alvard 1993). For example, it has been observed that Amazonian hunters typically do not hunt in depleted areas, which are usually areas close to settlements. They usually move on to new areas when the vicinity is getting depleted of prey. This is consistent with conservation, but is also consistent with optimal foraging models, which predict that hunters should not spend time in less profitable areas (Hames 1987). However, if hunters are simply maximising harvesting returns, they should always take prey from the depleted areas opportunistically as they pass through, which is what they have been shown to do (Alvard 1995). Furthermore, these models, based ultimately on fitness maximisation, stress that traditional populations must be seen as groups of individuals, where each person is trying to maximise individual fitness or the fitness of very close kin. Evolutionary models do not predict that such individuals will make sacrifices for the good of the wider group. A more likely explanation for the many examples of traditional foragers harvesting resources sustainably is limited technology, low population growth and distance from markets. Freehling and Marks (see Chapter 10) show how important these factors have been for the Valley Bisa people of Zambia. There are also examples of traditional communities co-operating to set rules limiting harvests in places where resource scarcity has become an issue (McCay & Acheson 1990), but this involves individual fitness maximisation as well; the members of a small, closed community are likely either to be relatives or to have long-term interactions with each other. Long-term interactions tend to promote co-operation among individuals because actions can be reciprocated.

More complex, but very useful, dynamic optimality models have been developed that can be used to analyse decisions when the state of the decision-maker y_i is important. These employ the technique of stochastic dynamic programming, which is a tool for finding a resource user's optimal strategy in a stochastic environment. It has been most widely used in behavioural ecology (Houston & MacNamara 1988; Mangel & Clark 1988). In human behavioural ecology, Mace (1993) used dynamic optimality to examine which mode of subsistence (farming, herding or agropastoralism) families would be likely to adopt to maximise household survival, depending on the resources available to them (Box 3.1). Fitness, as in number of descendants, can be used directly as a currency in some problems; Mace (1996, 1998) has used this approach to understand determinants of human family size in different environments, and suggests that the importance of inherited resources, combined with high costs of raising children, are likely to be key factors that will motivate people to have small families. Some use of stochastic dynamic models has been made in fisheries science (Walters 1978; Clark 1990) and in the analysis of pest control problems (Jaquette 1970; Shoemaker 1982). The technique has also been applied to terrestrial wildlife management problems (Reed 1974; Anderson 1975; Milner-Gulland 1997) and to conservation decision-making (Possingham 1996).

3.1.3 Game theory

Game theory becomes relevant when an individual decision-maker is not acting in isolation, but must make his or her decision in the context of the actions of others. It started as a branch of mathematics concerning how people play games (Luce & Raiffa 1957; Binmore 1992), and has been applied widely in the social sciences (Rapoport 1960). It is useful in the analysis of oligopolies, when there are only a few suppliers in a market, but it has also been used as the basis of other analyses of human behaviour towards natural resources.

Decision-making under conditions of uncertainty can be viewed as a special class of games – *games against nature*. In these games, the opponent is not a sentient being attempting to win, but is some external factor such as the climate. In this case, the state of nature is treated as an opponent which makes decisions that are completely independent of the player's decisions. The payoffs (in utiles, the unit of utility) are assumed to be *zero-sum* – a gain of 4 utiles for the player is assumed to be a loss of 4 for the opponent, so only the player's payoffs need to be listed. If there is a dominant decision that always gives the best result, then the choice of decision is clear. However, in many situations, dominance only rules out a few decisions, and the best decision is not clearcut. In these cases, psychological assumptions about the decision-maker's objectives need to be made. A common objective is to maximise the worst possible payout – given that things will turn out for the worst, which option is best? This objective is called *maximin*. Perhaps maximin is unduly pessimistic in a game against nature, as there is no reason to assume that things will always turn out badly. Against a real opponent, it is likely to be a more rational strategy, because in a zero-sum game, the payer wishes to pay out as little as possible given that the payee tries to get as much as possible. Therefore, the payer plays the mirror-image strategy to the payee, called *minimax*. A payoff matrix is a useful method of analysis:

a's decisions	b's decisions		min for a (payee)
	b_1	b_2	
a_1	4	2	2 (maximin)
a_2	3	1	1
max for b (payer)	4	2 (minimax)	

Here, decision a_1b_2 is the equilibrium solution of the game, where maximin = minimax. Most games don't resolve this easily, but go round in circles of bluff:

	b_1	b_2
a_1	0	2
a_2	3	−1

In this game, minimax leads to a_1b_2, but maximin leads to a_1b_1. As b knows that a will choose decision a_1, b chooses b_1, but knowing this, a

chooses a_2, so b chooses b_2, and so on. The game is solved by using a mixed strategy, in which each player chooses one or other decision randomly, with a fixed probability. The optimal strategy can be found relatively easily for two-person games (see Luce & Raiffa 1957). At the equilibrium, each player is playing their optimal strategy, because a player cannot improve his or her results by departing from the strategy, so long as the other player does not change their strategy. Similarly, if a player sticks to their optimal strategy, they cannot do worse if others change their strategies. This is a *Nash equilibrium*. The market equilibrium is also a Nash equilibrium – the best result that each supplier or consumer can achieve given the other people's supply and demand curves. The theory of games in evolution has used similar techniques to find *evolutionarily stable strategies*, which are also Nash equilibria (Maynard-Smith 1982).

Non-zero-sum games have been particularly studied in the social sciences, because they present the possibility of finding a co-operative solution. If the amount received by one person is not the same as the amount the other person must pay out, the interests of the players are not directly opposed. As long as communication is allowed, bargaining can take place. If utility is transferable, the players can maximise the total payoff, and then divide it between them. In the game below, a's payoffs are given in bold, b's in normal text:

	b_1	b_2
a_1	**9**/9	**2**/15
a_2	**10**/−10	**0**/−30

In this game, decision a_1b_1 gives the maximum total payoff, 18. b is in a much weaker position than a, because a can always threaten to choose a_2. Thus, the likely negotiated division is 10 for a, equivalent to a's maximum payoff, and the remainder (8) for b.

A typical, and famous, example of the problems involved in non-zero-sum games without communication or bargaining is the *prisoner's dilemma*. The game has two decisions – co-operate or defect. If both players defect, they both incur a negative payoff. If they both co-operate, they both receive a positive payoff. If one co-operates and the other defects, then the co-operator incurs a very negative payoff, and the defector a very high payoff. The values chosen for the payoffs are arbitrary, but follow these rules (where CD is the payout to player A if A co-operates and B defects, DC is A's payoff if A defects and B co-operates, and so on):

- The payouts are symmetrical between players.
- The values of the payoffs follow: CD < DD < CC < DC.
- The values follow: 2(CC) > (DC + CD).

The original story for this game was two prisoners held in separate cells, deciding whether to confess to a crime (defect) or keep quiet (co-operate):

		b	
		co-operate	defect
a	co-operate	**+5**/ +5	**−10**/ +10
	defect	**+10**/−10	**−5**/−5

If the game is played just once, *a* is better off defecting whatever *b* does; defecting produces a payoff of either +10 or −5. This is a better outcome than −10 or +5 from co-operating. *b* faces the same choice, and so makes the same decision, with the result that both defect. However, the total payoff is maximised by both co-operating (+10) and minimised by both defecting (−10). Individual rationality has precluded the jointly optimal decision of both co-operating, which could have been reached by bargaining. Instead, the worst outcome, in terms of total payoffs, has been reached. This result has parallels with the outcomes of resource harvesting under different market structures, as discussed in section 1.4. If a single decision-maker, such as the state, decides on the correct harvesting level for a resource, it can impose the decision that maximises the total benefits from harvesting. However, if the harvesters are left to make decisions independently, and have the choice of harvesting to maximise long-term sustainable yield from the resource, or of harvesting to maximise individual short-term gains, the prisoner's dilemma predicts that they will defect, and harvest to maximise individual short-term gains. The resource is over-harvested, reaching the open-access equilibrium. Analysing open access to a shared resource using the prisoner's dilemma has been very controversial, and we shall return to it in detail in section 3.2.

Axelrod (1984) showed that iterated prisoner's dilemma games can lead to stable strategies involving co-operation. A strategy called tit-for-tat does well in games iterated over time. Players start with co-operation and do what the other player did the previous round from then on. This allows for speedy retaliation and speedy forgiveness. In resource management, one important line of research has been to find strategies that ensure that the jointly optimal harvesting level is maintained over time (Box 3.2). Important considerations that determine when joint maximisation is likely to develop and be stable are:
• When there are few players involved, so that deals can easily be struck and compliance monitored.
• When there is communication between players.
• When the game between the same players is repeated over time, not just played once.
These considerations mirror those found by researchers working on

Box 3.2 Getting the optimal yield from a shared fish stock

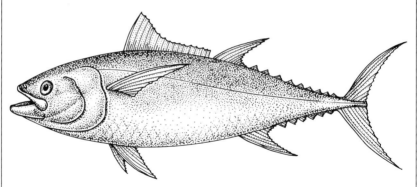

Fig. B3.2 Southern bluefin tuna (*Thunnus maccoyii*).

When the UN Law of the Sea came into force in 1982, many fisheries that were open access went into the private ownership of coastal states. Fish do not respect national boundaries, however, with the result that two or more countries often became owners of the same fish stock. For example, the Japanese and Australians both exploit the southern bluefin tuna (Kennedy 1987). When multiple ownership of the same fish stock occurs, a treaty needs to be devised that allows the countries to co-operate to obtain the maximum benefits from the fishery, rather than competing. Competition between the countries can result in a return to the open-access equilibrium, losing the advantage of the fishery being privately owned. Kaitala and Pohjola (1988) used game theory to explore a model of fisheries dynamics. They assumed that the two countries had come to an agreement to maximise fisheries yields, could monitor the agreement, and remembered previous years' catches. If the agreement is breached, the damaged party can retaliate. They found that a way to ensure that co-operative, profit-maximising harvesting is a long-term equilibrium solution is for both countries to threaten to return to competitive, non-cooperative harvesting in perpetuity if the agreement is ever breached. This threat is sufficiently credible to be believed, because non-cooperative harvesting is an equilibrium position too, and promises a loss of income large enough to discourage the countries from reneging.

If one country is a less efficient harvester than the other, then profits are maximised if only the efficient country harvests the resource. The country that doesn't harvest requires compensation of at least as much revenue as it would have received at the non-cooperative equilibrium, in order for it to gain from abiding by the agreement. This compensation can be paid as a one-off, effectively buying the less efficient harvester out of the fishery, or it can be paid as a proportion of the harvesting revenues each year. In the former case, once the non-harvesting country receives its one-off payment, it has no incentive to keep out of the fishery in perpetuity. If, on the other hand, it receives a flow of revenues from the fishery, neither party has an incentive to break the agreement, and a stable equilibrium results over time (Kaitala & Pohjola 1988).

the evolutionary stability of reciprocal altruism in animal populations (Trivers 1971). Solis Rivera and Edwards (see Chapter 7) discuss a community-based iguana-rearing project, in which the conditions required for co-operation between the participants are similar to those listed above.

3.1.4 Incentives for illegal exploitation

Many of the crucial questions about how to make resource use sustainable revolve around illegal exploitation. The options for a manager of a resource that is being illegally and unsustainably harvested include attempting to stop the resource use entirely, legalising the resource use and then trying to find ways to regulate it at sustainable levels, and attempting to contain illegal resource use at sustainable levels through law enforcement. The last option, though rather unsatisfactory, is the one that is usually arrived at *de facto*, when insufficient funds lead to imperfect enforcement. Game theory can be used to model the incentives for people to break the law, particularly when members of the same community are both law-breakers and law-enforcers. Mesterton-Gibbons (1993) describes a community irrigation scheme in which people take turns to guard the water supply, and members of the community have the opportunity to cheat by stealing extra water for their own crops. He uses game theory to show that whether stealing is optimal or not depends on how much extra benefit the stealers receive. If the probability of the supply failing is high enough, it is optimal to be trustworthy, even when there is no-one guarding the water supply.

Most theoretical work on law enforcement has taken the perspective of a resource manager trying to control poaching. A key component for understanding how different law enforcement policies work is the theory of how individual decision-makers choose whether or not to undertake an illegal activity. This has been studied in the economic theory of law (Becker 1968; Eide 1994). The theory was first developed to understand the behaviour of burglars in the USA. No empirical testing of its applicability to natural resource use has been carried out, but the results of these studies do have direct relevance to the problems of resource management.

In the economic theory of law, the law-breaker is seen as a rational utility maximiser in an uncertain world. The person's attitude to risk and uncertainty is a crucial determinant of the mix of activities undertaken: a risky high-return crime versus a safe, legitimate job. The amount of risk involved in the crime, and the amount of security in legitimate employment, varies with the type of crime and the social circumstances of the person involved. In general, a risk-averse person undertakes a greater proportion of legitimate activity than a risk-prone person. This is an application of portfolio theory (Ehrlich 1973; Copeland & Weston 1988). As Cook (1977) pointed out, illegal activity also has a threshold caused by respect for the law and social opprobrium, which is not easily quantified and varies with the type of crime. There has been strong resistance to the idea that law-breakers act in a rational economic way (Eide 1994), but poachers are perhaps more likely than most to act this way, either because they are already harvesting legally and are weighing up whether to take a bit extra, or because their social climate is relatively positive towards poaching, and they feel they have a right to harvest their local resources. Abbot and Mace (in press) describe how law enforcement aimed at regulating fuelwood gathering in Lake Malawi National Park came under strain for this reason (also see Box 2.2). Most women have no source of fuel available to them other than firewood from the park. Collecting wood from the park is legal, provided an inexpensive permit is purchased from park authorities. Due to the poverty of the women, the permits have to be extremely cheap. However, the women resented any expenditure; they felt they had an historical right to collect wood in the area, and as society in general agreed with the women's view, penalties for illegal wood collection were rarely imposed on transgressors. This meant that it was not economically rational for women to purchase permits, and they rarely did so. Eventually the park scaled down enforcement of the permit scheme.

Factors likely to reduce the crime rate are an increase in the perceived probability or severity of punishment, a decrease in the profits made from the crime, or an increase in the opportunity cost of crime through improved wages elsewhere. Studies seem to show a strong deterrent effect of the probabilities of being caught and convicted, but are contradictory as to whether the severity of the sentence has a deterrent effect. Ehrlich (1973) calculated the elasticity of crime rate to expenditure on law enforcement in his study area to be about -3, i.e. a 1% increase in direct spending on law enforcement leads to a 3% decrease in the crime rate, although this estimate has a large standard error attached to it. The elasticities that he found in different categories of crime were generally higher for the probability of capture than for the severity of the sentence, suggesting that expenditure on capturing criminals was more effective than money spent on more severe sentences. A study by Avio and Clark (1978) in Ontario showed that although the probability of apprehension had a strong effect on crime rate, the severity of the sentence had none. However, Cook (1977) cautions against taking these empirical studies too seriously, because most of them could have been strongly affected by measurement error.

It is the *perceived* severity of the sentence *before* the crime is committed that is the relevant factor affecting incentives to commit crimes. If a prison sentence is given, the person's discount rate and time-horizon (the distance into the future they look) affect the sentence's perceived severity. With a positive discount rate, 1 year in gaol with a probability of 0.2 is a worse option than 2 years with a probability of 0.1 (Cook 1977), because the second year in prison is valued less highly than the first. To a person with a short time-horizon, 10 years in prison may look exactly the same as 5 years. This discounting, together with the empirical evidence, suggests that with a limited budget, it could be better to concentrate on increasing the perceived probability of detection than to spend the same amount on keeping people in prison for long periods.

Natural resource users' attitudes to law enforcement have been little studied. Sutinen and Gauvin (1989) showed that the rate of violation of regulations by lobster fishermen in Massachusetts varied with the perceived probability of detection and conviction, as predicted by the theory. Freehling and Marks (see Chapter 10) show how hunter behaviour in the Luangwa Valley, Zambia, has changed as law enforcement has increased. They now tend to use less easily detected snares, rather than guns, and are more secretive in their consumption of meat. Simple bio-economic models of harvesting can be altered to incorporate the risk of capture and a fine into the costs of harvesting (Sutinen & Anderson 1985). Mazany *et al.* (1989) modelled the likely extent of illegal fishing in a situation where there was a legal fishing quota, with imperfect enforcement. People who fish above their quota face an expectation of a fine, expressed as the fine received multiplied by the perceived probability of receiving it. The poacher's short-term profit-maximisation (in an open-access situation) then becomes:

$$\max_E [pH - cE - \theta \mathfrak{I}] \tag{3.1}$$

where p is the price per unit of output, H; E is all the inputs to production, not just effort; c is the cost per unit input; and θ is the probability of receiving the fine $\mathfrak{I}$. θ is assumed to depend on the amount of input E. $\mathfrak{I}$ can be expressed as a function of either input or output. The profit-maximising condition found by Mazany *et al.* (1989) for Equation 3.1 was:

$$pH_E = c + H_E [\theta_H \mathfrak{I} + \mathfrak{I}_H \theta] \tag{3.2}$$

where H_E means the partial differential of H with respect to E. This is equivalent to the profit-maximising condition found for the standard model ($pH_E = c$, Equation 1.8), but with the addition of a term for the marginal change in the expected fine with a change in the output H. The long-term profit-maximisation of a monopoly hunter can be generalised in a similar way, by adding the expected fine as a second cost term in Equation 1.11. The mathematical analysis assumes that the decision-maker is risk-neutral, because it assumes that the cost of the fine is the expected monetary value of that fine. The actual set of incentives is much more complex, but this is probably an adequate approximation for a commercial fisher, who might weigh up a small breach of the regulations in purely monetary terms. The model also assumes that the actual fine and the probability of capture have equal weight in calculating the expected fine, implying that the policy-maker can increase the expected fine in two equally good ways – increase the probability of capture or increase the fine. As increasing the probability of capture is expensive, the strategy of lobbying for increased penalties has been a common reaction of wildlife authorities to unsustainable poaching (Leader-Williams & Milner-Gulland 1993). However, the socio-economic studies discussed above suggest that the best strategy to achieve effective law enforcement is to increase the perceived probability of detection (Box 3.3). This whole discussion is made more complex by the involvement of several authorities, with different priorities and budgetary arrangements. Crime-reduction initiatives taken by law enforcement officials may not be supported by the judiciary in sentencing. For example, in Zambia, concern about the loss of elephants and rhinos due to ivory and horn trafficking, led the government to introduce mandatory 5–15 year prison sentences for elephant and rhino poachers in 1982. After 1982, magistrates did tend to deliver more prison sentences to elephant and rhino offenders, but not all of them received prison sentences. Those that did receive prison sentences received only a few months. The maximum length given over the first 3 years of the new law was 36 months. The legislation that was required to increase the penalties was slow and difficult to enact, meeting much opposition. Once in place, it was widely ignored by the magistrates, and has failed to curb poaching. The rhino population of the Luangwa Valley declined rapidly to near-extinction over the same period as the new legislation was coming into force (Leader-Williams et al. 1990).

3.2 Natural resources as economic goods

3.2.1 Externalities

In section 1.6, the market equilibrium for normal economic goods was described as the level of production at which the total utility obtained from the good is maximised; this is called the 'socially efficient' level of production. However, the open-access equilibrium in natural resource harvesting involves both economic and biological over-harvesting, and utility is not maximised. What is it that makes people over-harvest, and what is special about natural resources as economic goods that makes their market equilibrium socially inefficient? A socially efficient market is one in which the supply curve represents accurately all the costs involved in producing the good, and the demand curve all the benefits from the good. If the supply or demand curves do not represent all the costs and benefits involved in the market, then the point at which they cross is not the socially efficient production level (Fig. 3.1). This is called *market failure*. The costs and benefits that are not represented in the supply and demand curves are called *externalities*. These costs or benefits are experienced by people other than those deciding the level of production or consumption of a good, so they are not taken into account in the market. External costs occur if the supply curve does not reflect all the costs of production; external benefits accrue if the demand curve does not reflect all the good's benefits. For example, Kenyan land-owners suffer externalities from wild-

Box 3.3 The effect of law enforcement on babirusa hunting

Fig. B3.3a A dealer's van, detained by law enforcement officials in Sulawesi, with meat from babirusas and other protected species. (Photo by Lynn Clayton.)

Law enforcement to prevent the illegal hunting of the protected babirusa wild pig in Sulawesi, Indonesia, has been sporadic, but was extremely effective when it was applied in the end-market. Government officials were notified that babirusas were being sold in the main wild meat market, and visited the market, as a result of which the number of babirusa sold openly declined. However, this enforcement was not sustained and the number of babirusas sold rose again to previous levels (Fig. B3.3b). The effect of this one law enforcement episode was strong and lasted into the medium term (more than a year) before fading. However, as this was the first episode of law enforcement babirusa dealers had experienced, it probably had a particularly strong effect on them, and the perceived probability of detection increased suddenly from near-zero to rather high. Once the law enforcement has been repeated (or not) their perceptions are likely to become more realistic (Clayton & Milner-Gulland 1998).

continued

Box 3.3 *Continued*

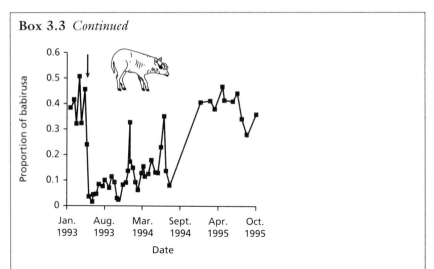

Fig. B3.3b Babirusas on sale in Langowan market, Sulawesi, Jan 1993–Oct 1995, shown as a proportion of the total number of wild pigs on sale. The remainder of the pigs on sale are unprotected Sulawesi wild pigs. The sudden drop in the proportion of babirusas on sale in May 1993 (marked by an arrow) coincided with a law enforcement action in the market.

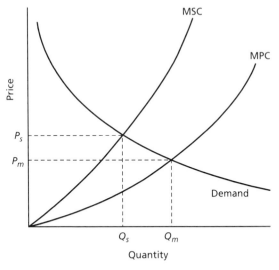

Fig. 3.1 The market model of supply and demand, modified to show the effect of external costs on the social efficiency of the market equilibrium. The private cost curve of the producer (MPC = marginal private costs) is the supply curve, leading to the socially inefficient market equilibrium, $P_m Q_m$. If external costs are taken into account, the supply curve becomes the marginal social cost (MSC) curve: MSC = MPC + MEC, where MEC = marginal external costs. In this case, the socially efficient equilibrium, $P_s Q_s$, is reached. The socially efficient equilibrium is at a lower level of production and a higher price than the market equilibrium.

life, which reduce the productivity of their land by damaging crops and property, killing people and livestock, spreading disease and competing for grazing. These externalities need to be taken into account when weighing up the costs and benefits of keeping wildlife in an area (see Chapter 11). Natural resource-based markets are particularly prone to externalities. The solutions generally proposed are:

• Define property rights. This can have two effects. People who own both the source of the external cost and the damaged property automatically 'internalise the externality' – they take it into account in their private cost–benefit analysis, and produce at the optimal level. Alternatively, people whose property is damaged can claim compensation, which would be difficult if they did not have clearly defined and enforceable rights.

• Government intervention and market regulation. The government decides what the socially optimal production level is, and intervenes to prevent market equilibrium being reached, and to ensure that the socially optimal level is achieved.

Valuing the non-market costs and benefits correctly would allow them to be incorporated into the supply and demand curves directly. But environmental costs and benefits are easily ignored, in comparison to mainstream economic costs and benefits. There are four major problems with environmental externalities:

• They are often *diffuse*. That is, their effects, good or bad, tend to be spread over many people. This is usually in contrast to the mainstream market effects. For example, if a sewage outfall is polluting a reef, many people are likely to bear relatively small external costs of pollution, which add up to a substantial amount overall (see Chapter 9). Yet most of the profits are clearly accounted for in one place.

• They are usually *hard to value* in monetary terms. For example, the mainstream economic costs and benefits of building a modern harbour are clearly identifiable. Price *et al.* (see Chapter 9) highlight how the full value of the loss of biodiversity in the area, due to the resultant degradation of the coral reef, is extremely hard to value, even if components of the loss (such as reduced coastal protection from the reef) are relatively easily valued. If these externalities cannot be expressed in monetary terms, they will not be included in monetary cost–benefit analyses.

• They are often *long term*. The profits from developments such as logging a forest are often relatively short term, while the environmental effects come into play gradually, after several years. Some effects, such as genetic erosion (see Chapter 6) may take many decades to come into effect. Even the costs that are easily valued monetarily, such as increased flood damage, are likely to be missed out of calculations if they occur in the future. Even if future costs and benefits are included, they are reduced in value by discounting.

- They are *poorly understood*. Even ecologists would feel less than confident in predicting the range and magnitude of effects of a harbour development on a coral reef in any but the broadest terms. In other cases, there may have been no ecological research done at all to establish how the ecosystem responds to perturbation. For example, Butynski and Kalina (see Chapter 12) show how virtually no scientific research has been done on the effects of human visits on gorillas, and yet tourists continue to come into contact with highly endangered gorilla populations. If these externalities cannot be quantified, then they cannot be internalised.

3.2.2 The 'Tragedy of the Commons'

Open-access resources are subject to external costs. Any resource that is subject to uncontrolled access by anyone who wishes to use it produces *reciprocal* externalities for the users. Rather than one person imposing externalities on others, as is often the case, all the users are imposing externalities on each other. The problem of reciprocal externalities in open-access resources was first described by Hardin (1968) as the 'Tragedy of the Commons'. He gave the example of a piece of common grazing land on which anyone could put their livestock. If a herder decided to increase his or her herd by one animal, the costs of this decision, incurred through the grass eaten by the extra animal, would be borne by all the herders equally, not just by the one who had increased his or her herd. The benefits would accrue only to that herder. The private benefits of increasing the herd decline gradually as the condition of the animals, and so their value, declines. These diffuse costs and private benefits mean that it is in every herder's interests to increase their private herd until the private benefits of an extra animal exactly match the herder's share of the costs of that extra animal. The density of animals that would have maximised the pasture's productivity is exceeded some time before the equilibrium is reached. This analysis of Hardin's Tragedy of the Commons exactly matches the description of the open-access harvesting equilibrium developed in section 1.4. It also has parallels with the prisoner's dilemma discussed in section 3.1.3. Each herder can 'defect' by adding to their herd, or 'co-operate' by showing restraint, and defection is the individually rational choice. If there are a large number of herders, with no possibility for communication, bargaining or exclusion of other people, then there is no obvious method of limiting resource use to the profit-maximising level. Hardin's model suggests that natural resources are inherently prone to over-exploitation, and there is very little that can be done to prevent this, other than privatisation or strong state control. In cases of international open access, such as the high seas or the atmosphere, international treaties are the only option. It is a depressing scenario, summed up in his words 'Ruin is the destination towards which all men rush, each pursuing his own best interest in a society that believes in the freedom of the commons. Freedom in the commons brings ruin to all' (Hardin 1968).

Hardin's model of the Tragedy of the Commons has provoked bitter arguments at the heart of the debate about sustainable use, with some anthropologists and economists lined up against what they see as the refusal of modellers to recognise the real-life complexities of systems of resource tenure and exploitation. Opponents of the concept of the Tragedy of the Commons, such as McCay and Acheson (1990), argue that the concept is deeply flawed because it fails to recognise that traditional societies in particular tend to have developed systems of sustainable resource use. They say that problems with unsustainable resource use arise only when outside interference is imposed on these cultural norms. Examples of systems disrupted in this way include the 'sasi' rules governing the harvesting of megapode eggs in Indonesia (see Chapter 5), and the totem status of beavers for the Khanti people in Russia (see Chapter 14). This argument has been useful in highlighting some important issues, but in most respects involves an attack on a straw man, because the models are acknowledged only to be relevant for open-access resources. Part of the problem seems to have been an unfortunate choice of example in Hardin's original exposition – the common grazing land of a village is rarely an open-access resource. Another major factor has been that political ideology has become involved, and the debate has implicitly become one about the privatisation of resources versus community ownership. For the system which he describes, Hardin's analysis is sound – the argument should be about the relative prevalence of open-access and other types of system in the real world. The insights from the prisoner's dilemma, and from biological research on the evolution of co-operation, are helpful in suggesting which societies might evolve to sustainable resource harvesting and joint maximisation, and which are likely to remain open-access harvesters. It is no surprise that the majority of the examples of successful communal resource management cited by McCay and Acheson (1990) are small, closed societies in which the same people have interacted over a long period.

One way in which the debate could be clarified would be the realisation that the *ownership* of a resource is fundamentally irrelevant to the outcome of exploitation. Privately owned, state, communal or genuinely unowned resources (rare as they are) can all be subject to open-access exploitation, or to profit-maximisation through an effective monopoly. It is the *control* of the resource that is the key issue. Stewart (1985) puts this argument clearly for forest resources. Similarly, in the Luangwa Valley, Zambia, a state-owned national park was exploited as a monopoly by poaching gangs, and the state-owned game management areas as an open-access resource by individual poachers (Milner-Gulland & Leader-Williams 1992). The communally owned resources used as examples by McCay and Acheson (1990) are protected from open-access exploitation because they are communally controlled, not because of their ownership. The political and legal environment has a large effect on a user's control of a resource; for example, Solis Rivera and Edwards (see Chapter 7) highlight how the future prospects of an iguana-rearing scheme would be improved by the formal licensing of the management programme by the Nicaraguan government. Even if a user has sole control of a resource at present, this may not be adequate to promote a long-term conservation ethic if control is not perceived to be secure. A user only has the incentive to conserve a resource if they feel that they or their descendants will reap the benefit of this moderation in the future. This is where ownership and control do intersect – legal ownership can be seen as a guarantee that control will continue in perpetuity.

The argument that it is important for resource users to have control over their resources if they are to conserve them can be expanded to include schemes involving the local community (see section 4.4). A resource such as a national park can be used directly by locals who poach in it or encroach upon it for grazing, agriculture or settlement. It can also be destroyed by neglect, which is the same as making a decision not to invest in the conservation of the resource (Swanson 1994). Local communities can disinvest in a national park by harbouring poachers and not co-operating with the authorities. Governments can do the same by not investing in the law enforcement and maintenance that the park requires. Norton-Griffiths (see Chapter 11) shows how this type of passive disinvestment has led to Kenya losing 44% of its wildlife over the last 18 years. Some schemes aim to encourage local people to invest in the conservation of their natural resources by giving them control over the resources (Child 1996). The extent to which these schemes are successful depends on the actual level of long-term, secure control that the local decision-makers feel they have. This depends partly on the political stability of the country and the local area, although a sustainable use scheme that is very successful on the local level can be ruined by large-scale external events. For example, the Gulf War reduced tourist revenues in a Caribbean marine park (see Chapter 9), and the Rwandan civil war halted the mountain gorilla programmes in the Virunga region (see Chapter 12).

Those who make decisions about the level of natural resource use are not necessarily the community leaders, who are often the ones targeted by development agencies, but are more likely to be individual hunters and herders. Freehling and Marks (see Chapter 10) discuss how ADMADE, a local involvement scheme in Zambia, has affected incentives to hunt. Provision of community-level services like schools and dispensaries, using the revenues from licenced hunting, has not stopped individual local hunters from hunting illegally, although increased enforcement has changed their hunting behaviour. As with law enforcement, it is the decision-makers that must be targeted, not the employees. Thus, arresting poaching gangs has much less effect on the incentives of an ivory dealer to continue to finance poaching than arresting the dealer. Similarly, babirusa dealers in Sulawesi are more susceptible to law enforcement than the hunters they employ (Box 3.3). Any attempt to re-organise incentives needs first to clarify the decision-making hierarchy in the area, and then to ensure that the incentives are tailored to that hierarchy.

3.2.3 Valuing natural resources

In section 3.2.1, four reasons were given why natural resources are particularly prone to market failure, and so to over-exploitation. Two of these involve the valuation of natural resources. These are the problem of valuing costs and benefits that occur in the future, and the problem of valuing non-monetary costs and benefits. Both of these problems prevent ecological costs and benefits being properly weighed against economic ones, but instead lead to them being undervalued.

Valuing natural resources over time

The correct valuation of costs and benefits that accrue at different points in time is important for any project that is intended to be financially sustainable in the long term. As shown in section 1.5, the present value of a continuous stream of revenues is calculated using the formula:

$$PV = \int_0^\infty A e^{-\delta t} \, dt \qquad\qquad (3.3)$$

where A is the revenue received at each point in time and δ is the discount rate. This expression for PV can be solved using the technique of definite integration (Bostock & Chandler 1981), to give a finite value of A/δ. Thus, if a payment of £100 is received from a resource in each period in perpetuity, then under a discount rate of 5%, this resource is worth £2000 in total. The value of the payments in each period declines rapidly; at $\delta = 0.05$ the value of the payments has already halved after 14 years. After 45 years, the value of the payments is only 10% of the original value (Fig. 1.7). The technique of discounting future costs and benefits is therefore not one that values long-term sustainability highly. This should ring alarm bells for those accustomed to ecological time-scales of damage and recovery.

The discount rate has three interrelated meanings in economics:

1   In the *capital market*, the discount rate (or interest rate) is the opportunity cost of investing in capital assets. This is related to the growth rate of the economy and the rate of time preference of investors (Price 1993). The discount rate on a particular asset also reflects the riskiness of the return on that asset. The discount rate in this sense is relevant to the decisions made by private companies that invest in resource use, such as fishing companies or game meat dealers and is the δ used in the models in section 1.5. It is also the δ used by Gullison (see Chapter 6) in discussing the incentives for commercial firms to over-harvest mahogany.

2   For *individuals*, the discount rate is that person's rate of time preference under the social and psychological conditions that he/she is experiencing. This discount rate determines whether the individual saves or spends, and is related to the individual's attitude to risk. It is used in decisions like a Bisa hunter's choice of whether to go hunting in a protected area, or whether to adopt different harvesting techniques (see Chapter 10). The discount rate was used in this sense in section 3.1.

3   In *investment appraisal*, the discount rate is chosen as a tool for assessing the relative merits of one project over another. Discounting is needed when comparing projects with different income and expenditure schedules over time, or comparing project performance to the performance of the general economy. For example, Gullison (see Chapter 6) shows how a government would weigh up different silvicultural regimes. The discount rate that is chosen may be the market rate, as in 1, or it may be an estimate of society's time preference rate. Society's time preference rate is presumed to be lower than the individual discount rate in 2, on the assumption that society takes a longer term view than an individual.

The first two meanings of the discount rate are of less concern here. Although governments can influence them, they exist independently of government policy with regard to sustainable use. Humans show time preference for consumption in the present over the future, and although the negative exponential is an inadequate model of human behaviour, it is simple and is a relatively good predictor in many cases. For example, discounting can be used to compare the time at which poachers would kill a dehorned rhino (whose horn grows back slowly) with the rotation period at which it would be most profitable for a rhino manager to dehorn, given that the manager is trying to maximise the returns from selling the horn (Milner-Gulland *et al.* 1992). It is the use of discounting in investment appraisal that has caused major disagreements among ecological economists, because the discount rate is chosen by the appraiser, and the rate that is chosen can have a major effect on the apparent profitability of projects based on environmental assets. This is because many projects involving the environment have large deferred costs or benefits (Box 3.4).

The knee-jerk reaction to the problem of choosing a discount rate for investment appraisal is to suggest that a low discount rate should be set for

Box 3.4 An example of the effects of discounting in forestry

Fig. B3.4 Logged forest in South-East Asia. (Photo by Kathy MacKinnon.)

This is a very simplistic analysis of the management options for a forest which is being harvested rapidly, following Price (1993). If harvesting continues, an output worth £25 000 per year is obtained for the next 10 years, followed by no further yield. Alternatively, a 40-year moratorium on timber harvesting will lead to the recovery of the forest, followed by a sustained yield of £50 000 per year in perpetuity. Which option is economically more worthwhile?

Clearance: $PV = \dfrac{A_1}{\delta}(1 - e^{-\delta t_1})$, where $A_1 = 25\,000$ and $t_1 = 10$

Moratorium: $PV = \dfrac{A_2}{\delta}(e^{-\delta t_2})$, where $A_2 = 50\,000$ and $t_2 = 40$

(PV = present value of the sustained yield from the forest)

The formulae used to calculate the values of the two options are obtained by the definite integration of Equation 3.3 between times 0 and 10 years for the clearance option, and between times 40 and ∞ for the moratorium option. The two options are equally good where:

$$\frac{A_1}{\delta}(1 - e^{-\delta t_1}) = \frac{A_2}{\delta}(e^{-\delta t_2})$$

Solving for δ shows that clearance is preferred at $\delta > 4.3\%$, which is a rather low discount rate. South American countries have a median discount rate of 11%, and some have discount rates up to 60% (see Chapter 6). Discount rates this high would strongly favour clearance.

environmentally sensitive projects, or for projects that have an important social function. This is not a useful response, firstly because the basic problem of assuming an exponential decline in value over time remains, and secondly because, as Pearce and Turner (1990) point out, high discount rates discourage investment. Investment tends to be bad for the environment because it uses resources and may lead to development of the countryside, so in some ways, a high discount rate can be good for the environment. It has even been suggested that a negative discount rate should be used for environmental projects. The snag here is that $\delta < 0$ intuitively means that the value of assets increases with time, giving the impracticable outcome that it is optimal to save constantly and never to spend. Several authors have suggested various alterations to the discounting procedure, including adding a sustainability constraint (Pearce & Turner 1990) and using discount rates which vary over time (Kula 1992). Price (1993) provides a comprehensive critique of discounting and the modifications suggested by others, and argues strongly for the use of methods which do not involve discounting at all, but deal explicitly with changes in value with time. The problem is that the negative exponential is a very simple and generally accepted model for change in value over time, and is therefore subject to an enormous inertia – rather like the logistic equation in population ecology. Also analogous to the logistic equation is the fact that people are constantly attempting to find ways to tweak the model for greater realism, which demonstrates its deep underlying problems. As the constituency using the negative exponential is orders of magnitude larger and more entrenched than that using the logistic equation, the challenge for those trying to get it abandoned is far greater.

One major issue in the treatment of time in economics is *intergenerational equity*. Using discounting as a convenient way to appraise the investments of the present generation is one matter; using it to decide on investments whose costs will be borne by future generations rather than ourselves is quite another (Norgaard & Howarth 1991). Lawrence H. Summers of the World Bank (cited in Price 1993) puts the standard position on the subject very clearly, arguing that the compelling needs of those alive today and the uncertainty of the future costs of global warming justify using the same discount rate for present-day sanitation projects and projects protecting against global warming (the latter will, of course, receive a very low priority under discounting):

> Do I sacrifice to help those in the future or help the 1 billion
> extremely poor people who share the planet earth with me today? I
> hold no greater grief for the people who will die 100 years from now
> from global warming than for people who will die tomorrow from bad
> water.

This argument applies with the same force to any project that aims to improve long-term sustainability at the expense of present-day development, for example, alleviating the long-term threat to the sustainability of life on the Maldives posed by sea-level rises (see Chapter 9).

Evolutionary theory is relevant to the question of how humans make decisions that affect future generations. It predicts that we are interested in future benefits only if they accrue to ourselves or our direct descendants. *Kin selection* theory (Hamilton 1964) provides a framework with which to make precise predictions about how we value benefits to the next generation compared with benefits to ourselves. Rogers (1994) uses this framework to give an evolutionary explanation for positive time preference and to estimate a long-term 'evolutionary discount function'. The marginal rate of substitution of preferences measures the rate at which present and future consumption can be exchanged without affecting utility. Humans, as evolved beings, should be indifferent between two choices which have an equal impact on our *Darwinian fitness*. Rogers uses this principle to define a marginal rate of substitution between benefits to oneself and benefits to future generations, which he uses to calculate an evolutionary discount function. Evolutionary discount rates can vary both between populations with different fertility and mortality schedules and between individuals of different ages; for example, young adults should discount the future more rapidly than their elders. This is because immediate benefits to a young person may have a large impact on their own reproductive success, whereas older people have more potential to increase their long-term reproductive success by helping their children or grandchildren to reproduce. For individuals in their thirties and beyond, or when benefits accrue 50 or more years into the future, Rogers calculates that a 'long-term evolutionary discount rate' of approximately 2% per year will apply in most cases when benefits accrue to descendants.

Valuing non-monetary benefits

Natural resources may produce revenues from the sale of products, but they are also likely to produce other, non-market, benefits. These might include a forest's function in protecting a watershed and so protecting fields from flash floods, or in providing a source of wild food in times of hardship. There may also be less tangible rewards such as the natural beauty and biodiversity of an area. If any conservation programme, be it based on sustainable use or total protection, is to be properly valued within the economy, then there is an argument for trying to give monetary value to these non-market benefits. If ecological assets are to be brought fully into the mainstream economy, there needs to be a single scale of measurement on which to compare ecological and economic values. Money appears to be the obvious scale to use. Other scales have been tried, such as energy or materials balance (Perrings 1987), but in order to be acceptable as an economic scale of measurement, a scale must be able to measure individual preferences, and therefore utility. Thus, the approach to the valuation of ecological assets has been to capture as much as possible of their monetary value to individuals. Whether all ecological value could ever be captured in this way depends on whether value is assumed only to be related to human needs and preferences, or whether ecosystems have intrinsic value irrespective of our uses for them. If the preferences of individuals are not an appropriate basis for ecological valuation, a single scale of measurement for combining all ecological and economic value is not achievable, because economics is based on individual preferences. Even if all ecological value is related to human needs, a single linear scale for expressing ecological value is probably unachievable, because of the many dimensions of ecological value (Gaston 1995).

There are several methods in use for estimating the non-market value of a resource. Environmental assets that are directly used by people are most amenable to the analysis of their non-market benefits, because some of the environmental values are represented indirectly in the market. Keith and Lyon (1985) valued national parks and the game within them in North America by estimating how much it cost people to travel to the parks; Brown (1989) valued elephant viewing by tourists in Kenyan parks using travel costs and direct questions on how much people would be prepared to pay to ensure elephants were present; Peters *et al.* (1989) and Hodson *et al.* (1995) valued the use of tropical forests as providers of non-timber products, such as traditional medicines, compared to their value for logging, using market values of these goods. One of the most illuminating studies is that of Norton-Griffiths (see Chapter 12), who calculated the current revenues to the Maasai from their land surrounding the Maasai Mara National Reserve in Kenya, and compared it to the potential revenues available if they were to develop the land as farmland. The present use of the land presents a substantial shortfall (US$26 million per year) over what could be achieved. This begs the question of whether that shortfall is more or less than the non-market value of the land as a provider of ecological services, and who should pay the costs of not developing it. It is clearly unreasonable to expect the Maasai to pay these costs. Norton-Griffiths and Southey (1995) calculated the shortfall in a similar way for Kenya's national parks, and obtained the figure of US$138 million per annum. Similarly, the switching value method calculates the monetary value of the alternative use of an environmental resource (such as using an area for agricultural production instead of conservation). If the environmental resource can be demonstrated to have a value greater than this, it should be retained. This is a more practical method than most, because there is no need to attempt a full valuation of the resource, with all the concomitant problems this brings (Saddler *et al.* 1980 cited in Pearce *et al.* 1990).

The final output of attempts to value ecological assets monetarily is an unquantifiable underestimate of the true value of the asset being valued. If the results of the incomplete cost–benefit analysis happen to favour environmental conservation, then this will be acceptable to conservationists. However, it is not particularly helpful to accept cost–benefit analyses that favour conservation and to argue that the calculated value of the environment is an underestimate, and so unacceptable, if the results happen to favour environmental destruction. Similarly, although estimates for the monetary value of environmental resources can be very high, these valuations are meaningless in terms of practical conservation action. Despite the difficulties, both philosophical and practical, in valuing environmental assets monetarily, there is great emphasis put by environmental economists on this research area. This is counter-productive if it means that other vital issues concerning resource use are ignored.

3.3 Summary

• Decision-making can be modelled within a static or dynamic framework, and the techniques used depend on the predictability of the decision's outcome. Economists assume that decision-makers aim to maximise utility; evolutionary biologists have pointed out that utility may be equivalent to Darwinian fitness. An individual's attitude to risk may depend on their current wealth and security. Desperate people are more likely to gamble on risky ventures.

• Models of human hunting and gathering behaviour, using the framework of optimal foraging, show that people do not generally act to conserve resources when they use them. Individuals are expected to maximise their own biological fitness, rather than making sacrifices for the good of the group. Rules governing resource use succeed only if they promote individual fitness.

• Game theory aids understanding of behaviour when outcomes depend on the behaviour of others as well as the decisions of the individual. The prisoner's dilemma game shows how one-off interactions are likely to lead to non-cooperative outcomes, whereas repeated interactions between a few people, who can communicate, may lead to co-operation.

• The incentives to break laws governing resource use depend on factors such as the expected reward from law-breaking, attitudes to risk and to the law, and the probability and severity of punishment. Increasing the probability of being caught has been shown to be a much stronger deterrent than increasing the severity of punishment.

• Natural resource-based markets are prone to externalities (costs and benefits that do not accrue to the decision-maker). They are tackled by defining property rights, valuing the externalities correctly, or through government intervention. Natural resource externalities are particularly intractable because they are diffuse, hard to value, long term and poorly understood.

• Open access to resources leads to reciprocal external costs. Open-access harvesters maximise short-run individual gains, rather than co-operating to maximise joint long-term gains. If decision-makers had control of the resources that they harvested, this might promote long-run sustainability.

• Valuation of costs and benefits over time is done in economics by discounting future gains and losses to their present value. Discounting leads to value declining exponentially with time into the future. Ecological costs and benefits that are slow to materialise will be valued less than immediate ones. Inter-generational equity is ignored by discounting.

• Monetary valuation of ecological goods and services is thought to be a way of improving decision-making concerning natural resource use. However, monetary valuation has major flaws that limit its practical usefulness. Even finding a single scale for ecological value is a fruitless exercise.

Chapter 4: Practical Considerations When Applying the Theory

There has been some disillusionment with the theory of resource harvesting and its applicability in the real world following the collapse of a number of fisheries while under management (Ludwig *et al.* 1993). This has led some to assume that the best option is to wipe the slate clean, rather than building on previous theory. While theoretical models may frequently appear to suggest clear recommendations for conservation action, in practice several issues tend to arise that make the implementation of such recommendations unsuccessful or inappropriate. In this chapter, we look at the practical issues involved in applying the theory to resource use. In particular, we show how the problems of data quality and uncertainty can be tackled within the theoretical framework of Chapters 1–3. The Revised Management Plan of the International Whaling Commission is used as an example of the modern approach to uncertainty in resource management. We discuss how to use whatever data are available to their best effect, while showing which data are essential to rational decision-making. The practicalities of conserving exploited species involve setting up a management structure, with both regulations controlling resource use and methods for enforcing those regulations. We address the policy issues surrounding the control of resource use, at the international, national and community levels. We conclude that, although there are areas that need to be addressed, the theory of resource exploitation as it stands is still a useful basis for assessing the sustainability of use (Rosenberg *et al.* 1993).

4.1 Uncertainty and sustainable use

Uncertainty is a major difficulty for the conservation and sustainable use of natural resources. It takes various forms, some more tractable than others. In section 2.1.3 we briefly discussed the effect of *environmental stochasticity* on sustainable harvesting, and how this kind of stochasticity can be included in population models. *Complexity* of ecological systems, and our lack of understanding of them, leads to uncertainty about the effects of our actions (section 2.2). Uncertainty in the technical sense of unknown probabilities being attached to states of nature can be dealt with in the framework of decison theory (section 3.1). Other forms of uncertainty arise from our attempts to get information out of systems. It is rarely possible to collect complete and accurate information about any aspect of a system, but estimates of parameters such as a population's size or range area are needed for management. The usual method for dealing with data collection under these circumstances is *sampling*. If populations are sampled using statistically valid procedures, an estimate of the sampling error involved can be obtained; this is a measure of the *precision* of the sampling procedure – how repeatable it is. Gunn (see Chapter 13) discusses how caribou population surveys are conducted to maximise precision. The important issues surrounding the correct sampling of biological populations are dealt with in Mead and Curnow (1983) and Caughley and Sinclair (1994). A less tractable problem is that of *accuracy* – how biased the measures are. If bias is severe, it needs to be acknowledged and dealt with. Bias can be introduced in various ways.

Sampling methodology. All sampling methods have potential bias problems. For example, aerial surveys are a common and useful method of estimating animal population size, but produce estimates that are biased towards underestimating population size, unless they are properly calibrated (Caughley 1974). Counts from vehicles are even worse, as they are biased in a much less predictable way. Road position depends on the geography of the area, so that counts from a road will tend to be from relatively dry and easy terrain at a relatively constant altitude. The presence of a road will also affect the behaviour of the organisms around it (Caughley & Sinclair 1994).

Deliberate misinformation. There is often no reason to assume that biological data about populations are intentionally biased, but this is not true of some types of data. Catch statistics, in particular, are often biased by under-reporting (Cannon, in press). It is usually in the interests of individual harvesters to conceal as much of their catch as possible in management regimes that limit offtake. This is even more of a problem when there is a substantial illegal harvest alongside the legal one, in which case the legal data can only be regarded as a minimum estimate of the harvest.

Excluded variables. If models are developed for resource management, there is always a judgement to be made about which variables are important to the dynamics of the system. Most variables will have to be excluded. This is not a problem if the influence of the unincluded variables is negligible (precision is not too low) and if the system does not show any trend with the unincluded variables (there is no bias). However, if there is a significant trend with the unincluded variables, then the model will produce biased predictions and will be a poor management tool. For example, environmental degradation might be gradually reducing the carrying capacity of an area over time, thus reducing the sustainable yield.

Modern approaches to resource management deal with uncertainty and the problems of data collection explicitly, rather than ignoring them and assuming that simple deterministic models are valid approximations to the behaviour of the system (Rosenberg & Restrepo 1994). One manifestation of this is the concept of *adaptive management*, in which managers are urged to treat their systems as dynamic, to be responsive to changes in the system, and to use their management as a way to learn about the system, often using Bayesian methods (Walters 1986). Information about the underlying dynamics of a particular population can be obtained by allowing it to go through a wide range of population sizes and offtake rates in a controlled way. Where possible, management actions should be treated as experiments, with proper controls (MacNab 1983). *Integrated fisheries management* is a similar approach, with a greater emphasis on socio-economics, that assesses the interactions between management objectives, the complexity of the social and ecological system, and the institutional structures surrounding the fishery, using dynamical models (McGlade 1989).

A contrasting approach is that of *no-take areas*, which are effectively reserves that act as refuges for the harvested species. This management measure takes a straightforward precautionary approach to complexity and uncertainty. Rather than relying entirely on developing an understanding of the dynamics of the system as a way of tackling uncertainty, the no-take area acts as a buffer against over-harvesting. No-take areas have been proposed particularly for reef fisheries (Roberts 1997), which are especially difficult to manage because harvesters catch many different species together (section 2.2). A strict no-hunting reserve should allow the less productive species to survive and continue to contribute to the harvest, as individuals migrate out of the protected area into the harvested area. Solis Rivera and

Edwards (see Chapter 7) describe how a no-take area of this sort is part of a harvesting scheme for iguanas in Nicaragua. The policy only works as a way of maintaining viable harvest levels if the protected area does act as a source for surrounding areas, i.e. if species' ranges are relatively localised, but with effective local migration along density gradients (Hatcher 1995). It would not work for migratory species such as squid or tuna. The approach is often popular with ecologists, because it should conserve the resource effectively within the confines of the reserve, whether or not it helps increase the harvested stock. Resource users are often less enthusiastic, because they find themselves banned from hunting in an area, while benefits in terms of increased yields are uncertain and slow to arrive (Valdes-Pizzini 1995).

4.1.1 The Revised Management Procedure of the International Whaling Commission

A particularly interesting approach to the problems of the sustainable management of renewable resources under uncertainty is the Revised Management Procedure (RMP) of the International Whaling Commission (IWC). The most important aspect of the RMP of the IWC is not the final management procedure itself, but the method by which it was chosen. The IWC was rather unsuccessful at preventing the over-exploitation of whale from its inception in 1946 to the moratorium on commercial whaling starting in 1986 (Cooke 1995). During the moratorium, the IWC's Scientific Committee worked to produce a management regime for the sustainable use of whales, which was accepted by the IWC in 1994. The development of the RMP broke new ground (IWC 1994; Cooke 1995; Kirkwood 1996). Due to pressure from some conservationists, the RMP has not been implemented, and commercial whaling remains closed for the foreseeable future.

   The first step in the method is to devise *explicit management objectives* for the system. The management objectives chosen by the IWC were (Kirkwood 1996):
1   Stability of catch limits (desirable for the orderly development of the whaling industry).
2   Acceptable risk that a stock is not depleted (at a certain level of probability) below some chosen level (fraction of carrying capacity), so that the risk of extinction of the stock is not seriously increased by exploitation.
3   Making possible the highest continuing yield from the stock.
These are quite general objectives that might be chosen for any harvested resource, although due to the special position of the IWC as a regulatory body, there is no mention of the overall profitability of the enterprise. Note that trade-offs need to be made – maximising yield is generally incompatible with yield stability and low risk of extinction, for example (section

1.3.2). It was decided that criterion 2 needed to be satisfied first, to ensure conservation of the stock, and that a management procedure could then maximise yield (criterion 3), subject to reasonable yield stability (criterion 1). Several possible management procedures were then tested to see how well they met these objectives. The best management procedure was chosen for its ability to give a robust and satisfactory performance in meeting the management objectives. This allowed some subjectivity in selecting the best management procedure, rather than trying to maximise utility over several complex criteria.

One way in which the method is different from adaptive management is that the IWC committee tested their procedures using computer simulation, rather than trying to test them on the actual populations to be conserved. The disadvantage of this is that the only true test of a management procedure is whether it actually leads to sustainable use in the real world. However, proper tests of a management procedure could be dangerous to the populations, take many years, and require several independent populations to give adequate statistical controls. Further, it is not always easy to see the exact effects of a management procedure, given that data-gathering from populations is subject to error (Cooke 1995). Using computer simulations means that the true status of the model population is known. This procedure is also known as the 'operating model' approach (Hilborn & Walters 1992). The process is illustrated in Fig. 4.1, which is a generalisation of the method used by the IWC Scientific Committee. The important steps are:

1   Data are generated for two aspects of the system – ecological and

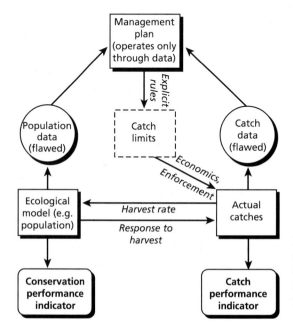

Fig. 4.1 A diagram showing the conceptual basis of the new method for developing management plans. (Inspired by Cooke 1995.)

economic. The ecological data are typically estimates of population abundance, but could also be population structure, or more complex ecological
measures if a multi-species system is being managed. The economic data
are typically offtake levels (e.g. number of individuals, biomass, age–sex
structure of harvest). Both data sources are biased and imprecise to a
degree that may be unknown to the managers, and the frequency of data
collection varies. There may or may not be records of historical harvesting
levels and population abundances.

2   The management procedure uses the population and catch data as
inputs to generate its explicit harvesting rules. Thus, the manager does not
know the true state of the system, but only the state of the system as
filtered through the data-gathering process, which is unreliable, as in the
real world.

3   The explicit rules from the management plan are expressed as some
form of catch limit. However, the economic situation and the manager's
success in enforcing the regulations determine the actual harvest level (see
sections 1.3–1.5 and 3.1.4).

4   The harvest rate has an impact on the ecological system (typically the
population being harvested), and the internal dynamics of the ecological
system determine the response to harvesting, and thus the offtake. The
ecological and catch data collected are the feedback to the management
procedure, which are used to determine the next set of catch limits.

5   There are two sets of unbiased performance indicators for the management procedure. The conservation performance criteria might be total
population size or population structure (age and sex, spatial structure or
sub-population size) each year. The catch performance criteria might be
catch size, quality (e.g. trophy size), or profitability each year. These annual
data can be aggregated over the period of a simulation to provide summary
data on variability, minimum values and long-term sustainability of yield.
These data are then used to evaluate the performance of the management
procedure against the original set of management objectives. Note that
these performance indicators are based on true, unbiased output from the
models, not filtered through simulated data collection procedures.

   The IWC Scientific Committee used a process similar to that illustrated
in Fig. 4.1 to test management procedures that the committee members
had devised. They developed a population model containing the best
available information about whale biology. They then carried out some
mild simulated trials of the chosen procedures to establish their baseline
performance. These trials typically involved 100 stochastic simulations of
100 years duration, for various starting population sizes. The procedures
could use the available historical data on population sizes and catches as
input, as well as the data produced as the model ran. Next, a set of severe
trials were run, to test model robustness. These trials showed what
happened when the assumptions about whale biology and data quality

broke down, including environmental deterioration, poor-quality data, epidemics and several independent whale sub-populations being exploited. If a procedure performed badly in certain cases, it could be modified and improved. The testing method is time-consuming and computer-intensive, but Kirkwood (1996) concluded that it would be feasible for government agencies and others responsible for important commercial fisheries. The method works even if the ecological models are not perfect representations of the true dynamics of the system, because the simulations can explore the whole range of plausible dynamics. As knowledge about the system improves, the procedure can be re-tested with the new model.

4.1.2 Insights gained from the RMP

The development of the RMP gave some insights into which type of management strategy is likely to lead to sustainable harvesting. Although these insights are still preliminary, were developed for whales, and have not been tested outside a computer, they are still worth repeating. They are based on more rigorous quantitative testing than any previous set of management recommendations, acknowledging the practical difficulties that managers face, and may be relevant to the management of exploited stocks in general. The list is adapted from Cooke (1995):
• Regular and direct surveys to determine abundance are crucial to successful management.
• Any other data about the exploited stocks are only marginally useful. Historical data on total offtake from the population are more useful than detailed data on population dynamics. Data on age structure, density-dependence functions, sex ratios, etc., are often seen as rather important for management. However, the IWC trials showed that regularly collected survey data on abundance are both *necessary* and *sufficient* for good management. One exception to this rule may be species that are hunted very selectively; for example, if hunters target only adult males. These populations may experience non-linear population crashes, and in this case data on population structure may also be necessary for good management.
• Safe management is only achievable by limiting catches to a small proportion of the total population size. Among the procedures tested, those that allowed higher catches in good years and tried to use trends in abundance as cues for raising and lowering harvest rates performed badly.
• Persistent bias in surveys of abundance was a problem for management strategies, while temporary biases were more easily overcome. This suggests that as survey methodology improves, it is much more important for managers to try to reduce bias than to try to maintain comparability with previous surveys.
• The management procedure that was finally adopted had a catch limit that was a function of the precision of the abundance data. It allowed

catches only when there was at least one estimate of abundance in existence. Given that estimates were available, the more precise the estimates (the narrower the confidence intervals), the higher the catch limit that was set. This meant that poor data led to lower catch limits, not higher risks to the stock. There was a built-in incentive for harvesters to pay for frequent high-quality surveys, so that their catch limits could be increased.

4.2 Data needs and availability

The previous section highlights some important issues concerning the kinds of data needed for sustainable resource use, and the problems created by inadequate data. *Data quality* is becoming an increasingly important issue for fisheries managers, who rely on catch data to make their stock assessments. Under-reporting of catches by fishers can lead to a vicious circle of stock assessments becoming more and more divorced from reality, whilst the fishers become more and more inclined to ignore the regulations (Cannon 1997). Stock assessments should not be totally reliant on data from users, but should use independently collected data as well. Information also needs to be up-to-date if it is to be useful in stock management; for example, the data used for management of the squid fishery in the Falkland Islands are collected weekly (see Chapter 8). Ideally, long-term and continuing statistically valid data on population sizes and numbers harvested should be used for analysing the sustainability of resource use. These data should be available for enough time before management starts to ensure that the system's dynamics are well described. For example, a single point estimate could easily be meaningless as a guide to population size, if the population in an area fluctuates seasonally or over a longer period. This could be due to climatic fluctuations or to migratory behaviour, as is the case for caribou and muskox in Canada (see Chapter 13).

In reality, it is necessary to base decisions on whatever data happen to be available. A *preliminary survey* of the resource before management starts is extremely important, in order to establish its baseline status, although in practice management is unlikely to be operating when a resource is first used. Knowledge of the resource's baseline status makes it possible to measure changes in the status of the resource against a meaningful standard. The baseline should not be taken as representing the point at which the resource should be maintained, as there is no reason to suppose that the status quo is optimal. For example, the present-day state of Caribbean reefs is radically different from their historical state, due to massive over-exploitation of turtles a few hundred years ago; turtles used to act as keystone species in the ecosystem. Comparing changes in reef condition to the condition of any reef in the Caribbean today is comparing against a highly altered ecosystem – it is like removing the large herbivores from the African savannahs and calling the result a 'baseline' (Jackson, 1997). People are becoming more aware of the issue of *shifting baselines*, in which we tend to assume that what we observe now is the norm. As successive generations do this, the norm tends to shift, often towards lower quality as ecosystems deteriorate (Pauly 1995; Sheppard 1995).

4.2.1 Population data

Cooke (1995) highlights direct estimates of population size as vital for successful management. However, direct population estimates are often hard to obtain, particularly for forest or marine species. It is often easier to use estimates of *relative abundance*, which give trends in population size over time, but not actual population sizes. These are unsatisfactory, as they cannot be used to calculate offtake levels in advance. However, they can be used as tools to monitor broad trends in population size (Box 4.1). Freehling and Marks (see Chapter 10) use data from hunters to estimate trends in relative abundances for large mammal species in Zambia over 30 years, while Gunn (see Chapter 13) uses aerial surveys to track annual trends in abundance of caribou and muskox.

The accuracy of a population survey depends on the technique used, and the conditions under which it is undertaken. For example, elephants in small national parks in savannah ecosystems can be relatively precisely counted using direct aerial surveys. However, the elephants of the West and Central African forests can only be counted using indirect methods, principally dung counts (Barnes 1989). Thus, the estimate of the total number of elephants in Africa includes relatively precise estimates for eastern and southern Africa, and very imprecise estimates for Central Africa, where up to 38% of the continent's elephants may live (Fig. 4.2; Said *et al.* 1995). Some of this variation also depends on the effort and money expended on surveys. If survey techniques cannot provide an accurate population estimate, this needs to be taken into account in the allowable offtake if the resource is to be conserved effectively, for example, by making offtake levels inversely dependent on the accuracy of population surveys.

Cooke (1995) emphasises that it is more important to improve *survey methodology* whenever possible than to try to keep surveys comparable. Keeping surveys comparable has the attraction of allowing relative trends in abundance to be monitored over time, but this is outweighed by the danger of persistent biases. Changes in survey methodology can, however, produce highly misleading apparent changes in abundance, so it is important to resist the temptation to maximise the number of data points by comparing old estimates with those obtained under new methods. For example, much has been made of the 94% decline in black rhino numbers from 65 000 in 1970 to 3800 in 1987. Yet the estimates of rhino numbers, upon which this massive decline is based, are dubious in quality. They could all be under-estimates (Western 1989b), or the first figure could be up to 100% over-estimated (N. Leader-Williams, pers. comm.) although this still gives a decline of 89% over the period. The opposite

---

Box 4.1 Using relative abundance data from game scouts

Fig. B4.1a Black rhino and calf, Tanzania. (Photo by Dave Currey/EIA.)

Game scouts can be very useful sources of data on relative abundances, for species like black rhinos, which are very hard to count from the air. As they carry out routine patrols, they record the number of rhinos and elephants (and other game species) they encounter, as well as recording information on the number of scouts in the group, the route taken and the date. In the Luangwa Valley, Zambia, data from game scouts were available both for the period 1947–69 and for 1979–85. When confounding factors such as time of year, number of scouts in the group and patrol length were factored out, the data gave estimates of trends in rhino and elephant abundances (Fig. B4.1b).

   These data suggest that there was no clear trend in rhino density over the period 1947–69. However, there was a catastrophic decline over the period 1979–85, caused by heavy poaching. The elephant data show a slower decline between 1979 and 1985, and also a rise in elephant numbers from 1947 to 1969, due to improved protection after setting up the national park. The recent rhino trend was confirmed by comparison with data on changes in actual rhino population sizes, collected in certain small and well-monitored areas of the Luangwa Valley. The two methods gave similar results. For elephants, aerial population surveys also tallied with the game scout estimates. The advantage of the game scout method is that the data are routinely collected anyway in the course of their duties, so that there is no extra cost involved in data collection, and large amounts of data are generated, with consistent monitoring over a long period (Leader-Williams *et al*. 1990).

continued

Box 4.1 *Continued*

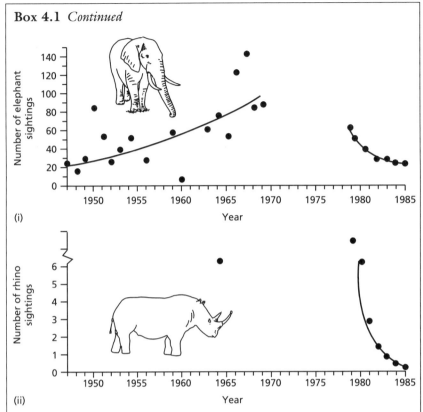

(i)

(ii)

Fig. B4.1b Trends in elephant and black rhino abundances, 1947–69 and 1979–85 (from Leader-Williams *et al.* 1990). The trends have been standardised to numbers of sightings made by a four-person group, patrolling for 7 days in January, using an amalgamation of the data for the North and South Luangwa National Parks. (i) Elephant sightings. (ii) Rhino sightings (only one datapoint is given for 1947–69, the mean for the 23 years, as there was no significant trend in sightings over time).

effect, of an apparently large increase in numbers as survey methods improve, is also common (see Box 4.2).

4.2.2 Harvest and trade data

Direct data on the number of individuals killed, or data from the trade in their products, can be quite extensive for commercially valuable species, particularly if their products are exported. For example, good data on the trade in ivory from Africa stretch back to 1814, and there are patchy data back to the 16th century (Parker 1979). Often, trade data are much more easily available than population data, so it is important to know how to use them to infer trends in population status. For example, the number of lynx and snowshoe hare furs received by the Hudson Bay Fur Company over the

Fig. 4.2 Estimates of the population size of the African elephant in 1995 (Said *et al.* 1995), divided by region and quality of the data. Although Central Africa holds up to 38% of Africa's elephants, only 3% of Central Africa's population estimates are categorised as definite, compared to 75% of Southern Africa's estimates.

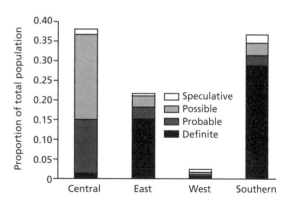

period 1840–1935 show such clear coupled oscillations that the data are used in ecology texts to demonstrate the dynamics of a predator–prey system (Fig. 4.3). If data on numbers killed are used as a proxy for the unexploited dynamics of a population in this way, the hunters must be taking a negligible proportion of the population, so that the hunting itself is not contributing to the population dynamics at all. Even using the data as an index of relative abundance requires that the hunting effort has not changed at all over the period, so that yields are directly related to population size. This is unlikely to be true in the case of the lynx and

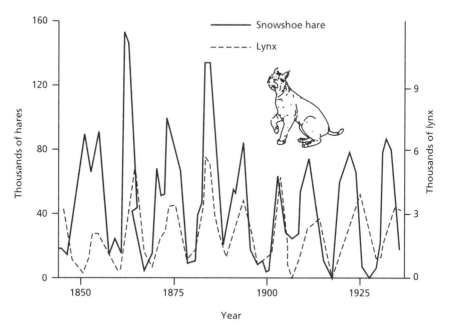

Fig. 4.3 The coupled oscillations of the lynx and snowshoe hare populations in Canada over the period 1840–1935. Data were collected from the number of pelts of each species taken to the Hudson Bay Fur Company each year. (From Begon *et al.* 1996a.)

Box 4.2 Estimating the population size of the Mongolian gazelle

The Mongolian gazelle (*Procapra guttorosa*) population has at first sight an enviably long-term set of population data. There is a 'baseline' data point from the 1940s, before large-scale legal hunting began, followed by regular estimates from 1975 to 1982. These seem to show that the population declined to about 20% of its former size between the 1940s and 1975, and since then, has remained roughly at the same level. However, in 1994, an aerial population survey was carried out, with radically different results – the population was estimated at more than double the size of the 1940s estimate (Table B4.1). This single result is not comparable with any of the earlier results, and the 1975–82 estimates are not comparable with the 1940s estimate either. Models show that, if the aerial survey is accurate, none of the other estimates could possibly be correct.

Year	Population size (1000s)	Source
1940s	1000	1
1975–6	180–200	2
1979	250	2
1980	180	2
1982	300–400	2
1994	2670	3

Table B4.1 Mongolian gazelle population size estimates from the 1940s to 1994, showing the effect of sampling technique on survey results.

Sources: 1 Bannikov (1954), based on extrapolation from range area and observed densities. 2 Lushchekina (1990), based on counts from a vehicle along a transect. 3 B. Lhagvasuren (pers. comm.), based on an aerial survey.

The Mongolian gazelle is an important game species, and is hunted both commercially and for subsistence. Official hunting quotas have been partly based on the population size estimates, and so have been low. Given the recent high levels of illegal harvesting, this has resulted in a redistribution of the profits from hunting rather than in the species being economically under-exploited. A major problem is that there is no evidence from population data about whether the Mongolian gazelle actually declined to 20% of its former abundance between the 1940s and 1975, although it is clear that its range area was reduced. Scientists believe that there has also been a decline in gazelle populations between 1982 and 1994, due to a large increase in poaching. Evidence for or against this has been masked by the change in survey technique. It would have been useful to have carried out a survey using the old method alongside the aerial survey, to give an idea of the degree of under-reporting from the old method, and so allow extrapolation of the population trend from previous surveys to the present day (Milner-Gulland & Lhagvasuren 1998).

snowshoe hare, because market conditions are unlikely to have been stable over such a long period. The problems with assuming that a non-declining CPUE (catch per unit effort) is a good indicator of sustainability were discussed in section 1.3.1.

Historical harvest data can be used to get an indication of past population trends without making restrictive assumptions about the relationship between yield and effort, if there is one good estimate of population size at the end of the data-series. It is then possible to model the population dynamics backwards, knowing the number of individuals remaining at the end, and the number that were removed from the population each year by harvesting. The more data there are on population size alongside the harvest data, the narrower the bounds of likely abundances are. A similar method is used in fisheries science, where it is known as virtual population analysis (Magnusson 1995). The method is widely applicable, although it has difficulties with error propagation. Historical harvest data have been used to show likely population trends of the African elephant population over the last 200 years (Milner-Gulland & Beddington 1993). The elephant population probably declined slowly over the 19th century, due both to habitat loss and hunting. There was a respite in the first half of this century due to the world wars. Since 1950, the population has declined at an accelerating rate, due principally to the ivory trade (Fig. 4.4).

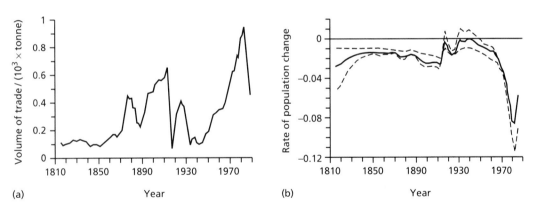

Fig. 4.4 An analysis of the likely trend in African elephant population size over the period 1814–1987, using data from the ivory trade. (a) Smoothed data showing the volume of ivory leaving Africa, 1814–1987. Note that there were two peaks of trade: one at the turn of the century chiefly supplying colonial markets in Europe, the second in the 1980s, supplying mainly Far Eastern markets. The ivory trade was halted by CITES in 1989. (b) The rate of elephant population change caused by the ivory trade and habitat loss combined. The solid line is the mean rate of change; the dashed lines are the bounds within which the modelled population size lies. The population declined at a rate of about 2% a year throughout the 19th century; then the population stabilised (rate of change around zero) before declining at an increasing rate from the 1950s onwards. (From Milner-Gulland & Beddington 1993.)

The example in Fig. 4.4 illustrates another point about trade data – note that Fig. 4.4(a) shows the volume of the ivory trade in tonnes, i.e. in terms of *biomass* removed, not numbers of individuals killed. In order to use harvest data to make inferences about population trends, the data need to be transformed into numbers of individuals killed. However, often the number of individuals is of lesser concern to traders than the total amount of the product that is sold, so only biomass is recorded. This makes the data much less useful. In the case of ivory, a mean tusk weight in the population must be assumed in order to convert biomass into numbers killed. Similarly, if the amount of meat is recorded, a mean mass of meat per individual must be assumed. These mean values can vary, particularly with the age and sex ratio of the population. Hunting itself causes changes in the age and sex ratio, as was discussed in section 2.1.2. This means that changes in volumes traded or in the size of trophies (tusks, pelts or carcasses) can be difficult to relate back to the dynamics of the hunted population (Box 4.3).

It is easy to jump to conclusions from ambiguous data, particularly to assume that populations are over-harvested because the size of trophies is declining. This is very tempting, because trade data are often all that are available for little-known, hunted populations. It is precautionary to assume that signals like declines in biomass are due to over-exploitation, and to use this as a reason to act to curb hunting. However, it is still important to remember that these data are only indirect indicators of trends in total population size, confounded by human decision-making, and by the dynamics of age and sex structure. They can easily be misleading.

4.2.3 Official socio-economic data

Ecological data are important for assessing the sustainability of resource use from an ecological perspective. Equally important, however, is sustainability from an economic and social perspective. Market data, information on the social and political situation, and on the institutional setting of resource use are needed to assess this. Governments routinely collect these data. Demographic data give information on human population movements, while economic data give mean incomes in various sectors of the economy, discount rates, and measures of people's consumption of goods. These aggregated data can be important for the analysis of internationally traded products like ivory, or for calculating the opportunity costs of hunting compared to the employment available in other sectors of the economy. Agricultural data on livestock holdings and productivity of farming are useful for comparing wildlife use to alternative land uses, as Norton-Griffiths (see Chapter 12) has done for Kenya. Often the data are available as time series, allowing wildlife population sizes to be correlated with factors in the economy such as changes in income or population

Box 4.3 Interpreting tusk weight distributions from the ivory trade

Fig. B4.3a Ivory seized by the Tanzanian government, Dar Es Salaam. (Photo by EIA.)

From 1986 to 1989, there was a quota system in force for ivory exports from countries that were members of CITES. Export permits from these countries were submitted to CITES, and the data were compiled into tusk weight distributions by the Wildlife Trade Monitoring Unit (WTMU) of IUCN. These distributions were of two types: in the majority of countries they were heavily skewed towards small tusks, with a long tail and a median tusk weight below 5 kg; however, three countries (Gabon, Congo, Zimbabwe) had a less skewed distribution, with a median tusk weight of 10 kg (Fig. B4.3b). A smaller sample of tusks from government ivory stores in various countries for the period 1979–81 had a similar distribution to that shown by the majority of countries from 1986 to 1989 (Fig. B4.3c). A distribution of tusk weights from an importing country (Singapore) is similar to the distribution for the majority of exporters (Fig. B4.3d).

These tusk weight distributions provide clues about the status of African elephant populations, remembering that old males have the heaviest, most valuable, tusks (Box 2.1). For example, the selective exporters could be assumed to have had healthier elephant populations, with more old males. The other distributions could represent heavily hunted populations, such that only smaller-tusked females and juveniles were left. There seems to be little difference between the tusk weight distribution from 1979 to 1981 and that from 1986 to 1989, suggesting that the tusk weight distribution had reached equilibrium. This can occur even when the population is declining rapidly. When populations producing trophies of age-related size are first hunted, the mean trophy weight tends to drop rapidly, due to reductions in the mean age of the population. The more selective the hunting is for large trophies, and the higher the hunting mortality, the faster the initial drop. After this the trophy size stabilises.

The explanations given above are supported by evidence from other sources. But it would be dangerous to draw conclusions about harvest rates in particular countries from tusk weight distributions alone. The

continued

Box 4.3 *Continued*

ivory market was fluid, so that tusks exported from one country could
easily have come from elephants shot in another (the extreme being
Burundi and Djibouti which exported ivory despite having no wild
elephants). Thus, the distributions of the 'selective' exporters could be
misleading as to the health of their elephant populations. There is
evidence (Barnes 1989) that small tusks left Gabon and Congo for sale in
other countries without CITES permits. Conversely, it seems that Sudan's
distribution was unrepresentatively biased towards small tusks because the
ivory store was systematically mined by people replacing valuable large
tusks with small ones (C. Huxley, pers. comm.). The distributions also
contain tusks from elephants shot over an indeterminate period of time,
so they say nothing about harvest levels in a particular year. Finally, mean
tusk weights in the trade depended on what was demanded by importers
as well as what what was supplied by exporters. The mean tusk weight in
importing countries declined sharply from 1979 to 1989, despite the
similarity of the distributions in exporting countries (Fig. B4.3c).
Singapore's distribution (Fig. B4.3d) is probably the best existing
snapshot of the illegal ivory trade in 1986. All tusks registered in
Singapore were legalised in that year when the country acceded to CITES,
causing the tusks to increase markedly in value. Dealers probably shipped
as many tusks as possible to Singapore in order to take advantage of this
windfall (Milner-Gulland & Mace 1991).

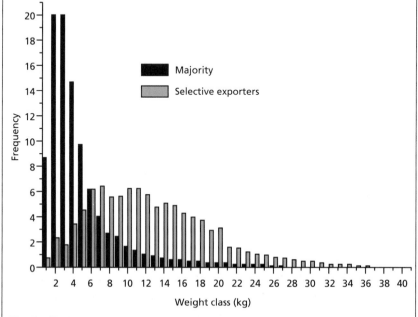

Fig. B4.3b A comparison of the 1986–89 tusk weight distribution of the majority of
African ivory exporting countries and the selective exporters. (Data from WTMU.)

continued

Box 4.3 *Continued*

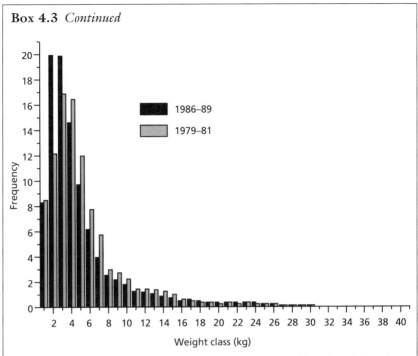

Fig. B4.3c A comparison of the tusk weight distribution for 1986–89 (majority of exporters only) and that for 1979–81. (All data from WTMU except that data from Kenya and Tanzania is from I. Douglas-Hamilton, pers. comm.)

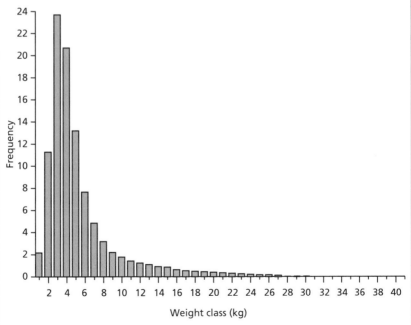

Fig. B4.3d The tusk weight distribution of the Singapore stockpile of 53 000 tusks, registered in 1986. (Data from WTMU.)

growth. There are, however, caveats about using these data:
• Official data sources may be prolific, but that doesn't mean they are of good quality. Some government statistics are not worth the paper they are written on.
• The data are usually heavily aggregated. If local incomes from hunting and agriculture in a remote area are to be compared, for example, there is little point in using aggregated statistics – locally-based income calculations are more relevant. Often prices for produce vary strongly between local, national and international markets. The appropriate level must be chosen to represent the incentives faced by the people who actually decide harvesting rates. For example, if the decision-makers are middlemen based in the town, then national data can be used for calculating opportunity costs, but if local people are the ones who decide the amount of resource use, national data will be less useful.
• Time series of incomes, prices and discount rates can be very useful for relating changes in harvesting rate to changes in social conditions. However, constructing an econometric model of the market for a good is technically difficult (Maddala 1989). It is necessary to convert time series of prices or incomes into *real* values rather than using the actual values experienced at the time. This involves dividing the data by the consumer price index or the GNP deflator to remove inflationary trends. Inflationary trends are misleading, as they affect the whole economy. For example, the actual price of rhino horn in Japan was relatively stable from 1951 to 1970. It started to rise slowly in 1970, then rose dramatically from 1978. Taking real prices still shows a dramatic price rise over the period 1978–80, but in the context of a much more volatile previous price, and no steady rise from 1970 (Fig. 4.5). These real prices can then be related to the quantity of horn entering Japan, together with other economic

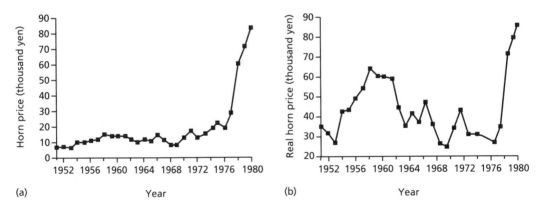

Fig. 4.5 (a) Actual rhino horn prices in Japan, 1951–80. (Data from Martin 1983.) These are the prices paid by consumers at the time. (b) Real rhino horn prices in Japan, 1951–80. (Consumer price index from IMF, 1981–88.) These prices are comparable over time, because inflation has been taken into account.

variables, in order to determine what drives demand for rhino horn. In this case, price did not significantly affect demand for rhino horn; GNP and changes in GNP were the main determinants of demand (Milner-Gulland 1993).

4.2.4 Ecosystem-level data

Ecosystems as entities are now receiving more attention from conservationists, who have traditionally been concerned with saving individual species, rather than protecting whole ecosystems and the processes within them.

The type of ecosystem-level data that is used for assessing sustainability depends chiefly on what is to be conserved. If a species is to be used sustainably, then data will be needed on the amount and quality of the habitat suitable for that species. If a whole ecosystem is to be used sustainably, or if conservation of an ecosystem is to be promoted by the use of a species from that ecosystem, then some measure of ecosystem health is needed.

For single species, the quantity of suitable habitat available may be relatively easily measured. The new IUCN criteria for *categories of threat* are used to classify species according to their danger of extinction (IUCN 1994). These new criteria explicitly recognise the size of the range area and habitat quality as important determinants of the threat status of a species. If a species is concentrated in very few areas, fragmented into many small areas, or if its habitat is declining in size or quality, it may qualify for threatened status. Habitat requirements are taxon-specific, so that large-bodied predators require large areas, while other species can survive indefinitely in very small areas (cave-dwelling endemics, for example). Calculations of the area requirements of different species can be done using the species richness of habitat remnants from the last ice age. They suggest that very few reserves are large enough to ensure that the populations of large carnivores within them can persist in isolation for 100 years, and none ensure persistence for 1000 years (Belovsky 1987). Although an adequate amount of suitable habitat is necessary for the conservation of a species, it is not sufficient, particularly if the species is exploited. In this case, there is no substitute for estimating actual population size to determine how fully occupied the habitat is. It is also important to be sure that the species is not intermittently reliant on particular habitat types that might be under threat. For example, the Soviet Virgin Lands project ploughed up most of the northern steppe of Kazakhstan in the 1950s. This appeared to make little difference to the saiga antelope population, but the ploughed area had been an important refuge for the species in severe drought years. The effect of this project on the saiga will only be seen under these extreme climatic conditions.

There is still a long way to go in terms of defining and measuring ecosystem health or quality. *Biodiversity* is one component of ecosystem quality that has received attention, although there are other components, such as buffering capacity, water-retention capability, connectance between components, productivity, and uniqueness, which may be of equal or greater importance, depending on the objectives of the conservation initiative. Even when biodiversity is chosen as a crude metric for ecosystem quality, the problems with measuring it are severe. Simple measures like species richness, which is a relatively crude approximation to biodiversity, are still fraught with technical difficulties (Gaston 1995). One practical method of monitoring the effects of use on an ecosystem is to use *indicator species*. Depending on the objectives of the manager, an indicator species may be chosen because it is particularly sensitive to human disturbance, or relies on some particularly important component of ecosystem function. Alternatively, the species may simply be easy to count, or be of particular interest. For example, the giant otters of Madre de Dios National Park, Peru, are both particularly interesting to tourists and particularly shy of humans. Monitoring the population of giant otters in the vicinity of tourist camps is a way of detecting whether tourists are causing unacceptably high levels of disturbance (Groom *et al.* 1991).

4.2.5 Making use of existing local knowledge

Outsiders, such as scientific researchers or development agents (be they governments, national or international NGOs), have frequently ignored the knowledge that rural people have of the natural resources in the area where they live. Where the natural resources are harvested by locals, either for subsistence or trade, then their knowledge of them is likely to be extensive. However, in many parts of the world, the individuals with this knowledge may be uneducated and illiterate; they are not generally accustomed to interacting with officials from government or international bodies (except as passive subjects of their laws). In the 1970s, development agencies began developing *participatory approaches* that could tap into this knowledge (Chambers 1983), and in the 1980s, a significant number of conservationists became aware that participation with local people is an essential part of effective conservation. The main thrust behind participatory approaches was to enable local communities to identify their needs, beliefs and opinions to the outsiders who were attempting to develop and/or conserve their area (section 4.4), but they are also a useful means of collecting data on natural resources.

Participatory research has been popular with anthropologists for decades. In its purest form, this involves the researcher living with a local community for an extended period, learning the language if not already a native speaker, and taking part in all aspects of daily life; in short, trying to live, as far as is possible, as if they were a member of that community. Recently, more of these data are being collected by local inhabitants themselves, working with or independently from outside researchers (Fig. 4.6). These are the approaches that produced the data on the Bisa people of the Luangwa Valley, Zambia, which have been analysed by Freehling and Marks (see Chapter 10). Whilst the knowledge gained from such an exercise can be very valuable, as Freehling and Marks show, this approach is considered too time consuming and expensive to fit in with the needs of many development and conservation initiatives. A family of measures known as rapid rural appraisal (RRA) was developed to try and gain as much of the relevant indigenous knowledge as possible in a short time. By 1990 it had become more popular to refer to these approaches as PRA (participatory rural appraisal) (FAO 1989; IIED 1994), and lately PLA (participation, learning and action) is the acronym of choice in many development agencies.

With respect to data on natural resources, methods that can be used to enable rural, possibly illiterate, people to impart their knowledge include semi-structured interviews both with groups and with key informants (such as those specialising in the harvest of a particular resource). Thus, consensus views, individual opinions and knowledge can be gained. The participatory review of an iguana-rearing project in Nicaragua described by Solis Rivera and Edwards (see Chapter 7) included using drawings, charts, games and theatre to facilitate discussion. Through such interviews it is much easier to gain information from a range of different sectors of society, and thus to get information the local leaders might not have. For example, nearly all fuelwood and most non-timber forest products are collected by women, especially poor women, who may well not be represented in any political groups or committees, if such groups exist at all. The lack of participation by women in the decision-making of the

Fig. 4.6 Interviewing a Bisa elder on lineages and local history, 1973. (Photo by Stuart Marks.)

Nicaraguan iguana-rearing project was identified as one cause of the problems it encountered (see Chapter 7). Mapping the location of resources (possibly just by drawing in the sand) and walking transects with informants are often useful parts of such discussions. Ranking can be used to establish the relative quantities or importance of different resources. Oral histories and descriptions of daily or seasonal routines are often very informative. Thus, data can be collected on which resources are harvested and how, which are the most important and the areas where these are found, seasonal changes, and historical trends, including any history of conflict in the area over these resources. If researchers are planning to make use of questionnaire surveys, such knowledge could be an essential prerequisite, both with respect to designing appropriate questions and to alerting the researcher to possible biases in the answers obtained (Moser & Kalton 1971; de Vaus 1996).

PRA will not give enough information to form the basis of a fully fledged community-based conservation project; nor should the data obtained be considered as some kind of baseline against which future progress or declines can be measured. However, PRA is a good starting point for almost any interaction between the local community and development or conservation agencies, or researchers who do not already have extensive involvement in an area.

Sometimes external agencies do have a greater capacity to assess the status of a harvested population than does the local community, but that knowledge is either not shared or not believed by locals, making the data less useful. Gunn (see Chapter 13) describes a situation where information was usefully exchanged between local hunters and government scientists regarding caribou and muskox in Canada's Northwest Territories. Bad relations between these two groups had meant that hunters did not believe government figures. Co-operation between the groups led to learning in both directions. The government had extensive monitoring tools available to it, such as aerial surveys. When hunters were involved in the formal population surveys, they were more inclined to believe, and better able to make use of, the information obtained.

4.3 Policies for regulating resource use

The analysis of regulatory instruments tends to focus implicitly on management by national bodies external to the users, such as government agencies. In fact, regulation of resource use can occur at several levels: within a community, at the national, regional, or international level. Regulatory effort needs to be concentrated at the appropriate level, both ecologically and socio-politically. Ecologically, it would be pointless to regulate harvests of highly migratory tuna at the village level. Socio-politically, externally imposed regulation might just exacerbate locals' alienation from sustainable resource management; Freehling and Marks (see Chapter 10) show how local Bisa in Zambia are alienated from the ADMADE programme set up to promote sustainable resource management. Within a country, the way in which particular regulatory instruments act is generally independent of the scale at which they are applied, but international agreements and small local communities do require separate analysis.

4.3.1 International agreements

Many conservation problems are problems of more than one country, and measures taken in one region may be ineffective if similar measures are not taken elsewhere. The international trade in endangered species is one such case. As environmental problems become more global, international agreements will increase in conservation importance. International law is different from domestic law, in that there is no higher authority to enforce it, should any country decide not to comply. Thus, international regulatory agreements are based on the voluntary co-operation of signatories. The incentives to sign up to international agreements on environmental issues (such as the Convention on Biological Diversity, now ratified by more than 160 countries) include that they are popular with the domestic electorate, or that there are political advantages to being seen as a responsible and co-operative member of the international community. International agreements are not likely to be effective if countries have a strong national interest in not co-operating, or are not interested in how they are perceived by the international community. Although international agreements can have positive effects for the environment (e.g. the unprecedented success of the Montreal Protocol for reducing CFC use), too often they are weakened by domestic interests (e.g. the USA failing to sign the Convention on Biodiversity).

The Convention on International Trade in Endangered Species (CITES) is an important tool for the conservation of internationally traded endangered species. It recognises three categories (or appendices) in which species can be listed. Appendix I effectively bans all commercial trade, although there can be exemptions, for example, for sport hunting trophies. Appendix II requires the parties to the Convention to monitor trade in the species concerned, and to issue import or export permits only if they are satisfied that the trade does not constitute a threat to the survival of the species. Any state can list their own population of a species on Appendix III, which requires the state to issue a permit for export, and importing states to check the origin of the product, and to require an import permit if it comes from the listing state. Most, but not all, countries are parties to CITES, notable exceptions being Taiwan and South Korea.

CITES came to public prominence in 1989, when the parties voted to ban the international trade in ivory. Zimbabwe, and some other states that traded in ivory, exercised their right not to comply with the agreement, due to their national interest. In this particular case, the only countries that did not comply were those countries wishing to sell ivory, whilst all the major importers of ivory (USA, EU, Japan) did sign up to the agreement; so this international ban was broadly successful at reducing the quantity of tusks traded and the number of elephants killed. This illustrates how control of buyers can often be a more effective conservation tool than attempting direct control of harvesters, particularly if the harvested re-source is in remote areas that are almost impossible to police. Although CITES regulates the trade in hundreds of species, the ivory issue has caused a re-examination of its role and success. CITES has an ambiguous status, as do other regulatory organisations like the International Whaling Commission and the International Tropical Timber Organisation: is it primarily a conservation tool or a trade agreement to prevent over-exploitation of valuable resources? The two roles overlap, but when countries that neither export nor import the products concerned take an interest for purely conservation reasons, conflict tends to arise. CITES may be in danger of becoming unstable, as some nations become more vocal in their demands for a stronger focus on resource use rather than conservation (Sharp 1997). In 1997, pressure from three Southern African countries (Zimbabwe, Namibia, Botswana) was successful at getting CITES to allow them to sell their ivory to a single buyer (Japan). Some governments and conservation-ists opposed this measure due to fears that it will increase the incentives for poachers and traders to seek legal and illegal means of getting their ivory into the legal international trade.

Another question that has arisen is whether CITES has been sufficiently successful at conserving threatened species even to justify its existence. The continual upgrading of species from one level of protection to higher and higher levels attests to the ineffectiveness of most listings in the face of increasing market pressure (Box 4.4), although populations of a few species have been downlisted because of successful sustainable use schemes (e.g. the vicuña; Rabinovich et al. 1991). In the case of the elephant, there have been clear conservation benefits from the Appendix I listing in terms of reduced poaching (although the earlier Appendix II listing only served to document a drastic population decline, not to limit it). Rhinos, on the other hand, have been listed on Appendix I since the inception of CITES in 1973. Since then, poaching for the international trade has continued to exterminate rhino populations in the wild. In this case, the major buyers (Yemen, China) were either not signatories or had no interest in comply-ing with CITES, so trade was virtually unaffected by the ban. MacKinnon (see Chapter 5) shows that CITES listings were counter-productive for two endangered parrot species. The parrots are endemic to the Tanimbar Islands of Indonesia. Despite a very restricted range, the parrots are

Box 4.4 The international trade in bears and bear parts

Fig. B4.4a A European brown bear, *Ursos arctos*. (Photo by Debbie Banks/ EIA.)

Bears illustrate many of the difficulties involved in attempting to control the international trade in wildlife products. There are eight species of bear, distributed world-wide. Some of these species, particularly the Asian ones, are highly endangered; others, such as the North American brown bear populations, are still relatively healthy. Bears are threatened by habitat destruction and fragmentation, the latter being particularly important because of the large area that is required to support a bear population. They are also threatened by the international trade in their products, particularly gallbladders, for traditional Chinese medicine.

Fig. B4.4b Bear gallbladders and tiger penises on sale in a North Thailand village. (Photo by Adrian Barnett/EIA.)

The bear trade is clearly an international problem. Although Asian populations suffered first, the demand for bear products has led to poaching on all populations, including those in North American national parks. Bears illustrate the phenomenon of gradual uplisting on the CITES appendices, a sign of CITES' failure to control the trade. Table B4.2 shows the gradual increase in the number of populations of the most widespread species, *Ursus arctos*, listed on CITES appendices.

continued

Box 4.4 *Continued*

The trade in bear parts, like the ivory and rhino horn trades, is fuelled by demand in Asian countries, which is increasing as the wealth of those countries increases. High prices are unlikely to put consumers off for long. The trade reaches to remote areas, such as parts of the former Soviet Union and Mongolia, and to areas where law enforcement is relatively good, such as the USA. The collapse of the Soviet Union has also played a part in the increasing trade, by exposing previously secure bear populations to hunting (see Chapter 14). A major problem for the legal trade is the impossibility of telling species apart from their gall bladders alone. Even DNA testing (which requires fresh, unprocessed gallbladders) can only pinpoint the continent of origin. As happened for ivory in the 1980s, the legal trade in some species under Appendix II has provided cover for illegal traders. In 1997, proposals to put all European and Asian populations of *U. arctos* onto Appendix I, and to declare a moratorium on the international trade in the gallbladders of all Ursidae, were unsuccessful (the Environmental Investigation Agency, pers. comm.).

Table B4.2 The listing of *Ursus arctos* populations on CITES appendices. (Data from the EIA.)

Population	Common name	Appendix	Year
U. arctos, Italy	European brown bear	II	1975
U. arctos, N. America	Brown bear, grizzly	II	1975
U. arctos nelsoni (Mexico)	Mexican brown bear	I	1975
U. arctos pruinosus (Bhutan, China, Mongolia)	Tibetan blue bear	I	1975
U. arctos isabellinus	Red bear/Himalayan brown bear	I	1979
U. arctos, All Europe	European brown bear	II	1983
U. arctos, All Asia (incl. Iran, Iraq, Syria, Turkey)	Brown bear	II*	1990
Ursidae spp.	All bears	II*	1992

* Except if already listed on Appendix I.

common on these islands. Crop-raiding individuals were caught and sold by farmers until CITES legislation prevented export. Now they are still trapped, but not sold, which helps neither the farmers nor the birds. The Arabian oryx (which is listed on Appendix I, as it is highly endangered in the wild) is breeding well in American zoos, who sell surplus stock within the USA but cannot export the animals. There are potential clients in Arabia, who would like to keep the species as exotics on their private land; demand for such animals is thought to be contributing to poaching in Oman, where the species was reintroduced after being hunted to extinc-tion. Vested interests frequently disrupt attempts to list species in cases where there is a good chance that importing countries would otherwise comply. Recent attempts to list bigleaf mahogany on CITES Appendix II (monitored trade) have been overturned because of opposition from various countries involved in the lucrative trade (see Chapter 6).

Part of CITES' recent re-examination of its mission has been to redefine the criteria for a species to be listed on its appendices. They have adapted the IUCN criteria for assessing the threatened status of a species (section 4.2.4), in the hope of providing a more objective measure of whether a species is threatened by trade. Whether scientific advice continues to be outweighed by politics and vested interests remains to be seen. CITES listing is not the only way to conserve species, and it is not always the best way. Listing a threatened species on a CITES Appendix is only likely to limit trade if:

• The species is actually threatened by international trade. Species threatened by trade within a country, or species that are threatened but not traded, are not helped by CITES, and only clog up the appendices, reducing the attention given to the species that are traded internationally.

• The CITES listing will have a conservation benefit. For example, listing a species that is traded only between non-members of CITES, or that has products which are indistinguishable from those of a common species, is unlikely to influence the trade. A way of distinguishing the products of the threatened species must be found first. This is a problem for many species that are used in Chinese medicines, because the medicinal name can cover several different species, some threatened, some not, and it can be hard to tell the species apart, especially after processing (see Box 4.4).

A major problem for environmental agreements in the future will be the extent to which they conflict with other treaties, in particular the *General Agreement on Tariffs and Trade* (GATT), which has now evolved into the *World Trade Organisation* (WTO). Trade sanctions have been the only credible threat to countries that have not abided by international treaties such as CITES. However, these are probably now illegal under the WTO, weakening the powers of signatory nations to influence those who renege on environmental agreements. This area of international law is still largely undefined, but it seems likely that when treaties contradict one another, the WTO will take precedence. For example, Gullison (see Chapter 6) highlights the fact that schemes to certify timber from sustainably managed forests are likely to be illegal under WTO, unless certification is required for all timber entering a country, regardless of origin.

4.3.2 Economic instruments for regulation within a country

In this section we consider the policies that a resource manager (such as a government department) might implement to control commercial resource use within a country. These include standards, transferable permits, taxes and property rights. Within a particular regime, there will also be

detailed decisions to make, particularly about how monitoring and enforcement of the regulations will be carried out. These regulations are
usually discussed in terms of their ability to achieve *optimal* resource use. A
major tenet of the economic theory of resource regulation is that there is
an economically optimal level of resource *over-use* (use that generates
negative externalities), and that this is non-zero (Fig. 4.7; Field 1994).
Different people have different views of optimality – as Norton-Griffiths
points out (see Chapter 11), some ranchers may hate wildlife, and view the
optimal amount of wildlife on their land as zero; others receive large
non-monetary benefits from conserving wildlife and have high-density
wildlife as their optimum. $Q*$ stands for the level of resource use that the
regulator has decided to aim for, whether this is the economically optimal
level or some other level. There are several criteria that can be used to
assess the success of a regulatory measure (adapted from Field 1994):

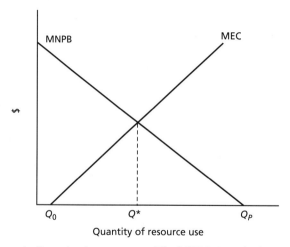

Fig. 4.7 Economically optimal resource use. The MNPB (marginal net private
benefits) curve shows how a resource user's profits per extra unit of resource used
decline as the amount used increases. The profit-maximising level of resource use is
where the MNPB curve crosses the x-axis, because here the area under the curve
(which is total profits) is maximised. This point is shown as Q_P. If a regulatory
authority made a resource user reduce the amount of resource used from Q_P to $Q*$,
this would lead to a reduction in profits, and would increase the marginal costs of
further reductions in resource use. From the perspective of the community, the ideal
situation would be no external costs from resource use. This is at the point Q_0. Q_0 is
not usually zero, but the socially optimal harvest rate. As harvests increase, the
marginal external costs increase (i.e. the external cost of each extra unit of resource
use increases). These external costs might be associated with the size of a harvested
population being lower than is socially optimal. This is shown by the MEC (marginal
external costs) curve. Where the externalities imposed on the community equal the
costs to users of complying with regulations, the optimal level of externality has been
achieved, assuming that the costs to users are of equal weight to costs to the
community in general. It is this point, $Q*$, that an economically optimising regulator
aims for.

- *Efficiency*: the ability of the measure to move output levels to Q^*, the optimal level of resource use. This is usually thought of as the major criterion for judging a policy, but in practice should take second place to practical considerations such as robustness and enforceability.
- *Cost-effectiveness*: the maximum possible benefit should be achieved for the money spent. The policy is good value for money.
- *Equity*: the policy has a fair distribution of costs and benefits, both among resource users and between resource users and the rest of society. It is a moral question for the policy maker to decide how much weight is given to the distributional effects of a policy.
- *Enforceability*: there is no point in a policy that is impossible (or very expensive) to enforce, regardless of how efficient it might be under perfect compliance.
- *Innovation*: a policy that gives an incentive to resource users to innovate, and find new and better technology, is a good policy in pollution control. However, in harvesting, the use of new and better technology often leads to progressive over-harvesting (section 1.4).
- *Moral message*: some policies send perverse signals about the acceptability of over-exploitation. They may then be politically unacceptable, however well they work in other ways.
- *Robustness*: this relates to the practicalities of monitoring and controlling resource use. Some regulatory measures work better than others under conditions of uncertainty and poor data quality. In particular, in order to calculate the curves in Fig. 4.7 the population size and dynamics of a harvested resource need to be known. In most cases, this knowledge is lacking, and so the ability of a mechanism to protect the stock against inadvertent over-harvesting due to a miscalculation of stock size is crucial for conservation.

Enforceability is particularly important for the successful regulation of resource use, but is often overlooked. Only if the breach of a regulation is associated with a credible penalty is there any reason to comply. To be effective, the perceived cost of a breach to the private resource user must be greater than or equal to the cost to the user of complying with the regulation. As discussed in section 3.1.4, this means that the probability of detection must be perceived as non-zero. This imposes *enforcement costs* on the resource manager. The resource manager's costs are of two kinds – the fixed costs of administering and monitoring resource use, and the variable costs of enforcing the regulations, depending on the required probability of detection of non-compliers. Enforcement therefore has an opportunity cost – money spent on anti-poaching patrols could have been spent on other conservation measures. There is a trade-off for the resource manager between enforcement costs and the benefits of increased compliance. The regulator's costs must be added to the user's costs of compliance to obtain the true costs of regulation, because the more reduction in resource use

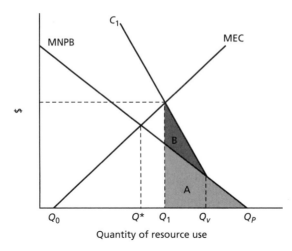

Fig. 4.8 The optimal level of resource use under costly enforcement of the regulations. The optimal level of harvesting under costless enforcement is Q^*. Under costly enforcement, the enforcement costs are added to the MNPB curve to give curve C_1, because reducing resource use is increasingly costly both for the complier (MNPB curve) and the regulator (C_1-MNPB). Note that the first stage of reduction in use, from Q_P to Q_V, is voluntary compliance due to the user's social conscience, or wish to be seen to be co-operating, which does not incur enforcement costs. The new socially optimal level of resource use, Q_1, is higher than the level without enforcement costs. The shaded areas show the costs of compliance, accruing to different people – area A to the resource users, area B to the regulatory authority.

there is, the more costly it is, both to the enforcer and to the complier (Fig. 4.8). This extra cost means that the economically optimal level of resource use increases from Q^* to Q_1 (Fig. 4.8) – the higher the enforcement costs for a policy, the higher is the optimal level of resource use, and so the more externalities are imposed on society by this resource use. Different types of enforcement have different cost structures, as well as different benefits, all of which must be weighed up in deciding on an appropriate enforcement strategy (e.g. is it better to use foot patrols or to invest in an aircraft, a satellite tracking system, or inspection of markets for illegal produce?).

Some common regulatory instruments

Standards. Standards are also called 'command-and-control' policies. They are very commonly used in natural resource management, despite the fact that they are generally thought to be economically unsatisfactory. Some reasons for continued use include their simplicity of implementation and, in some cases, their direct link with the status of the exploited population. There are three main kinds:

1   Standards for the use of the whole resource, for example total allowable catch (TAC) standards. The harvest is stopped once the total catch has

reached the allowable level for the season. These standards cannot be directly enforced on individual users, but only on the industry as a whole. They are usually used in conjunction with other policies, such as transferable permits, and are rarely used alone. Constant escapement policies, such as that used for the Falkland Islands squid fishery, are inverted TAC policies, in that the number escaping capture is set, rather than the number to be killed (see Chapter 8). Policies involving taking a constant proportion of the population or a constant number of individuals also involve a TAC, which is then divided up in some way among the users.

2   Output standards for individual users. These include individual non-transferable quotas in fisheries. Individual users are given a quota for their resource use which they must not exceed. This quota may be a percentage of the TAC or a number of individuals that they can harvest. Gunn (see Chapter 13) describes a system of individual quotas given to communities for caribou and muskox harvests.

3   Technology standards. These standards are not directly related to the output of the users, but to the inputs. The users are required to use a particular type of technology, but otherwise can use the resource as much as they wish. Common restrictions include net size, engine size, weapon type. MacKinnon (see Chapter 5) gives examples of the use of technology standards in the forests of Irian Jaya, where hunters were required to use only traditional weapons.

It is useful to note that a *total ban* on harvesting is a standard of type 1, but with the TAC set at zero. Thus, the comments made about standards in general apply to bans as well. However, total bans can be useful because they make enforcement easier, as they remove the need for regulators to distinguish between goods from legal and illegal sources. This was the major argument in favour of the ivory ban agreed by CITES in 1989 (Pagel & Mace 1991).

• The *efficiency* of standards depends on whether they are set at the correct level. If set at Q^*, they are as efficient as any other policy instrument.

• A lack of *cost-effectiveness* is the main reason why standards are unpopular with economists. Standards do not satisfy the *equi-marginal principle*, which states that when there are several producers with different production costs (e.g. different boat sizes), the profit-maximising level of production is when the marginal costs of production are equal. This means that the cost of producing one extra unit is the same for each producer. Standards are not equi-marginal unless a different one is devised for each individual user, based on the user's cost curve. This would be a difficult job, so standards are usually applied uniformly. Using a TAC policy alone, not backed up by regulations that affect individuals, is a particularly unhelpful policy, as it leads to continuing open-access behaviour by resource users. Each user ignores their effect on stock size and the date of closure, and profits are dissipated in the scramble to take as much as possible of the resource before the season is closed. The only difference is that, instead of the stock size reaching the open-access equilibrium size N_∞, it reaches the target escapement level N^*, at which point the season is closed to allow the stock to recover. The resource is conserved, but the economic performance of the harvesters is likely to be highly sub-optimal, with too much money going into unnecessary equipment and gear improvements, raising the total costs of harvesting. An example is the US Pacific halibut fishery in which the industry has become so over-capitalised that the TAC is reached rapidly, and the annual season is now only a few days long (Hilborn & Walters 1992). However, recent research has suggested that simple standards-based regulatory measures, that may appear economically sub-optimal, can actually be very helpful in lowering information and transactions costs (the costs of reaching and enforcing agreements), paving the way for the future introduction of more economically effective regulatory regimes (Wilson & Lent 1994).

• *Equity* is less of an issue for standards than for other methods. Each user must meet the same criteria, and the basis for standards is generally transparent.

• Incentives for *innovation* are perverse in the case of technology standards. A technology standard leads the users to innovate to circumvent the standard, and increase their output. Thus, an engine size restriction may lead to users investing in better nets or fish location devices. The technology standards need to be constantly expanded to prevent resource over-exploitation reappearing. For this reason, technology standards are generally not effective regulatory tools. The other types of standard have no clear effect on incentives to innovate.

• *Enforcement* is important for standards. Frequently, the stricter the standard, the more costly enforcement is, because there is a greater incentive to breach the standard. Thus, a lax standard properly enforced might actually lead to more overall reduction in resource use than a strict standard that is too expensive to enforce properly. The one exception to this is a complete ban on hunting, which is the strictest standard, but straightforward to enforce, as it is clear when someone is breaking the law. In general, as there is no complicated administration involved in setting a standard, an agency can simply set the standard and choose the level of enforcement that it can afford, without the basic structure of the policy being weakened. This lack of prior commitment to expensive enforcement programmes is a feature that makes standards attractive to managers.

• The *moral message* sent is an all-or-nothing one. The user has the right to exploit the resource at any level up to and including the limit, but no right to exploit beyond the limit. This is a slightly perverse message. There is no concept of the user paying society for the right to use the resource.

• *Robustness* varies, as there are several different kinds of standard. Those that are directly related to stock size, such as setting a proportional hunting mortality, are generally good for the conservation of the stock, as discussed in section 1.3.2. However, the difference between the discussion in section 1.3.2 and the present one is that a market with many independent users is now being considered, rather than a single user. Standards can run into problems when the management authority has different priorities from the harvester. For example, a constant escapement policy without any control on individual effort might ensure stock safety, but it would be economically inefficient. Standards need to be set according to ecologically meaningful boundaries, rather than administrative zones, to ensure they do not unintentionally lead to over-exploitation in certain areas.

Transferable permits. These are also known as individual transferable quotas (ITQs). They are currently used by the British government to regulate milk production by the dairy industry. They have become very popular in theory in recent years, although problems in their implementation are now beginning to surface (Box 4.5). The policy involves setting up a market in permits to use a unit of the resource. Once established, this market is allowed to run without intervention. A resource user will buy an extra unit of quota if the marginal profit from the extra output is greater than the price of the quota. Conversely, a user who is producing at a level Q is facing an opportunity cost from each unit of quota that he/she is using that could be sold instead. The market in quotas has a demand curve that is equivalent to the users' MNPB curve, and a vertical supply curve because the manager has released a fixed number of quotas for sale (Fig. 4.9a). If there are new

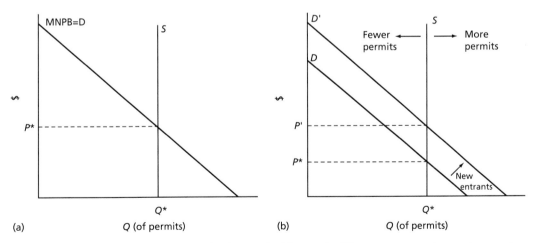

Fig. 4.9 (a) The market in transferable permits. The manager sets the supply curve by releasing Q^* permits. The price, P^*, is set automatically by the demand for permits. (b) The effect of new entrants and changes in the number of permits on the market. New entrants cause the demand curve to move up, and so increase the permit price to P'. Releasing more permits shifts the supply curve to the right and so decreases the price; withdrawing permits does the opposite.

entrants into the industry, then the demand curve shifts upwards, which increases the permit price (Fig. 4.9b). The market needs to be competitive, with many users, so that an individual user cannot influence permit prices. There is great flexibility in the system, because the issuing authority can buy and release permits as it chooses, like a central bank controlling the money supply. It would even be possible for environmental groups to buy up permits to lower use levels. The government might not be in favour of this, however, if it feels that it has taken the views of the whole of society into account by setting $Q*$ as the socially optimal level of resource use.

- *Efficiency*: permits are efficient if the number issued is set at the correct level, $Q*$.
- *Cost-effectiveness*: permits are cost-effective due to the equi-marginal principle. A user with high harvesting costs would rather sell permits, whereas one with lower costs would choose to buy more permits. The key feature here is that the permits must be tradeable.
- *Equity*: the profits from resource use accrue to the original permit holders, rather than to the management authority. These original holders can keep them, loan them to others at the market price $P*$, or sell them at their value in perpetuity $P*/\delta$ (see section 3.2.3). A few privileged people may grow rich from the ownership of these exploitation rights. This suggests that permits should be sold to the resource users at full price when the market is first set up. A major problem is the *allocation* of the original permits (Box 4.5). Allocation only occurs once, and since an allocated permit is a valuable commodity that can be sold on the permit market, it is important to allocate the permits fairly if user resentment is not to damage the policy. Allocation might be by the size of the user's operation, or by the amount of resource use currently occurring (thus penalising the less damaging low level users). Using recent data to allocate permits is dangerous, because users can increase their operations in the short term in order to gain more permits. Use levels over a long period before permits were mooted would be better. Other methods include auctioning of permits (so that only rich users can afford them), selling permits at a unit price, or giving them away by lottery.
- *Innovation* to increase harvesting efficiency is encouraged by ITQs.
- *Enforcement* requires that the managing authority knows who owns each permit, and how much of the resource each person is using. The ownership of the permits is no problem, as there is a strong incentive for self-reporting; if someone buys the right to use a unit of resource, they will inform the authorities of that right. Monitoring has problems because of the strong incentive to under-report output, but there is some incentive for monitoring each other. This comes because anyone who cheats by not buying enough permits is lowering the market price of the permits, and this reduces the value of the permits held by the other resource users.

Box 4.5 The experience of ITQs in New Zealand fisheries

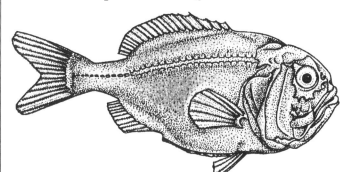

Fig. B4.5 An orange roughy (*Hoplostethus atlanticus*). This species is fished in New Zealand under the ITQ system, and has been the subject of recent concern. The species is long-lived, and easily overharvested, and its quota has been heavily disputed.

On economic grounds, ITQs are potentially the best available option for resource regulation. They are a relatively new idea, and are becoming more widely used. However, not all experiences of ITQs have been happy; there has been particular criticism of the New Zealand system (Duncan 1995). ITQs were introduced there in 1986, covering the main commercial species. The ITQ unit is a percentage of the TAC. The TAC is varied from year to year on scientific advice. An individual holder cannot have more than 20% of the TAC for a given inshore area, or 35% for a deep sea fishery, and pays the government an annual administration fee. The original allocation was free, and was made on the basis of past catch history. The government then bought back a proportion of the quota as a way of reducing catch to more sustainable levels, while compensating fishers who agreed to fish less than previously.

There have been some problems with setting the TACs, which can be difficult to calculate with enough accuracy to prevent political interests from overriding scientific advice. In fact, in 1990, the four largest fisheries had such high TACs that the fishers could not catch their quota. The orange roughy quota, in particular, has been heavily disputed, and the stock appears to be over-harvested. TACs have also been regularly exceeded, legally and illegally. There have been high-profile prosecutions for large-scale poaching (for example, rock lobster poaching is equivalent to more than a third of the legal commercial catch). Both these problems are common to all forms of resource regulation, and the criticism of ITQs on this basis is more because they are not the panacea they were hoped to be, than because they are worse than other methods of regulation.

The two problems that are intrinsic to ITQs are the initial allocation of quotas, and big business taking over the fishery, to the detriment of small-scale local fishers. They are interlinked, because the initial allocation was

continued

> **Box 4.5** *Continued*
>
> based on past catch history, was allocated to the boat owner rather than the crew, and was biased against part-time fishers. The free allocation of quota is difficult to accomplish without causing resentment among those who do not receive it, but requiring payment can exclude poorer people. Since the initial allocation, the concentration of quota in large companies has been such that the top three companies own more than 60% of the total quota. There are laws aimed at preventing such accumulation, but they seem to be lax and to have loop-holes. This concentration of profits from fishing has provoked great resentment among the smaller-scale fishers. It could be defended on economic grounds as leading to increased efficiency and economies of scale, but is not socially equitable.

- The *moral message* of permits is that people are buying the right to use resources.
- *Robustness*: permits are convenient for managers, because the optimal permit price P^* is determined automatically by the market, so the only parameter that needs to be set is Q^*. However, as with all regulatory instruments, setting Q^* is fraught with the usual ecological uncertainty. One problem is that the optimal level of resource use might not be constant throughout an administrative region, but if the permits are freely tradeable, the price of the permits is constant. Permit sales need to be zoned in ecologically meaningful ways; the permit cost may need to vary from area to area, and it should be impossible to trade a permit for a resilient population for a permit for a vulnerable population. Another problem is that under an ITQ system, fishers have an incentive to maximise their revenue per unit of quota, leading to the discarding of lower value fish and the landing of only high value fish. This is a particular problem in multi-species fisheries where several species are caught together but have different market values. Large quantities of low value species may be caught but discarded at sea, thus not appearing in fisheries statistics (Arnason 1994). Some other potential problems are illustrated in Box 4.5.

Taxes. Taxes involve levying a charge per unit of resource used. Taxes are often advocated by economists, but are rarely used in renewable resource management, suggesting that there are serious practical problems in their implementation. For example, Norton-Griffiths (see Chapter 11) discusses the drawbacks of a land use tax to influence how Kenyan rangeland farmers use their land. Taxes on resource use are called Pigovian taxes after Pigou, who first proposed them (Pearce & Turner 1990). The idea is to impose a tax of optimal size, t^*, on each unit of Q, so that the MNPB curve is lowered far enough that it crosses the x-axis at Q^*. Q^* then becomes the optimal level of output for the profit-maximising user, who

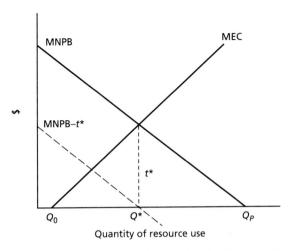

Fig. 4.10 A Pigovian tax. A tax *t* is charged for each unit of output. The users then automatically produce at the point where the curve MNPB-*t* crosses the *x*-axis, i.e. where MNPB = *t*, because this has become their profit-maximising point rather than Q_P. The optimal Pigovian tax *t** is equal to the marginal external cost of resource use at the optimal level of resource use *Q**. The regulator needs to know the user's MNPB function and the shape of the MEC curve to estimate *Q** and *t**. This might be difficult, as users are not usually willing to reveal their profit curves. However, an approximation to the theoretical optimum is usually adequate.

automatically produces at this level (Fig. 4.10). The major advantage economists see in taxes is that once the tax rate has been set by the regulator, the decision about how much to produce is retained by the individual resource users, rather than being centralised to the regulatory authority. Taxes are theoretically equivalent to ITQs, because the ITQ managing authority can fix the total number of quotas so that the price of one unit of quota *P** is equal to the optimal tax rate *t**. Thus, we highlight the differences below, rather than reiterating features that are common to both ITQs and taxes.

- *Equity*: a tax is often seen as unfair to resource users, because it gives the impression of penalising the users twice; at *Q**, they are still paying taxes, and they have incurred the loss of profit from moving from Q_P to *Q**. Should users continue to pay taxes when they are operating at the optimal level? The answer depends on our view of the property rights to the resource – it is entirely fair if the user is seen as paying society for the use of environmental services. Fishers are completely opposed to taxes; they move from an open-access situation where they received zero profits to the profit-maximising situation, but they are still receiving zero profits, while the state receives the tax revenues.

- *Enforcement* can be tricky. Taxes give a strong incentive to cheat by under-reporting the amount of resource use. This is particularly true if the tax is perceived as unfair.

- The *moral message* is that users are paying for the right to use the resource.
- *Robustness*: taxes require a lot of information if they are to be set at the correct level. Both Q^* and t^* need to be set. Any changes in market conditions (such as new entrants or inflation) require the regulator to adjust the tax level. Another potential problem with taxes is that if the user's MNPB curve is elastic to quantity harvested, then users will respond to small changes in the tax with large changes in resource use. This makes it hard to control resource use accurately. Finally, a regulatory instrument that is resented by users is not generally helpful for conservation, as violations of the law tend to be tolerated.

Granting property rights. Property rights have been put forward as the way to solve the problem of the 'Tragedy of the Commons', and are the dominant means of resource management in Western countries. It is argued that if clear property rights to a resource are assigned to someone, then the optimal level of resource use will be reached automatically, with no need for further intervention by a resource regulator. An owner can manage a resource to its maximum value, whereas an open-access resource is prone to over-harvest. That owner could be an individual or a community. For example, the community of Cosigüina, Nicaragua, may be more likely to succeed in managing their iguana-rearing enterprise sustainably if their co-operative is officially recognised (see Chapter 7). In the theoretical analysis of the effects of property rights, who gets the rights is irrelevant economically, but is an ethical question. It could be the users, the sufferers of the externalities, individuals, communities or the state. The users and externality sufferers could even be the same people.

The *Coase theorem* states that once property rights are assigned, Q^* is reached by bargaining (Pearce & Turner 1990; Fig. 4.11). If the Coase theorem held, this would seem to negate the need for any intervention to solve resource use problems, as long as someone is granted property rights over the resource. However, there are several requirements for the process to work:

- Property rights must be well defined, enforceable and transferable. Thus the owner must be able to prevent others from doing what they like on their property, and must be able to sell the property, so that long-term damage to the property is reflected in a reduction in its sale value.
- The negotiation process must be efficient and competitive. Transactions costs must not be too high (as they will be if many people are involved).
- All the value associated with the resource must be capturable by the bargainers. This includes the non-monetary values as well as the market value. Sufferers, in particular, may not be able to express their non-market values in the bargaining process. If there is a lot of free-riding, the full value of the good is unlikely to be expressed.

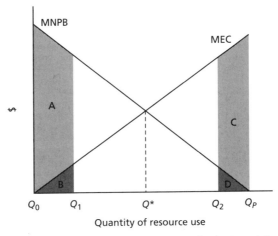

Fig. 4.11 Reaching the optimal level of resource use, Q^*, by bargaining. If the externality sufferer owns the resource, the level of use starts at Q_0, where there are no externalities. A move from Q_0 to Q_1 would cost the sufferer area B, but the user would gain area A + B. So it is in the user's interests to offer to pay an amount greater than B but less than A + B as compensation if the sufferer agrees to the level of resource use increasing from Q_0 to Q_1 and it is in the sufferer's interests to accept the offer. If the resource user owns the resource, the resource starts at Q_p, the profit-maximising point. In the same way, compensation paid by the sufferer to the user (greater than D and less than C + D) will induce both the user and the sufferer to move to Q_2. This process continues until Q^* is reached and neither party wishes to change the level of use.

There are other considerations that determine how effective property rights are. The point at which the profit curve of the resource user and the marginal external cost curve cross is only the optimal level of resource use if the users are in perfect competition. This is because the optimal result occurs when the price of the good fully reflects the costs of its production, both private (MPC) and social (MEC): P = MPC + MEC. Thus, for this optimal point to be reached by bargaining, the users' profit (MNPB) curve must equal P – MPC, so that MNPB = MEC is equivalent to P – MPC = MEC. In a monopoly, the price is not the same as the marginal revenue, and so P – MPC is not the firm's profit curve, and the incorrect equilibrium is reached by bargaining (Pearce & Turner 1990). Similarly, if there are any inequalities between the sufferers and the users, so that one side can threaten the other, the efficient solution cannot be reached. For example, if the damaged parties are future generations, they cannot bargain effectively, and are bargained for by proxies, weakening their position. In general, there seem to be very few real-world situations in which property rights and bargaining actually solve resource management problems. This seems logical, because if such a solution were easily reached, there would be few cases of resource over-exploitation remaining.

Other regulatory instruments

Other regulatory instruments exist, that work in more limited circumstances than the ones discussed above, or that have flaws that make them unsuitable for use in most cases. These include limited entry; voluntary restraint; subsidies; and liability laws.

Limited entry may be imposed if the industry has too many users. The optimal number of users is calculated, and the numbers are reduced to the appropriate level. The problem is, however, that each user is still exerting open-access effort, not the optimal effort. As the stock recovers and they start to make profits, the users can afford to invest in new technology, and their overall harvesting capacity may increase. This then leads to more restrictions being imposed. This is a policy (like technology standards) which addresses a superficial symptom, and fails to deal with the underlying causes of over-exploitation. However, it does work in certain circumstances where entry can be effectively controlled, such as the Falkland Islands squid fishery (see Chapter 8).

Voluntary restraint has been useful to those politically opposed to legislation for resource management, and is also very cheap, as there is no requirement for enforcement. It can be the only option when laws are unenforceable. Users are asked to restrain themselves from over-exploitation. Often moral suasion is used, by appealing to users' moral values and civic duty. Taking litter home and taking bottles to bottle-banks for recycling are examples. Butynski and Kalina (see Chapter 12) describe the system of voluntary restraint in which tourists who feel ill are asked not to visit habituated gorilla groups – this is unlikely to be an effective protection for the gorillas when tourists have travelled great distances and spent large sums to visit them. In general, some restraint may occur, as shown in Fig. 4.8, but it is very unlikely that Q^* will be reached by voluntary means in large communities, due to the problems of small individual benefits (frequently limited to social praise) and free-riding (people reaping the benefit of others' restraint while continuing to over-exploit). Free-riding tends to undermine the long-term effectiveness of the policy, because compliers become cynical as violators receive no concrete penalty. Moral suasion is useful as a supplement to other regulatory instruments, as a positive public climate reduces violation rates. Of course, if regulations are seen as unjust, moral suasion may act the other way, encouraging violations. For example, local communities living around African game parks may support poachers (Gibson & Marks 1995), just as fishing communities may see 'quota-busters' as Robin Hood figures.

Subsidies involve using a carrot rather than a stick to get users to Q^* but should have the same effect as taxes on individual users. They involve giving a subsidy of s^* for every unit of resource not used by an individual, thus imposing an opportunity cost of s^* per unit on those who continue to use the resource. The optimum then becomes the point at which $s*$ equals the cost of reducing resource use. For example, Kenyan landowners could be given a subsidy for keeping wildlife on their land (see Chapter 11). The problem arises because, although the incentives are correct for individual users, the money that is being put into the industry means that the industry becomes more profitable. This encourages new users to enter the industry, and so the total level of resource use is likely to increase. To avoid this, taxes and subsidies can be combined, with the tax raising the money for a future refund for those operating at the correct level.

Liability laws attempt to internalise externalities by making resource users liable for the damage they do to the environment (the 'polluter pays principle'). They are taken to court and required to compensate the damaged parties. Evidence given in court is used to calculate $Q*$, and the victims must either prove liability by the user or, less strictly, negligence. If the court is able to calculate the externalities correctly, then the resource user automatically reduces use to $Q*$, where the marginal compensation (MEC curve) equals the marginal profit from resource use (MNPB curve, Fig. 4.7). Liability laws are common (e.g. for oil spills), and are popular with governments because once the legislation is in place, they need do no more. As was discussed in section 3.2.1, this approach has grave problems. The first is *burden of proof*, which involves a damaged person establishing a direct causal link between their losses and the resource user's activities. This is difficult, for the very reasons why natural resources are subject to market failure – costs are usually diffuse, long term, hard to value and poorly understood. The second is *transactions costs*, the costs of reaching and enforcing agreements. These can be very high when the process involves legal action. Liability laws are only likely to work as a way of finding $Q*$ if there are few people involved, there are clear causal linkages, and damage is easy to measure.

4.4 Community-based conservation

4.4.1 The origins of community-based conservation

Community-based conservation is historically the oldest form of conservation; but traditional rules governing who is allowed to harvest resources, or areas that are meant to be left unexploited (often sacred areas), have not in general been successful at resisting modern economic and political pressures. The measures that replaced traditional systems, or were put in place where no traditional systems existed, were largely devised by outsiders, such as governments. Lately there has been more emphasis on community-based initiatives. The idea has emerged in response to two different thrusts, one ethical and one practical. Ethically, there is growing uneasiness, both locally and internationally, with the notion that the local population can be overruled, ignored and frequently forced to relocate, to meet the needs of wildlife conservation – an ideal that they do not necessarily share. Practically, conservation initiatives that have ignored local people are failing.

One of the most important tools in the conservation of terrestrial ecosystems has been the setting up of protected areas. Under colonial or other undemocratic regimes, this frequently involved excluding local people that may have used that area either to live in permanently or to visit occasionally to harvest resources. Even today, people are still evicted from prospective wildlife parks on which their economy was based, for example pastoralists were excluded from the Mukomazi reserve in Tanzania in 1988. Protected areas have been successful in conservation terms – Norton-Griffiths (see Chapter 11) shows that loss of wildlife from protected areas in Kenya is far less than from outside protected areas. However, more than 90% of terrestrial ecosystems are outside protected areas (WCMC 1992). Furthermore, a protected area cannot normally exist as an ecological island. Animals frequently migrate in and out of a protected area, making them particularly vulnerable to a hostile local population. People living on the boundaries of such reserves may also suffer damage to their farms from the wildlife. Whilst these costs are paid by those living around the parks, locals are frequently not allowed to venture into the parks; the benefits of wildlife conservation are normally reaped by outsiders (tourists, film-makers, wealthy élites). It is therefore not surprising that protected areas are frequently resented by local people, to the extent that their very usefulness as reserves is threatened (Bell 1984).

This imbalance in costs and benefits has two important consequences for conservation. First, as democratic notions spread to more and more countries in the world, protected areas will not be supported by governments if local populations do not support them. Second, local people will not comply with the legal restrictions imposed upon them, unless draconian enforcement is introduced. For example, a shoot-to-kill anti-poaching policy has been in force in several African wildlife parks over the last decade, without any noticeable effect on poaching rates (Leader-Williams & Milner-Gulland 1993). In some cases, wildlife has even been killed by local people primarily to make a political point. Western (1994) describes an epidemic of poaching in Amboseli National Park in Kenya after government authorities failed to honour a local agreement. As the human population grows, hunting technology advances and the need to earn money increases, the pressures on protected areas have increased whilst the policing of these unpopular reserves has fallen down the list of priorities of poor governments. Many parks are protected areas in name only.

The idea of community-based conservation grew after the importance of local participation in rural development projects was recognised in the 1970s. The hope was that, if conservation initiatives were structured in such a way that the benefits of conservation accrued to local people, then they would support and even enforce and manage measures to conserve their natural resources. The means by which conservation and development initiatives have been linked are many and varied. Thus, case studies, such as those presented in Part 3, are important in understanding the breadth and efficacy of community-based conservation.

4.4.2 Community-based conservation in practice

Community-based conservation projects are generally initiated in areas where there is stress on people, wildlife, or both, causing a need to be identified. They are frequently set up on the boundaries of fully protected areas, sometimes in a buffer zone between the reserve and some more ecologically destructive type of land use.

In some cases, traditional land-use systems have proved themselves sustainable over centuries, for example, Maasai and wildlife have coexisted successfully in the Ngorongoro Crater area in Tanzania (Homewood & Rogers 1991). In such cases, conservation efforts aim to protect such systems from being disrupted by the numerous political and economic changes brought about by coexistence with modern market economies. However, once these changes have begun, the clock cannot be turned back. Once a logging road has opened an area of forest up to outsiders, for example, it is extremely difficult to prevent settlement, agricultural encroachment and increased hunting pressure. More generally, the effects of improved communications on previously isolated communities can be hard to predict. A closer linkage to the mainstream economy can encourage commercial hunting by reducing transport costs to markets and allowing people access to improved technology, but it can also decrease people's reliance on game meat for subsistence and raise the opportunity costs of hunting (Ayres *et al.* 1991; section 1.4).

In other cases, new development projects have been set up in areas where resources are threatened by human use, in the hope that creating new means of income generation will relieve the pressure on threatened resources. However, developers have frequently not considered that people are most likely to adopt the new income earning opportunities in addition to the old ones, rather than instead of them. Further problems arise if people migrate into the area specifically to benefit from the new options for earning income, where they actually increase the threat to the resource that the development was intended to conserve; Oates (1995) describes one such unfortunate sequence of events as development projects were established in the Okumo Forest Reserve in Nigeria. More successful efforts are those in which the economic developments are more closely integrated with the conservation measures, where the aim is that the success of one is dependent on the success of the other. A range of *integrated conservation and developments projects* are now being established. MacKinnon (see Chapter 5) gives examples of such projects in South-East Asian forests. Sometimes, totally protected areas can be negotiated, with local agreement, as part of a wider land-use project. Amboseli National Park in Kenya has such an agreement, where profits from tourism in the protected park are fed back into the surrounding community in various ways (Western 1994). This is partly responsible for the benefits to wildlife conservation in the areas described by Norton-Griffiths (see Chapter 11). It should be remembered that a wildlife park in Kenya (in this case, within easy reach of the capital) has far greater opportunities for tourist revenue than most protected areas, but the success of the project is still dependent on constant negotiation.

In some cases, projects are designed actively to encourage the use of wildlife as a market product – a notion based on the idea that such use will increase local incentives to conserve both that species and its habitat. 'Use it or lose it' is a snappy slogan that is frequently repeated in this context, but it is a dangerously simplistic chain of logic, which ignores most of the theoretical and practical complexities that we outline in this book. The economic conditions which might generate such incentives for conservation are fairly restrictive (section 3.2.2). These conditions only really apply when the wildlife resource concerned can be managed in a way rather similar to a farm or ranch. The iguanas reared in a project described by Solis Rivera and Edwards (see Chapter 7) are one such resource.

Most community-based conservation initiatives are recent and in poorer countries. However, in other countries, especially those with a democratic tradition, conservation has for some time had at least some element that is community-based. For example, in England and Wales, 11 national parks were established in 1952. They are in areas that have been inhabited for centuries, and most of the land is farmed and in private ownership. There was never any question of relocating the inhabitants; nor would it be particularly desirable from a conservation perspective, as many of the species are adapted to live in a farmed environment. Conservation initiatives in these parks are thus aimed at preserving more traditional land-use practices, which are not the most economically efficient ones. A combination of planning restrictions and economic incentives are the main conservation tools. Subsidies for environmentally friendly farming are available, especially in regions designated as environmentally sensitive areas, and there have been local successes in encouraging farmers to promote nature conservation on their land (Statham 1994). Political support for these subsidies derives from the pleasure that a large portion of the country's population gains from visiting these beautiful areas for recreation. Unfortunately, larger subsidies are available for environmentally damaging farming methods, through the Common Agricultural Policy of the European Union. This has proved the more powerful influence in most areas, resulting in a dismal record of nature conservation in Britain as a whole. A report on the status of farmland birds in Britain shows dramatic decreases in most species over the last 20 to 30 years (Campbell & Cooke 1997). For example, formerly common species such as the tree sparrow (*Passer montanus*) declined by 89% on farmland over the period, with the turtle dove (*Streptopelia turtur*), the bullfinch (*Pyrrhula pyrrhula*) and the song thrush (*Turdus philomelos*) all declining by more than 70%, and the lapwing (*Vanellus vanellus*) and skylark (*Alauda arvensis*) by around 60%. These declines are linked to the intensification of farming methods, with the greatly increased use of pesticides which destroy prey species being a major contributing factor. These are practices that have been heavily subsidised by the EU over the relevant period, highlighting the power of economic incentives to land users.

4.4.3 Rights and regulation

Granting the legal right to exploit a resource to local users is generally considered an essential part of community-based conservation. Land that is owned by government, but not effectively managed by it, is frequently treated as an open-access resource, with serious consequences for conservation. The transfer of land to individual or community ownership can change that situation. Because those with power are always reluctant to relinquish it, this is frequently done only in a very limited way, if at all. National governments, by definition, will always continue to hold some interest and influence. Perhaps as a reaction to coercive and undemocratic conservation policies, combined with a generally poor record of governments at conserving public resources, many have hailed the devolution of property rights as a panacea, a magic solution to conservation problems, but it has yet to prove itself as such (Little 1994).

There is no particular theoretical reason to suppose that devolved property rights automatically lead to more sustainable use. As shown in section 4.3.2, national governments and local users might converge on a similar economically optimal level of exploitation, although the circumstances in which this occurs are rather limited. The granting of rights to local users may reduce inequalities in the bargaining power of different stakeholders, and lead to more non-market values being introduced into the equation. Whether these changes move a situation nearer to or further from a solution which will benefit biodiversity depends on the values of those concerned, and the legal and institutional framework. The case of the Ache, an indigenous group in Paraguay (see Box 1.2) gives an example where granting exploitation rights to a local group led to the immediate felling and sale of trees on the land.

The regulatory aspects of community-based conservation have not been given much attention. This is partly due to a belief that communities will regulate themselves. However, threats to conservation do not disappear, either from within or from outside the community, just because land-rights, or other community-based measures, are introduced.

Where communities have a strong tradition of managing use in a way that has proved itself sustainable, then self-regulation may suffice. The exclusion of other groups is frequently an important part of such systems. However, governments have a tendency to overrule this, due to a political desire to play down indigenous sub-divisions within a nation state or even a group of nation states in union. For example, the Common Fisheries Policy of the European Union allows fishermen from all EU states access to each others' traditional fishing grounds. Sometimes, corrupt or unregulated government officials exploit resources that were being used sustainably by local communities, either for their own personal gain or for the benefit of more powerful political or economic lobbies. Freehling and Marks (see Chapter 10) show how the valuable game species of the Luangwa Valley, Zambia, are managed for the benefit of wealthy hunters, rather than the Bisa hunters who traditionally used them. Community-based private tenure systems may reduce these risks, although it is unrealistic to suppose that these can function without the strong support of the state. A strong system of legal enforcement may be necessary to prevent illegal exploitation by outsiders. Prior to the international ban on ivory trading in 1989, military-style forces were required to defend many of the regions where elephants lived – a level of enforcement that African governments were frequently unable to provide. Local communities cannot be expected to be able to resist such commercial poaching pressures without help. Some indigenous forest-dwelling communities, from Amazonia to Cameroon, may have used sustainable harvesting systems throughout their history, but are usually powerless in the face of encroaching agriculture and commercial forestry. Even more extreme threats from external forces have recently been faced in Central African national parks, when the rule of law itself disintegrated in a state of war (Box 4.6).

Threats from within communities derive from the fact that communities are, of course, made up of individuals, whose personal costs and benefits are likely to differ from those of the community as a whole. When populations are small and communities close-knit, it may be possible for harvesting systems to be devised and enforced within the community, perhaps informally. Fear of reprobation by the rest of the community may be sufficient to deter transgressors. Such a local system would have the advantage that knowledge of the harvested resource, and of the behaviour of those harvesting it, is likely to be better than that of an external manager. However, such conditions are easily threatened by population growth, immigration, increased financial incentives, easier technological means to over-harvest, or a weakening of community authority. MacKinnon (see Chapter 5) discusses the traditional 'sasi' rights system that existed for the harvesting of

Box 4.6 War and conservation in Central Africa

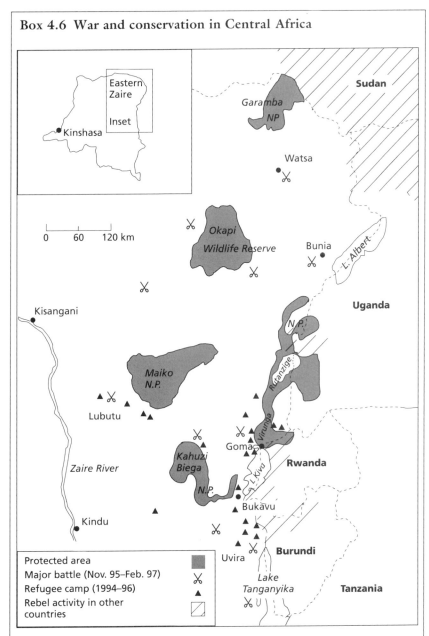

Fig. B4.6 Major concentrations of refugees and armed rebel activity in eastern Zaire (now DRC) and adjacent countries. Sources include observations from field staff, UN reports, and Internet news services. Not all sites of refugee camps or skirmishes are shown. (From Hart & Hart 1997.)

The recent and ongoing conflicts in Rwanda and Zaire (now the Democratic Republic of Congo) threaten several national parks of major

continued

Box 4.6 *Continued*

international importance, including several World Heritage sites and the region containing the only population of mountain gorillas (see Chapter 12; Fig. B4.6). The ability of staff to withstand the huge pressures of war and maintain some degree of protection for the parks leads to lessons for the future of conservation in areas of political instability. Two stories from Western conservationists involved with projects in the region (both working for the American NGO, the Wildlife Conservation Society) illustrate this. Although the authors consider that their experiences are success stories, it is still too early to say exactly how much damage the wars have done to the wildlife of the area. Early indications are that, despite these positive stories, the overall damage has been severe (see Chapter 12).

Hart and Hart (1997) contrast the work of development, conservation and humanitarian organisations. Development organisations generally aim to pass projects over to national governments once they become self-sufficient, whereas a strength of pure conservation programmes is that they often have an international link, which staff can rely on when governments disintegrate. Humanitarian organisations cannot be relied upon to conserve natural resources in a crisis situation, because their priority is to save lives in the short term. In Zaire, refugee camps were situated very close to national parks, and most international aid for conservation was withdrawn, leaving a minimal, unpaid staff of Zairean nationals. Their headquarters were looted and equipment was destroyed. Despite this, the staff continued to defend their parks, and foreign-trained Zairean conservationists based at Virunga National Park persuaded the rebels who took over the park to help in restarting anti-poaching patrols. Hart and Hart suggest that the best preparation for political breakdown is to have well-trained staff, and strongly site-based conservation programmes, supported by an independent international conservation organisation that can continue to support them at times of need.

Fimbel and Fimbel (1997) describe the effect of the war in Rwanda on a conservation project in Nyungwe Forest Reserve, which had emphasised training and empowering of local staff, with the intention that one day the reserve would be managed entirely by Rwandans. When war came, the expatriate project leaders were evacuated, then the senior Rwandan project staff either fled or were killed, leaving junior staff recruited from local villages in charge. They continued with daily patrols in the park, and other project activities, throughout the war, despite the lack of salaries or communication with senior staff. This conservation project is still the only one present in the Nyungwe Forest Reserve, despite the war finishing in late 1994. This is in contrast to before the war, when there were five large internationally funded projects there. The lesson that they draw is that what leads to continued conservation efforts at times of stress is the commitment of locally-based individuals, together with independent support from an international NGO.

megapode eggs in Indonesia. The 'sasi' system has broken down in many areas for the above reasons, but remains effective on the Molluccan island of Haraku. Where community management breaks down, or where it never existed, then all the issues of setting and enforcing levels of sustainable harvests, discussed in section 4.3.2, will be as relevant to community-based systems as they are to national ones. The need for formal estimates of sustainable harvests, and for regulating harvesting within these limits, is frequently not appreciated. Sometimes it is deliberately ignored, because suspicion of the technical fix, based on many conspicuous failures, is not an uncommon view amongst development workers.

In many cases, there are no local political groupings appropriate to manage a community-based conservation project. Whilst efforts can be made to encourage such groups, a community-based project engineered entirely by outsiders is not, by definition, community-based. However, in practice, projects with almost any level of community involvement are frequently referred to as community-based. Many of the problems that typically threaten the sustainability of any rural development project can arise with these projects. If projects are set up by ecologists or conservationists with no former experience of rural development projects, then the risk of such pitfalls is great. Similarly, development workers with no knowledge of how to obtain or interpret ecological data may threaten the ecological sustainability of such projects. If a crucial part of the project is run by NGO staff, the project will collapse after the departure of the NGO. If local political structures are weak, locally influential individuals who may claim to represent the community, in fact take a disproportionate part of the benefit; Solis Rivera and Edwards (see Chapter 7) describe how one family was initially taking too great a share of benefits from an iguana-rearing project in a Nicaraguan community.

Gunn (see Chapter 13) describes a system of *co-management* (i.e. a set of formal agreements) between the government of the North-West Territories of Canada and indigenous groups. Land rights, including the right to manage the exploitation of wildlife, have recently been transferred to the local population. The community can draw on the technical expertise of government scientists for advice on ecologically sustainable harvests and other matters.

Improved relationships between governments, professional managers and conservationists on the one hand and local populations on the other is one of the successes of community-based projects that can already be seen. Whether this will be followed by conservation benefits is hard to judge, as most initiatives are recent. The development of successful projects usually requires a great many political and technical issues to be successfully resolved. Whilst community-based conservation is sometimes hailed as a new paradigm – a label that is not inappropriate – it is certainly not the case that it can be considered a quick fix.

4.5 Summary

• Natural resource managers face various kinds of uncertainty, caused by environmental stochasticity, complexity of ecological systems and sampling error. Bias is a particularly serious problem, which can be caused by sampling methodology, deliberate misinformation, and excluded variables. Uncertainty can be tackled in various ways, including adaptive management and setting aside reserves.

• The Revised Management Procedure of the International Whaling Commission approaches uncertainty using computer simulations. The performance of a management strategy is tested under various extreme scenarios, to see how it reacts to problems such as biased and unreliable data, or unexpected biological or environmental problems. The tests show that regular estimates of population size are the most important data needed for good management. A precautionary strategy is to make harvesting levels dependent on data quality.

• Population data can give an index of a harvested resource's relative or actual abundance. Data on relative abundance are usually easier to collect and can show population trends, but are of limited use for managing hunted species. Survey methodology can distort population estimates enormously. Estimates that incorporate a measure of data quality are useful. Baseline surveys are vital, but can be misleading due to the shifting baseline syndrome.

• Harvest and trade data can be extensive and long term for commercially important species. The data can be used to estimate population sizes, but care is required, particularly if the data are expressed as biomass rather than number of individuals. Trade data represent both the status of the resource and demand for the product; neither are likely to remain static over time.

• Official data can give useful information on social and economic conditions. However, the data are usually aggregated, and can be inaccurate. Economic data need to be converted to real values by removing inflationary trends.

• The biological data required depend on what is to be conserved. For single species, measures of habitat quality and extent can be useful. Biodiversity is one of many measures of ecosystem health. Indicator species can be used to simplify monitoring.

• Existing local knowledge can be extremely valuable, but skill is needed to collect it. Participatory approaches and interviews can be used, for example, to map resource locations. Co-operation between scientists and local users can improve understanding in both directions.

• Regulations for controlling resource use can be applied at the international, national and local levels. The appropriate level depends on the ecological and social system being regulated. International agreements between sovereign states are difficult to enforce. CITES has suffered from a lack of clarity of objectives, and political differences between signatories. Recent trade agreements under GATT threaten to conflict with previous environmental agreements.

• Regulatory instruments at the national level can be judged by their economic efficiency, cost-effectiveness, equity, enforceability, innovation, moral message and robustness. Enforceability and robustness are particularly important in practice. Standards are widespread and popular because of their simplicity and robustness, despite their economic inefficiency. Taxes, by contrast, are rarely used for exploited species, despite their economic efficiency, because they are unpopular and require a lot of information. Transferable permits have had only limited success. Their problems include the initial allocation and their effects on users' social structure. Other regulatory instruments are less generally applicable.

• Community-based conservation includes traditional harvesting rules in small communities. These are rarely robust to external influences. Local communities tend to bear the costs of conservation without receiving the benefits, a situation that must be changed. Integrated conservation and development projects may work under limited circumstances. Examples of the successful integration of the needs of human and non-human residents of an area can be found.

• Granting property rights to local users is sometimes hailed as the answer to the conservation of exploited species. However, it is unlikely to be fully implemented in practice. There is no theoretical reason why granting property rights will necessarily lead to resource conservation.

• Local communities may be able to regulate their own resource use, but cannot be expected to protect their resources from large-scale external threats such as wars, commercial poachers, and industrial resource use. Even without the external threats, informal community regulation may be weakened by population growth, technology, social and economic changes. Then more formal regulatory structures will be required, such as co-management agreements.

Part 3
Case Studies

Man in Bwindi-Impenetrable National Park, Uganda. (Photo by Tom Butynski.)

Introduction

Part 3 consists of 10 case studies, in which conservation practitioners discuss in detail the issues surrounding the sustainability of use in the systems in which they work. These detailed case studies show how the theoretical tools presented in Part 2 are used in practice, and highlight some of the complexities of real-life conservation. The case studies are intentionally diverse, in their geographical locations, the taxa that are used, and most importantly in the perspectives of the authors. Two cases involve harvesting by large commercial ventures and three concern exploitation primarily by local communities. Two are cases where the predominant use of the resource is for tourism, and two other studies concern the history of an exploited ecosystem; one is largely anthropological and the other political. One case study considers the problems of resource management from an economic perspective. The case studies represent the broad range of approaches to sustainable use that can be found among conservationists today.

Kathy MacKinnon is a professional ecologist working for the World Bank. The World Bank is an organisation that is predominantly concerned with development issues, but in recent years it has moved towards the idea that the conservation of a country's natural resources is an important component of sustainable development. Kathy MacKinnon discusses a range of examples of the interaction between development projects and conservation, several of which are projects that have been specifically set up to link sustainable use with conservation. The examples are taken from the species-rich forests of South-East Asia, with a particular emphasis on non-timber forest products like seeds, gums and fruits. For this reason, her case study demonstrates the importance of the complex ecosystem-level effects of harvesting that are discussed in section 2.2 of Part 2. Several of her examples show how involving local communities in conservation efforts (particularly in buffer zones around national parks) can have a positive effect, as discussed in section 4.4 of Part 2. She also sounds a note of caution about the effects of conservation policies – she describes a situation where an international trade ban (as discussed in section 4.3.1) was counter-productive.

R.E. Gullison has spent several years researching the ecology and management of mahogany in Bolivia. He shows that the history of mahogany exploitation over the last few hundred years has involved harvesting an area until it is unprofitable and then moving on. The ecology of mahogany, and particularly its slow growth rate, means that sustainable use is unlikely ever to be profitable. Its ecology also means that it is unsuitable as a flagship species for the sustainable use of a forest ecosystem. He shows that international agreements for regulating the mahogany trade have been ineffectual so far, and are likely to remain so. He uses the concepts of harvesting over time and net present value, discussed in section 1.5 of Part 2, in developing his arguments.

Vivienne Solis Rivera and *Stephen Edwards* are professional conservationists working for IUCN's Sustainable Use Initiative. This initiative has been involved in running pilot sustainable use projects, to build knowledge about use as a tool for conservation. They approach the issue from a development perspective. Rather than discussing how best to conserve natural resources, they concentrate on how using wildlife sustainably can help very poor people to improve their standard of living. Thus wildlife use is a tool for rural development, with conservation of the resource being a prerequisite for sustainability. This is an approach to conservation that has become popular with IUCN in recent years. From first-hand experience, they discuss the potential pitfalls of wildlife-based development projects. They emphasise that these projects are dynamic and unpredictable, so that their future sustainability is impossible to determine. Their case study project concerns iguana rearing in Nicaragua and links with the discussion of community-based management in section 4.4 of Part 2.

Sophie des Clers has been involved in managing the Falkland Islands squid fishery for a number of years. She is concerned with the practicalities of resource management, in the context of managing a commercial fishery for profit whilst ensuring the survival of the harvested stock. This is the tradition of fisheries management within which much of the theoretical background discussed in Part 2 has been developed. The managers of this new fishery are able to manage the resource purely in the interests of long-term profit and conservation of the stock, with no reference to any historical rights of the harvesters. In this situation, the best management tools for the system can be used, as discussed in sections 1.3.2 and 4.3.2 of Part 2. Sophie des Clers shows that the scientists and managers running this fishery have succeeded in using it sustainably for more than 10 years, and she contrasts the reasons for this success with the poor record of commercial fishery management in Europe, where the rights of harvesters are politically paramount.

Andrew Price, *Callum Roberts* and *Julie Hawkins* are researchers and consultants who work on coastal zone management throughout the world. They look at reef-based tourism in their case study. This presents a very different set of issues to those discussed in other case studies, because the use is not targeted at a particular species, but affects the whole ecosystem (see section 2.2 of Part 2). Reefs are coastal ecosystems, and as such are indirectly affected by the land-based development associated with tourism, as well as being directly affected by use. They discuss the linkages between various sectors of the economy (such as tourism, construction and waste management), and how they combine to affect the conservation status of reefs. The two areas that they study, the island of Saba and the Maldives, are in different oceans and face different problems, but are linked by the current low levels of ecological damage that they have sustained. This means that there is still time to implement conservation measures that could prevent over-exploitation of the islands' natural resources.

Joel Freehling and *Stuart Marks* are anthropologists. Stuart Marks has studied and lived with the Bisa people of Zambia for 30 years. The authors look at the relationship between communities and wildlife, taking an anthropological perspective. They discuss the incentives that people face to act in particular ways, the changing social structure of the community, and the effect of external influences (such as government policies) on people's relationships to wildlife. The case study illustrates the theory of how people make decisions, discussed in section 3.1 of Part 2, and community-based conservation, discussed in section 4.4. It also shows the value of long-term, in-depth studies, as discussed in section 4.2.5. The Bisa live within the area covered by ADMADE, one of several Southern African projects that aim to give local people the benefits of the wildlife on their land, and so promote conservation. Those who set up these projects are convinced of their effectiveness (Child 1996). Others attack the projects on conservation grounds (HSUS/IHS 1997). A new set of insights can be obtained by looking at the effects of these projects from the perspective of local individuals, as Freehling and Marks have done. They trace the effects of several different administrations' conservation policies on wildlife populations in the Luangwa Valley area of Zambia, and on the Valley Bisa people who use the wildlife. None of the policies have been successful, either in conserving wildlife or in improving the lives of local people.

M. Norton-Griffiths discusses conservation policy from the perspective of an economist. He is a rare economist – one with 20 years previous experience as a field ecologist, working in Africa for UNEP. Economics can be a powerful tool for explaining resource users' incentives, and why conservation policies that do not take into account economic incentives are doomed to failure. M. Norton-Griffiths' perspective is similar in many ways to that of Joel Freehling and Stuart Marks, but they emphasise the importance of livelihood and the complexity of human motivations, while he emphasises the monetary decisions that people make. Section 3.2 of Part 2 gives the theoretical background to this approach. The models and concepts that M. Norton-Griffiths uses are introduced in section 1.6. In his case study, he shows that the incentives for Kenyan landowners to develop the agricultural potential of their land outweigh those to conserve their wildlife. For this reason, Kenyan wildlife has suffered drastic declines in recent years. He discusses the possibilities for reversing this trend by using regulatory instruments such as those outlined in section 4.3.2 of Part 2.

Tom Butynski and *Jan Kalina* are professional zoologists. They have worked directly on gorilla conservation, and demonstrate that, although gorilla tourism has a strong conservation image, it is founded on a very weak research base, and presents a substantial threat to gorilla survival. The major direct threat is disease transmission between humans and gorillas. Indirect threats stem from the reliance on revenue-earning as the justification for gorillas and their habitat being worthy of conservation, when tourism is such an unstable and locally unprofitable form of use. Little of the revenue is returned to local people, and events such as the recent conflicts in Central Africa can ruin the tourism market. Section 1.6.1 of Part 2 explains the instability of tourism using elasticities, and sections 3.2 and 4.4 discuss the effects of local incentives on the success of sustainable use ventures.

Anne Gunn is an ecologist in the Wildlife and Fisheries Division of the Canadian government. She discusses the commercial and subsistence harvesting of individual species from the perspective of a resource manager; thus, she has much the same perspective as Sophie des Clers. However, the system within which she works is rather different; harvesting of caribou and muskox in the tundra of northern Canada is predominantly by aboriginal groups. She describes the population dynamics of the hunted species, and how the hunting quotas are set, using annual census data. Data collection is discussed in section 4.2, and setting harvest quotas in section 1.3.2. The interactions between the hunted herbivores, their forage and the unpredictability of the climate are important determinants of the sustainable level of harvest; the theoretical background to this is presented in Chapter 2 of Part 2. Political changes are causing changes in resource management. Previously government agencies were responsible for setting and enforcing hunting quotas; now all interested parties co-operate in co-management, with the aboriginal groups having a major say in resource management. Although it is too early to comment on its success, this is an innovative form of community-based management (discussed in section 4.4).

Leonid Baskin is a scientific researcher in the Institute of Ecology and Evolution of the Russian Academy of Sciences. His case study takes an historical look at the fortunes of game mammals in the former Soviet Union. He describes how the Russian Empire of the 19th century over-harvested many valuable game species. The advent of the Soviet Union had a positive effect on game mammal populations, partly because political ideology favoured conservation and sustainable use, and partly because local people were excluded from using wildlife resources. His case study illustrates how politics can have a strong influence on conservation.

The Soviet Union was unparalleled in its successful sustainable use of wildlife, and yet it broke most of the rules for sustainable use set out in section 4.4 and the other case studies. He also emphasises the importance of the ecology of a species in determining its resilience to harvesting, as discussed in section 1.3 and Chapter 2 of Part 2. Historical analyses of wildlife use are rather uncommon, but are valuable in bringing out the important factors that affect sustainability in the long term.

Following the case studies, Part 4 of the book, 'Making Conservation Work', will bring out some of the lessons from these case studies, as well as highlighting other major issues in the conservation of biological resources. The prospects for conserving exploited species are discussed in the light of theory and practical experience.

Chapter 5: Sustainable Use as a Conservation Tool in the Forests of South-East Asia

Kathy MacKinnon

5.1 Introduction

Less than 10% of the world's terrestrial ecosystems lie within protected areas (WCMC 1992). Even though these protected areas are the cornerstones of any national or global conservation efforts, on their own they will be insufficient to protect biodiversity representatively across all ecosystems. In South-East Asia very few countries approach this 10% target, and in most countries the most species-rich ecosystems are poorly represented in the protected area network (MacKinnon *et al*. 1986). Lands outside, and even within some categories of protected area, will come under increasing levels of exploitation. It is therefore imperative to explore all possible opportunities to conserve biodiversity, natural habitats and species richness within the production landscape. In this context, the concept of sustainable use is attractive as a conservation tool.

Sustainable use can be promoted as a tool for conservation in three contexts:
- Wise and sustainable harvesting of resources at one site reduces the need to extend the area under exploitation, or to spread exploitation to another resource as the original resource is depleted.
- Sustainable use of biological resources in buffer zones can provide alternative and sustainable livelihoods that reduce destructive and exploitative practices in conservation areas.
- Sustainable use (i.e. harvesting at low levels, tourism) can justify the conservation of areas of natural habitat rather than their conversion to other forms of land use.

All of these scenarios will require the active and willing co-operation of local communities. Whether they will help to conserve biodiversity, however, may still depend on the way biological resources are valued, on the balance between benefits that accrue to local communities or to central government and/or big business interests, and on issues of ownership. Moreover, access to new technologies and new markets may put additional pressures upon these resources to the point where harvesting is no longer sustainable (Kartawinata *et al*. 1984). With many different types of resource extraction under many different socio-economic conditions, sustainable use may be impossible (Robinson 1993; Southgate 1996).

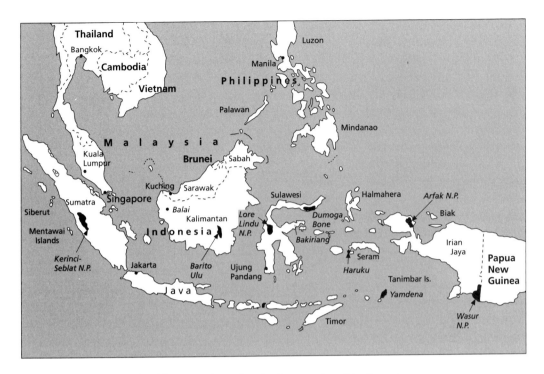

Fig. 5.1 The countries and main sites mentioned in the text.

Throughout South-East Asia vast areas of species-rich tropical forests and other habitats are threatened by logging, agricultural encroachment and conversion. At the same time, populations of vulnerable plant and animal species are being pushed towards local extinction due to habitat loss, over-exploitation, hunting and pollution. Local communities, national and local governments, NGOs and donor agencies are wrestling with the difficult task of reconciling the legitimate needs of local communities and conservation. Innovative solutions are being tried and tested. The following case studies from Indonesia, and elsewhere in South-East Asia, illustrate some of the conflicts and trade-offs between utilisation and conservation and the challenges of promoting sustainable use as a conservation tool. Project sites are shown in Fig. 5.1.

5.2 Profiting from plants: harvesting non-timber forest products

Non-timber forest products, including wild game and fish, may have a value to local and national economies far in excess of the value of standing timber (Caldecott 1988a; Peters *et al.* 1989). Plant products can provide valuable foods, palm oils, fodder for livestock, fibre, fuel, beverages, horticultural plants, antioxidants, sources of chlorophyll, enzymes, food colourings, sweeteners, spices, vitamins and medicinal plants (Duke 1992). Throughout South-East Asia many non-timber forest resources have been harvested for hundreds of years, both for subsistence and sale (Burkill 1935; de Beer & McDermott 1989; MacKinnon *et al.* 1996).

Because of this history of human use, it is often assumed that commercial harvesting of non-timber plant products has little or no ecological impact on tropical forests. In fact there is a very high probability that intensive resource extraction will gradually lead to depletion of these resources over time. Sustainable harvesting depends on the selection of species, resource and sites, and adjusting harvest levels to allow regeneration and growth of the species being harvested. Ecology and forest management are the keys to sustainable resource exploitation in tropical forests; it is critical to understand what happens to the plant populations being exploited and what levels of harvesting they can sustain (Peters 1996a).

Ecological constraints to sustainable harvesting of non-timber plant products in tropical forests include: high diversity and low population density of key plant species; irregularity of flowering and fruiting; dependence on animals for pollination and seed dispersal; high mortality and low success rate during seedling establishment, and the sensitivity of population structure to changes in levels of natural regeneration. Tropical forests are highly species rich; this means that although there are a large number of species, each species may be represented by only a few individuals. Within forest areas, however, there are mosaics of different habitats, some more uniform and less diverse in species composition than others (Whitmore 1984). Sustainable harvesting is more likely to succeed in habitats where there is relatively low species diversity and a high density of commercial species (Reining & Heinzman 1992).

The overall ecological impact of harvesting non-timber forest products will depend on the floristic composition of the forest, the nature and intensity of harvesting, and the species or type of resource being harvested. Most non-timber forest products can be categorised into three groups: fruits and seeds, plant exudates and vegetative structures. The ecological impacts of harvesting these three groups of products will differ.

Ecological constraints on sustainable harvesting of fruits and seeds

When the resources being harvested are fruits and seeds, the unpredictability of flowering and fruiting in tropical trees is a key issue in sustainability. Very few forest species produce reliable fruit crops at well-defined intervals. This is well illustrated by the dipterocarps, some of the most important timber trees in South-East Asian forests. At irregular intervals from 2 to 10 years, many dipterocarp species will start to flower more or less simultaneously; this mass flowering is followed by high levels of fruit production, known as mast fruiting. In an intense mast year almost every dipterocarp and up to 80% of all canopy trees may burst into flower. Such irregular fruiting patterns make harvest predictability difficult. However, illipe nuts or *tengkawang*, the fruits of *Shorea*, a common dipterocarp in the forests of Borneo, are an important source of cash for some Dayak communities in good harvest years (Chin 1985). Illipe nuts contain an oil that is used in the manufacture of chocolate, soap, candles and cosmetics. In 1987, a mast year, Indonesia earned more then US$8 million from exports of illipe nuts and oil (de Beer & McDermott 1989); the province of West Kalimantan alone exported over 13 000 tons of illipe nuts worth US$5 million. While most of the illipe trade depends on wild harvests, the crop is so valuable that some Bornean communities have developed small *Shorea* plantations in East Kalimantan (Seibert 1988), West Kalimantan (Peters 1996b) and Sarawak (Chin 1985).

Many tropical trees rely on animals for pollination and/or fruit dispersal. Thus, if there are no pollinators there will be no fruits, no seedlings, no harvest and no profits. In some cases the number and availability of pollinators may depend on land-use patterns far away from the harvest site and beyond the control of the resource harvesters. In Peninsular Malaysia the cave-dwelling fruit bats *Eonycteris spelea* are the exclusive pollinators of durian trees. These bats roost in limestone caves and feed preferentially on the flowers of the coastal mangrove *Sonneratia alba*, but pollinate the durian flowers they visit on their long nightly flights *en route* to the mangroves (Start & Marshall 1976). Destruction of either the limestone caves or the mangroves on which the bats depend could lead to reductions in the bat populations and a consequent drop in durian pollination and fruit production. As durians are a valuable and much-prized fruit throughout South-East Asia, a fall in durian production would have both economic and gastronomic consequences.

Frugivores are important seed dispersers but they are also major seed predators. In South-East Asia over 90% of canopy trees produce fruit adapted for consumption and dispersal by animals (MacKinnon *et al.* 1996). Bats, birds, primates, pigs, fish and other vertebrates are important seed dispersers and the distribution and abundance of seedlings produced by forest trees are frequently controlled by such animal dispersers. In forests where these key dispersers have been hunted to extinction, or to very low densities, the number of regenerating seedlings will also be reduced. The impact of hunting on vertebrate populations in Amazonia and Borneo, the 'empty forest' phenomenon, has been well documented (Redford 1992; Robinson & Bennett, in press). Conversely, heavy human harvesting of fruits and seeds may impact on animal populations by reducing the available food supply. Herds of bearded pig *Sus barbatus* are a valuable source of meat for rural communities in Sarawak (Caldecott

1988a), yet clearance of forests for logging, and damage and over-harvesting of food trees, has led to a decline in pig populations.

During fruit dispersal a large number of seeds may be lost to predation; over 98% of some forest species are lost to predators, especially rodents, weevils, beetles, ants and other insects (Peters 1996a). Only a small fraction of seeds will survive seed predation, germination and seedling predation and grow into canopy trees which will bear fruit. Even if the parent tree is left undamaged, commercial collectors of fruits and seeds are increasing the levels of predation on fruits and seeds and competing with other ground-foraging animals. The net result is likely to be an increase in the total percentage of fruits and seeds lost or destroyed and a reduction in seedling recruitment. Sustainable resource use hinges on a species' ability to establish new seedlings while being subjected to repeated and intensive harvesting. By increasing levels of predation, and reducing the number of seedlings available for recruitment, over-harvesting of fruits and seeds can eventually lead to changes in forest structure and species demography. A situation may develop where adult trees still occur in the forest but there is no reservoir of young seedlings and saplings to replace them – the 'living dead' phenomenon described by Janzen (1988). Even where regeneration does occur, heavy harvesting and the selective removal of only the best fruit over time may affect the genetic composition of the exploited population to leave a fruit tree population dominated by trees which produce less fruit of lower economic value.

It can be expected that some species and populations will be more susceptible to over-exploitation than others. In general, forest species that occur at high densities, exhibit high regeneration and are pollinated by either insects or wind should be able to tolerate more harvesting. It is interesting to note that neither wild durian in Borneo nor brazil nuts in Amazonia, both heavily harvested, fit this pattern. Instead both species occur at low densities and have obligate relationships with seed or fruit dispersers, characteristics that make them more susceptible to over-exploitation (Peters 1996a).

Gums, latexes and resins

Throughout South-East Asia gums, latexes and resins are collected for domestic and commercial use. Jelutong, the latex of the tree *Dyera costulata*, was particularly important as a source of rubber prior to the establishment of plantations of Brazilian rubber *Hevea brasiliensis* through-out South-East Asia. Other locally and commercially important tree exudates include gutta percha from *Palaquium*, damar from various dipterocarps and resins from *Agathis borneoensis* and *A. dammara*. Gaharu, a resinous product of *Aquilaria* trees where the heartwood is infected by fungus, is collected in Bornean forests for its fragrant incense wood.

In theory the tapping of latexes, resins and gums need not disturb the forest canopy, kill the exploited tree or remove seeds from the site, so should be sustainable. In practice exploitation of plant exudates can be very destructive. Several species of *Dipterocarpus* are exploited for their oleo-resin or damar which is valued as a source of varnish, for its use in caulking and as a perfume base. In Borneo and West Malaysia this damar is routinely collected by chopping holes in the parent tree and lighting a fire in the hole to make the resin flow. This sequence of boxing and firing can seriously weaken and eventually kill the tree. Similarly the collection of gaharu (incense wood) in most of South-East Asia is accomplished by felling the parent *Aquilaria* trees. While many collectors believe in a relationship between the occurrence of gaharu and outward signs of decay, it can also be found in apparently healthy boles. As a consequence, collectors routinely fell all *Aquilaria* trees that they find, including healthy trees with no fungal infection, so that collection is invariably wasteful. Nevertheless, as gaharu fetches prices of up to US$500 per kilogram the wasteful harvest continues until the search effort for rare and widely separated trees is no longer worthwhile.

Rattans and other vegetative structures

A diverse array of plant parts are used for building materials, utensils, fibre, tools, foods, fish poisons and medicinal use. The plant parts exploited may be roots, stems, bark, leaves or apical buds (growing tips). The plants will either be killed during collection or will survive the harvesting and later regenerate the stem, leaf or other structure that was removed. Sustainability depends on the form and ecology of the plant being harvested, as well as the intensity of harvesting.

The current harvesting of rattans in South-East Asia is unsustainable (Peluso 1983; Caldecott 1988b). Rattans are climbing, spiny palms (subfamily: Calamoidae) found in South-East Asian forests, with the largest concentration of species found in Borneo (151 species) and Peninsular Malaysia (104 species) (Dransfield 1988). At least 20 of these species are widely sought as a source of cane for manufacturing furniture, woven mats, baskets and other household goods (Fig. 5.2). Rattan is harvested by cutting the plant at the base and pulling the entire spiny stem and leaves out of the canopy. The impact of harvesting depends on the growth form of the specific rattan. Large cane rattans possess a single stem which does not resprout after cutting; harvesting kills these individuals. Intensive harvesting has drastically reduced the abundance of these solitary rattans. For instance, the valuable manau rattan *Calamus manan*, prized for furniture, has been virtually wiped out in Kerinci National Park in Sumatra. On the other hand, smaller-caned rattans are typically multi-stemmed and can resprout after cutting if sufficient time is allowed

(a)

(b)

Fig. 5.2 Rattan harvesting. (a) A rattan collector's camp. (Photo by K. MacKinnon.) (b) Baskets made from rattan, reeds, sedges and pandans on sale in Indonesia. (Photo by W. Giesen.)

between harvests. Unfortunately the rising demand for smaller canes has also led to over-exploitation in many parts of South-East Asia, with collectors cutting stems too young or close to the ground.

Re-harvestable rattan species such as *Calamus caesius* and *Korthalsia* can be cultivated. In the alluvial soils along the banks of the Barito river in Central Kalimantan there are long-established plantations of *Calamus trachycoleus* and *C. caesius* (Dransfield 1974). Ripe rattan fruits are collected during the fruiting season, crushed, cleaned and allowed to germinate. Seedlings are tended in shaded nurseries and by 14 months may have a cane 1 m tall with seven to eight leaves and one sucker. At this stage the seedling is transplanted to a garden cleared of undergrowth and with a fairly open canopy. *Calamus trachycoleus* grows rapidly with little care other than occasional clearing of the canopy but *C. caesius* needs constant clearing around the cane buds for maximum harvest. A first harvest can be cut 7–10 years after planting and every 2 years thereafter. Mature canes may be 15–20 m long after 10 years with 10 or more canes in a colony. One hectare can produce 10.5 tonnes of wet rattan per hectare (about 6 tonnes of dry rattan) and these yields could be improved by more intensive harvesting. Elsewhere in Central and East Kalimantan these short-stemmed rattans are planted in old rubber gardens and agricultural fields left to fallow as part of the swidden agriculture cycle (MacKinnon *et al.* 1996). By the time the fallow fields are brought back under cultivation the rattans can be harvested for the first time. If the demand for rattan drops the plants can be left growing until the market improves again (Weinstock 1983).

Indonesia produces more than 75% of the world's rattan supply. Exports quadrupled between 1968 and 1988, with rattan earning more foreign exchange than any other forest product except logs (Peluso 1986). Nearly half of these exports came from Kalimantan, the majority from East Kalimantan, with the greatest part of the stock derived from the wild. In 1989 the government of Indonesia banned the export of unprocessed rattan in an attempt to conserve stocks and retain more added value from processing in Indonesia. Initially this led to falling prices for collectors, which reduced pressure on wild stocks but also gave less incentive to farmers to cultivate rattan in their old fields. In the long run the sustainability of rattan harvests will depend on improved regulation of the trade, greater investment in rattan as a plantation and buffer zone crop, and incentives to small farmers to grow rattan in managed community forests and regenerating fallow fields.

Throughout South-East Asia much of the current exploitation of non-timber forest products is not sustainable. Some species, because of their reproductive biology, regeneration and growth strategies, or population structure, are inherently more able to withstand the impact of regular harvesting than others. Choosing the 'ideal' forest resource for harvesting will depend on the species' life cycle characteristics, type of resource produced, density and abundance in different forest types, and the size-class distribution of the population. An ideal species might show one or more of the following characteristics: occur at high densities in certain forest types, fruit annually, be pollinated by a common generalist species or wind, be a primary species adapted for growth under closed canopy, produce a resource such as a fruit or seed where harvesting does not kill the parent plant, and show prolific natural regeneration (Peters 1996a). Even with species chosen for these characteristics, sustainability of harvests will depend on the ability of the resource managers to monitor the impacts of harvesting on regeneration and their willingness to adjust harvesting levels accordingly.

Managed forests in West Kalimantan

How do local communities manage forest resources? Many indigenous groups in the tropics have developed effective systems for manipulating the distribution and abundance of important forest resources. Often these agroforestry systems involve intensive management of regrowth and secondary vegetation after forest clearance for agriculture (Dove 1985; Denevan & Padoch 1988). Elsewhere, indigenous communities implement silvicultural management in areas of primary forest under their control. Such forest management systems usually focus on a variety of different species and resources. Thus, Dayak groups in West Kalimantan manage their community forests to selectively favour the regeneration and growth of nine different species of illipe nuts (Padoch & Peters 1993) as well as trees which provide fruits, latex, resins and sometimes timber. This has the additional benefit of providing some security against sudden fluctuations in local market conditions with harvesting focused on products that will fetch a good price that season.

The Daret of Balai are a Dayak group in West Kalimantan that manage hill dipterocarp forests which contain a variety of useful plant resources, including illipe nuts, rattans, bamboo, sugar palms *Arenga pinnata* and construction woods such as ironwood *Eusideroxylon zwageri*. Edible fruits, including durian *Durio zibethinus*, rambutan *Nephelium* spp., langsat *Lansium* sp. and mangosteen *Garcinia mangostana*, are common. The distribution and abundance of these resources, both wild and introduced, have been enhanced by the Daret through selective weeding, enrichment planting and occasional low-intensity harvesting of poles (Padoch 1992; Padoch & Peters 1993).

Selective weeding is done around durian and illipe trees during the harvest season to facilitate the location and collection of fruits. Seedlings and saplings of valuable species are spared while less desirable species are removed. Enrichment planting is sometimes deliberate but often occurs as a result of casual actions, such as the discarding of fruit seeds as people walk through the forest. Other species such as rattans, medicinal plants and construction timbers are carefully transplanted. The resulting forest is a mixture of wild, transplanted and haphazard plantings, i.e. a modified natural forest where certain useful native species are encouraged. Many of these species are shade-tolerant primary forest species that require only minimal canopy openings for establishment and growth. These canopy openings are provided by natural treefalls and occasional felling of trees for timber.

This system of community forest management works with minimum labour investment. Although most of the forest products are for household use or subsistence, valuable crops such as durians and illipe nuts are collected and sold to local traders. In 1991, during the peak of the fruiting season 10 000 durians were transported for sale from four Daret villages in one day (Peters 1996a), and over the month-long season durian marketing provided a substantial income. Although village management modifies the natural species composition and structure of the forest this sort of agroforestry system conserves native species and maintains populations of useful species in a species-rich and ecologically sustainable system. The access to non-timber forest products for subsistence and income encourages villagers to manage and maintain the forests.

5.3 Harvesting wildlife resources

Even where harvesting of non-timber forest products appears sustainable and forests seem intact and healthy, those forests may be empty of larger mammals and birds, which are some of the main predators, frugivores and seed dispersers. Wildlife surveys conducted in Sarawak hill forests illustrate the impact of hunting (Bennett *et al.*, in press). In areas of high hunting pressure the number of species of primates, hornbills, and birds in general were lower than in areas with little or no hunting (Fig. 5.3). Populations of gibbons, langurs and macaques were greatly reduced by hunting, often to the point of local extinction. Continued hunting at current levels will drive animals to extinction, disrupt forest ecology and deprive local people of a major source of protein. With loss of these key animal species forest ecosystems could start to collapse and decay. From a conservation viewpoint the message is clear: to preserve forest diversity it will be necessary not only to protect large areas of forest, but also to direct hunting efforts away from more vulnerable species, such as primates, to species with higher reproductive outputs, such as pigs and deer, which are more able to withstand the offtake.

Throughout South-East Asia, animals and animal products are harvested from wild populations: fish, game, river terrapins and their eggs are collected for food and trade; snakes, crocodiles and monitor lizards are collected for their skins and marine turtle populations are subjected to intense harvesting for meat, shells and eggs. The primitive and valuable arowana fish *Scleropages formosus*, collected for the aquarium trade, has been so heavily harvested that populations have declined dramatically in Kalimantan rivers (Giesen 1986). For some commercially important

Fig. 5.3 An orang-utan skull and a hornbill casque; trophies from Dayak hunting expeditions in Borneo. (Photo by K. MacKinnon.)

species, such as crocodiles, there is open acknowledgement that harvesting from the wild may not be sustainable; monitoring mechanisms have been put in place to regulate the trade and commercial crocodile farms have been established. These strategies have been more effective in some Asian countries than others, and while these measures may ensure sustainable harvests, they may do little to promote protection of wild populations and their habitats. Just as for plant products, the sustainability of wildlife harvests depends on the species and resource being harvested, the species' life history, the intensity of harvesting and, perhaps most important, local conditions that regulate harvesting levels.

Parrots and corn in Tanimbar: sustainable harvesting of a CITES species

An understanding of species ecology and the likely impact of harvesting, both on the total population and different age groups, can give some indication of whether harvesting is likely to be sustainable. It is especially important to monitor the impact of harvesting on rare or endangered species and species with restricted distributions. The Convention on

International Trade in Endangered Species (CITES) provides mechanisms to monitor and restrict trade in endangered and threatened species.

Two parrot species are endemic to the southern islands of the Banda Sea in the province of Maluku, Indonesia. *Cacatua goffini* is endemic to the Tanimbar islands while *Eos reticulata* occurs in the Tanimbar islands and the small islands of Babar and Damar further west. Concern over the number of birds of these restricted-range species being traded internationally led to *C. goffini* being placed on Appendix I of CITES at the Eighth Meeting of the Conference of the Parties in Tokyo in 1992. Indonesia proposed a zero catch quota for *E. reticulata* pending the results of field surveys on population numbers.

In 1993 the Indonesian Conservation Department (PHPA) and the international NGO Birdlife International surveyed Yamdena, the largest of the Tanimbar islands, to assess the status of the two parrot species and to evaluate the damage caused to local farmers by *C. goffini* preying on maize fields. The populations of *C. goffini* and *E. reticulata* on Yamdena were estimated to be $255\,000 \pm 36\,000$ and $220\,000 \pm 52\,000$ respectively (Cahyadin *et al.* 1994). Minimum exports of live birds reported to CITES in the seven years up to 1989 were between 8651 and 14 234 (average 11 349) for *C. goffini* and 1888 to 7703 (average 3198) for *E. reticulata*. Because the birds are sold only for the international trade, numbers of birds traded give some approximation of numbers caught, though there will be some losses in captivity. Past annual catch levels, estimated at less than 5% of the Yamdena population of *C. goffini* and less than 2% of *E. reticulata* on Yamdena, appear unlikely to have caused population declines in these species on the island. *C. goffini* is caught only when it raids crops and the main offenders are juvenile birds, which are the part of the population likely to suffer the highest natural mortality prior to birds finding territories. Flocks of *C. goffini* damage an estimated 1.7% of the island's maize crop annually (Cahyadin *et al.* 1994). Overall this loss seems insignificant, but for the individual farmer it can be a disaster to have flocks of parrots clean out his maize fields. In the past, capture and sale of *C. goffini* compensated for this loss and provided valuable extra cash income.

The harvesting levels of parrots on Yamdena seem sustainable yet trade has been stopped due to international concern over the parrots' status in the wild – a situation that prevails even after publication of the PHPA/ Birdlife International survey. A large number of captured birds could not be sold on, there is little or no domestic market for these species and some birds were returned, at considerable expense, to Tanimbar for release back to the wild, with funding from international organisations. Parrots still continue to raid the maize fields and are still trapped to stop their predations but now have no cash value so are exterminated. This case illustrates how even a rare and restricted-range bird can be sufficiently numerous to be a pest locally. Furthermore, it re-emphasises the need for a good understanding of species ecology and monitoring of wild populations of traded birds. With such monitoring in place it seems likely that both parrot species on Yamdena could sustain healthy populations and tolerate sustainable harvesting to meet the current limited demands of the international bird markets. Allowing local farmers to capture and sell crop-raiding birds would not only compensate them for crop losses but could help foster support for the proposed Yamdena reserve where the main parrot population feeds and breeds.

Megapodes and maleos: maintaining the supply of golden eggs

The impact of local traditions and harvesting regimes on sustainability is illustrated by comparison of megapode egg harvesting at three different sites, namely Dumoga-Bone National Park and Bakiriang in Sulawesi and the Moluccan island of Haruku (Argeloo & Dekker 1996). Megapodes are birds which use external heat sources such as volcanically- or sun-heated soils or rotting vegetation to incubate their eggs. Two of the 15 species of megapodes which occur in Indonesia, maleo birds *Macrocephalon maleo* in Sulawesi and the Moluccan megapode *Eulipoa wallacei* on Haruku, lay their eggs in communal laying grounds. Each egg is equivalent in volume to about five hen's eggs. Megapode eggs are harvested by local collectors for food and sale.

Traditionally, levels of harvesting at megapode beaches were controlled by local customs, including contracting of egg-collecting rights and closed seasons. Loss of these traditions plus increasing human population and immigration and improved road access to nesting grounds have resulted in over-harvesting and serious reductions in maleo populations in North Sulawesi. The maleo population at Bakiriang in Central Sulawesi is also threatened by increasing human population and the loss of traditional management customs (Watling 1983). On Haruku, however, the Moluccan megapode population is still being managed under the customary rights or 'sasi' system. This system allows the adjustment of harvesting levels to maintain population levels, either through community-enforced reductions in harvest quotas or compensation for the harvesting rights (Argeloo & Dekker 1996). In North and Central Sulawesi, however, stricter law enforcement seems to be the only option to protect megapode nesting grounds. The different threats and opportunities for megapode management at different sites demonstrate that conservation and use strategies must be appropriate to local situations.

Butterflies, oils, conservation and traditional use in Irian Jaya

Involvement of local communities and endorsement of their traditional management systems is working effectively for conservation in the Arfak

Mountains Nature Reserve in the Bird's Head of Irian Jaya. The area was gazetted as a nature reserve because of its biological importance. It harbours at least 110 species of mammals (53 of them New Guinea endemics) and 320 species of birds, half of the avifauna of Irian Jaya (Craven & de Fretes 1987). Prior to gazettement all the land was traditionally owned by the Hatam people and it was obvious that trying to stop them from collecting resources in the reserve would be counter-productive. Instead the World Wide Fund for Nature (WWF) worked with local Hatam villagers to develop co-operative management agreements that would enable villagers to continue their traditional lifestyles but engage them as guardians of the area against outsiders (Mandosir & Stark 1993).

The reserve and adjacent outlying lands were divided into 16 nature reserve management areas (NRMAs). The size and boundaries of each NRMA were defined by the extent to which each collective group of landowners were willing to work together. A committee of influential people, such as village heads and church leaders, were assigned to manage each NRMA in accordance with tribal customs and community decisions. The committee was responsible for identifying the official landowners and overseeing the correct marking of the boundary. The Hatam were allowed to retain enough land outside the reserve for future subsistence needs.

The management system developed with the Hatam works because the boundary falls under multiple jurisdiction and allows rapid identification of violators as either landowners or outsiders. No one is allowed to establish permanent houses or gardens in the reserve but the indigenous people are allowed to collect firewood and timber for home use and to hunt with traditional weapons such as bows and arrows. Members of one community may not take forest resources belonging to another community without permission of the owners. Fires may be built for cooking and comfort but not to aid hunting. The regulations allow the continuation of the Hatam's traditional lifestyles but outsiders face much stricter regulations. They are not allowed to hunt, to make temporary shelters from forest materials or to remove plants, trees or animals. Infringements are initially dealt with by the committees which have government-sanctioned powers to enforce reserve regulations. Usually violations cease after warnings and fines at the community level but options exist for the committees to pass the matter higher to the reserve management authority (PHPA) or to the district government officer (*camat*) if necessary. The local communities have taken their responsibilities seriously and have seized the opportunity to play an active role in protecting their own traditional lands and resources.

As well as promoting local stewardship of the reserve, WWF worked with local communities to provide alternative income-generating activities to reduce the need to extend gardens within the reserve. One such project involves butterfly farming and sustainable harvesting of the Ornithoptera birdswing butterflies for which the area is renowned (Mandosir & Stark 1993). Gardens of swallowtail food plants such as the *Aristolochia* vine have been established in secondary forest areas outside the reserve. Wild butterflies lay their eggs on the vine; larvae feed and pupate high on the vines. The villagers harvest the live pupae which are then sold to a marketing centre. No adult butterflies can be caught or sold. Since only a proportion of the live pupae are found and harvested, wild populations are continually being replenished. With careful control of collection and marketing by the committees and a local NGO, the butterfly farming should be sustainable. This activity is directly linked to protection of the nature reserve, where wild butterflies spend most of their lives, yet it provides local people with a cash 'crop' that is light to transport and yields high returns (from US$1.50 to US$60 per pupa according to species) without damaging the natural forest. Initial problems in marketing practices are now being smoothed out, a crucial step to ensuring that villagers see a quick return on their protection and collecting efforts. Similar butterfly-farming initiatives are operating successfully in the highlands of Papua New Guinea (Hutton 1985) and in buffer zones around Lore Lindu National Park in Sulawesi.

WWF have subsequently begun another project to strengthen local community involvement in park management in Wasur National Park (412 000 ha) in South-east Irian Jaya. The land within the park boundaries is owned traditionally by 2000 members of the Kanum, Marind, Marori and Yei tribes who use it for shifting gardens. Another 65 000 people live around the fringes, many of them subsistence farmers. The need for cash to buy household essentials has led these local communities to engage in small-scale logging, hunting and selling land. The WWF project is working with local community groups and government agencies to recognise the park as a traditional use area where indigenous people and long-term residents are allowed to continue their traditional agricultural and hunting activities (Craven & Wardoyo 1993).

Part of the management strategy has been to stop illegal hunting with firearms by outsiders while allowing local people to continue hunting traditionally with bows and arrows. Each clan and family has a traditional range where they hunt, garden and carry out rituals. In 1992 local villagers were able to earn US$3750 over 3 months from selling deer hunted with bows and arrows. The deer are an introduced species and keeping their numbers in check probably helps native wildlife; the income generated alleviates rural poverty and wins support for protection of the park. Other income-generating options are also being investigated; they include eco-tourism and the extraction of essential oils from the leaves of the native paperbark trees *Melaleuca*. Paperbark occurs in pure stands, especially in degraded freshwater swamp forests, and regenerates new shoots quickly after harvesting. Cajeput oil, extracted from the leaves, is used as insect repellent and in the manufacture of soap, throat pastilles and ointments. The involvement of local villagers in the protection, management and controlled exploitation of common natural resources is proving an effective tool for conservation (Craven & Wardoyo 1993).

The Irian Jaya programmes have worked well, through a mixture of acknowledgement and extension of traditional rights and customs, and the provision of small-scale economic activities. They offer some simple lessons and ingredients for success: close consultation with local people; identification of key players; understanding of community needs; provision of alternative income-generating or social benefits which come 'on stream' quickly; strict enforcement of agreed boundaries and regulations, with the communities themselves engaged in enforcement and guarding the reserve; employment opportunities for local people; and flexibility to adapt management strategies to resource availability, local needs and situations. It could be argued that the situation in Irian Jaya was predisposed to conservation since the Irianese tribes live close to the land and are dependent on natural resources. As community aspirations and opportunities change, other management solutions may be needed. It is crucial to ensure that economic incentives or any other benefits are seen to be linked to the park and conservation. Too many supposedly integrated conservation and development projects have turned out to be simply rural development projects with vague or no obvious links to conservation (Wells & Brandon 1992).

5.4 Logging forests to save them: an NGO dilemma

Elsewhere NGOs are engaging in the challenge of more sustainable forestry. In Papua New Guinea (PNG), international NGOs are entering into partnerships with commercial logging companies to bid for forest management agreements and timber concessions. Until recently PNG's forests have remained relatively intact, but as forest resources are depleted in Asia many logging companies have expanded their operations into PNG. The Ministry of Forestry awards the actual timber concession but local landowners own the timber rights and receive royalties on the timber extracted. The intent behind the NGO-backed initiatives is to encourage more responsible 'green and clean' logging companies to bid for concessions with the expectation that not only will the landowners get a better deal but that the companies will log and manage their concession in a more environmentally friendly and ecologically sustainable manner.

The logging company gets access to a new resource base, the landowners get royalties, and ultimately the buyers have the comfort of buying 'green' timber – certified according to the standards of the Forest Stewardship Council. It is hoped that the involvement of the NGOs will ensure that there is better monitoring of logging impact and improved forestry management, so that their private sector partners go much further than merely meeting the requirements of the existing, but rarely enforced, guidelines of the PNG Forestry Code. The real challenge for the NGOs will be to see how far they can push the conservation agenda and logging best practice, within the bounds of company profitability. There could be opportunities for the establishment of conservation zones to protect high biodiversity areas within the concession (Marsh & Sinun 1992), for low impact logging activities to leave habitat corridors between logging blocks, and for the creation of conservation revenues and trust funds. There is a real opportunity for demonstrating to landowners that business does not have to go on as usual and that logging can be done in a more ecologically and socially sustainable way. Building on their concession experiences, the NGOs could seize the opportunity to work with partner NGOs in market countries in Asia to promote a public constituency for more responsible forestry practice and green consumerism.

What are the conflicts and contradictions inherent in the principle of logging a forest to save it? The logging company, the landowners, the government, the NGO, even the consumers benefit, but what are the real benefits for biodiversity? Bad logging becomes better logging but logging, however 'green', will change forest structure and ecological succession. Logging roads open new landscapes to settlement, agriculture and hunting, leading to habitat loss and increased predation pressure on certain forest species. Even where forests may remain relatively intact, any extractive use or manipulation of forest ecosystems is likely to lead to some loss of biodiversity, and the species that are most likely to be lost first are those that are rare and local in their distribution. Through their involvement in the PNG forestry initiatives the NGOs are seeking to minimise those biodiversity losses. Depending on levels and frequency of logging, many primary forest animals and birds can survive in selectively logged forests (Johns 1988). Forests under sustainable logging regimes can thus effectively extend the conservation estate as a supplement, but not an alternative, to fully protected areas (Johns 1992)

5.5 Bringing biodiversity into the mainstream

Conservationists are caught in a painful dilemma. With increasing pressure on land and natural resources in South-East Asia it is probably unrealistic to hope that many more large areas will be designated purely for conservation, even though that may be environmentally the most appropriate form of land use. Indeed, in many parts of the tropics there is increasing emphasis on establishing protected areas as multiple use areas, aiming to conserve habitats and wildlife while allowing local communities to harvest forest products, including meat. The assumption is that such resource extraction is sustainable. In fact research in Sarawak and Amazonia has shown that hunting, at least of some species, is unsustainable even when it is only done for subsistence (Redford 1990; Bennett *et al.*, in press). Similarly, throughout South-East Asia many plant products such as rattans, gaharu and ironwood are also being over-exploited; to encourage these activities within protected areas (Siebert 1988) can only exacerbate the problem (Dransfield 1992). It is doubtful that current levels of utilisation can continue without destroying the resource base; that would be a disaster for the forests, for the wildlife and for the local people. Forest-dependent communities will lose access to wild meat, other forest products, potential tourism revenues, reliable water supplies and cultural benefits. It is important to get policy-makers to understand that conservation and use are not always compatible, and that the best strategy is to designate different areas for strict protection or other uses and to manage those areas according to their very different objectives.

Conservationists must look for innovative opportunities for conservation and more sustainable use in the production landscape but they need to do this honestly and critically. A forest can be managed for extraction of forest products, logging, swidden agriculture or converted for intensive agriculture. These represent increasing levels of utilisation; all are conceivably sustainable but each has a different and increasing impact on biological diversity (Robinson 1993). Even extraction of non-timber forest products, the lowest level of forest manipulation, will eventually lead to loss of forest species (Anderson 1990). Moreover, higher prices for fruits or other non-timber forest products will not necessarily induce better management of renewable natural resources nor greater benefits to rural households (Browder 1992). Ultimately the contribution of extractivism to rainforest conservation may turn out to be very limited indeed (Southgate *et al.* 1996). In a similar way ecotourism may generate revenues and support for biodiversity conservation and benefits to rural communities, but such benefits are not automatic; they will only come about as the result of clear objectives, planning and management to attain these specific goals (Brandon 1996).

Many of the countries of South-East Asia support rich, and often unique, biodiversity which is part of their natural heritage and resource base. Many sectors of the economy are dependent, directly or indirectly, on the diversity of natural ecosystems and the species and environmental functions that they protect. Conservation of biodiversity is crucial to the sustainability of sectors as diverse as forestry, agriculture and fisheries; health care; science; industry and tourism. Some of the poorest and least advantaged rural communities are the most dependent on biological resources. Conservation and sustainable use of the region's biodiversity will be critical to sustainable development. The case studies described above have illustrated some of the challenges and trade-offs between conservation and sustainable use at the local level. A far greater challenge will be to persuade policy-makers, governments, the scientific community, NGOs and local communities to work together to bring biodiversity issues, options and concerns into the broader picture of sustainable sectoral and regional development, promoting conservation within the production landscape (World Bank 1995, 1996).

Chapter 6: Will Bigleaf Mahogany Be Conserved Through Sustainable Use?

R.E. Gullison

6.1 Historical use of mahogany

The neotropical mahoganies (*Swietenia* sp., Meliaceae) have the longest commercial history of any neotropical tree, with inter-continental trade dating back more than 400 years (Lamb 1966). The rot-resistance and fine working quality of the wood makes it ideal for boat building. Upon arrival in the New World, the Spanish were quick to recognise this value, and used mahogany extensively in the construction of their ships, including those of the Spanish Armada. By the middle of the 17th century the English too were incorporating mahogany into their naval fleet, where it soon became the preferred wood. Mahogany was adopted widely by English cabinet makers in the 18th century, and its use defined an entire period in furniture design. Mahogany continues to be a premier wood for both the ship and furniture trades, with its main markets in the UK and the USA (Rodan *et al*. 1992).

The trade has been maintained not through management for sustainable production, but by shifting sources of supply (Lamb 1966; Rodan *et al*. 1992), increasing the efficiency of extraction, and by harvesting a wider range of qualities and sizes (Snook 1996). As early as 1735 mahogany was becoming scarce in Jamaica, and British timber producers began shifting their operations to Central America (Lamb 1966). This trend has continued, and the Caribbean mahogany, *S. mahogani*, and the Pacific Coast species, *S. humilis* (ranges shown in Fig. 6.1), now contribute negligibly to international trade. Virtually all international trade is of bigleaf mahogany (*S. macrophylla* King) originating from South America. In this region there is much concern about the sustainability of extraction, as shown by the President of Brazil's recently decreed moratorium on all new mahogany logging (Rainforest Action Network 1996).

The long commercial history of mahogany and the mounting concerns about its conservation status mean that it presents an excellent opportunity to examine the prospects for conserving neotropical tree species through sustainable use. This chapter begins with an overview of the ecology of mahogany, and the problems it poses for management, and then discusses whether recent conservation and management initiatives are likely to prevent *S. macrophylla* from sharing the fate of its congeners.

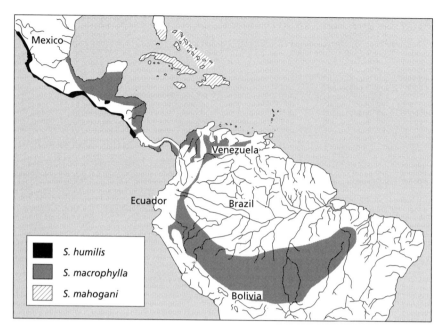

Fig. 6.1 Ranges of the three neotropical mahogany species. (Adapted from Lamb 1966)

6.2 Ecology of mahogany

Bigleaf mahogany grows in a wide variety of forest types, ranging from sub-tropical and tropical dry forests to moist forest formations (Lamb 1966). It has a very wide distribution, occurring as far north as the Gulf Coast of Mexico, and as far south as Bolivia and Brazil (Fig. 6.1). It is an emergent tree, attaining heights of up to 50 m. The large trees may be heavily buttressed. Seeds are winged, wind-dispersed, and are released near the beginning of the wet season. The density of commercial sized trees (60 cm diameter at breast height (dbh)) varies widely, from a reported low of 0.2–0.3 trees ha^{-1} in some Bolivian forests (Gullison *et al.* 1996), to a high of 22 trees ha^{-1} documented for some Central American stands (Lamb 1966).

The ecology of bigleaf mahogany is unusual. It is classified as a climax species (*sensu* Swaine & Whitmore 1988) that is light demanding in its younger stages. It is present as adults but does not generally regenerate in closed forests. Rather, it requires large disturbances to create the conditions favourable for the growth of its seedlings and saplings. In its northern range, the agents of disturbance are fire and hurricanes (Lamb 1966; Snook 1993). Adult mahogany trees survive these disturbances better than other species, and are subsequently able to disperse their seeds to open sites. Light levels are high because most of the larger size classes of trees are destroyed, and the destruction of the seed and seedling banks means there is little competition from other species. Once established, mahogany trees are capable of maintaining emergent status for hundreds of years.

The ecology of Amazonian mahogany is more poorly understood than that of Central American populations. In Bolivia, mahogany has recently been shown to regenerate after hydrological disturbances, either after erosion or in areas of flood-killed forest caused by logjams (Gullison *et al.* 1996). In *terra firme* forest in eastern Amazonia, the disturbance agents that led to the existing populations have not been conclusively identified. A drier climate in the past may have permitted forest fires, explaining the occurrence of mahogany where currently neither flooding nor fire are prevalent (Snook 1996). Another possibility is that mahogany is maintained in these forests by large blowdowns that have been recently documented (Nelson *et al.* 1994).

There is growing evidence that pre-hispanic cultures have had a strong influence on current forest composition, and this factor complicates interpretation of the ecological processes that have structured existing mahogany populations. In Central America and Mexico for example, it appears that past cultures have purposely increased the abundance of useful trees such as mahogany and chicle (*Manilkara zapota*) (Gomez-Pompa & Kaus 1990). Similarly, in Bolivia Gullison *et al.* (1996) found evidence of ceremonial sites, roads, and raised agricultural beds underneath or near most of the mahogany stands they studied. In the latter case it is not known whether previous cultures actively managed these species, or whether the high density of mahogany in these areas is simply due to the fact that suitable regeneration sites were created by the abandonment of agricultural fields.

The natural disturbances that create conditions favourable for regeneration are infrequent. As a result, mahogany populations are typically composed of one or a few cohorts of trees, with few younger individuals present where trees are of commercial size (Snook 1996; Fig. 6.2a). If a long time has elapsed since the last disturbance, the population can be composed almost entirely of old trees. The unpredictability and long time period between disturbance events appear to have shaped a life history strategy of longevity, achieved by the allocation of photosynthates to secondary compounds and to structural support (Snook 1993; Gullison *et al.* 1996). In addition, the onset of fecundity is delayed, with high rates of capsule production not occurring until the trees reach 80 cm in diameter. This may take 150 years or more.

6.3 Problems in managing mahogany

Mahogany is typically managed with a minimum cutting diameter limit,

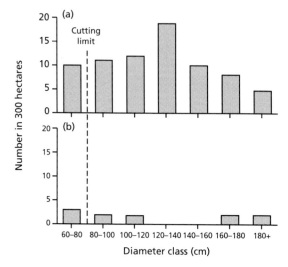

Fig. 6.2 The size distribution of mahogany trees from 300 ha in the Chimanes Forest, Bolivia. (From Gullison 1995.) (a) Before harvest. The 'humped' or unimodal size distribution that is characteristic of mahogany populations is evident. This is a result of regeneration occurring only after large infrequent disturbance events. (b) After harvest. Timber harvest is regulated with two rules. Trees smaller than 80 cm diameter cannot be cut, and 10% of the trees of > 80 cm diameter must be left as seed trees. In this case, seven of the 10 pre-commercial sized trees were illegally cut. The second rule was inadvertently followed because some trees were missed by the tree spotters, and others that were rotted but alive were left standing. Even so, logging resulted in a six-fold reduction in density.

which specifies the size above which most or all of the trees can be legally harvested. For example, the minimum diameter limit is 80 cm in Bolivia (Gullison *et al.* 1996), and as low as 40 cm in Mexico (Snook 1991). However, this type of management is inappropriate for the ecology of mahogany. The rarity of regeneration events means that in some cases virtually the entire population can be above the cutting limit and vulnerable to harvest (e.g. Fig. 6.2). Furthermore, the low density of mahogany trees means that logging does not create much of a disturbance to the forest (Gullison & Hardner 1993), and there is little or no recruitment of seedlings that survive the removal of the parent trees (Quevedo 1986; Verissimo *et al.* 1995; Gullison *et al.* 1996; Fig. 6.3).

An additional conservation concern is that unregulated logging may cause genetic erosion by concentrating mortality on the largest and best formed individuals. Newton *et al.* (1996) report results from provenance trials that demonstrate enough heritability in height growth and pest resistance to support claims that selective logging will cause genetic erosion. The amount of genetic erosion in natural populations is, however, unknown because the relative importance of genetic and environmental

(a)

(b)

Fig. 6.3 Logging mahogany in Bolivia. (a) Cut logs. (b) Mechanised removal of the logs from the forest. (Photos by Ted Gullison.)

factors in determining phenotypic variation in form and pest resistance has not been measured.

The potential for genetic erosion will be higher the greater the relative reproductive value of commercial sized trees, because logging will cause a greater reduction in effective population size, and trees will have reproduced less before they are harvested. Two factors increase the reproductive value of commercial sized mahogany trees compared to faster maturing species. The first is the delayed onset of fecundity. Mahogany trees below the minimum diameter of 80 cm produce relatively few seeds, and fecundity peaks when the trees are about 110 cm diameter (Gullison *et al.* 1996). The long time periods between disturbance events also maintains the reproductive value of the older larger trees, because younger trees may have had no opportunity to reproduce. These two factors are likely to be related, the delayed onset of fecundity being an adaptation to the long return time between disturbance events.

A second possible genetic impact is that logging may increase the inbreeding rates of logged populations by lowering the density of reproductive trees. The low density at which mahogany often occurs before harvest can be drastically reduced by logging. For example, logging reduced the density of the trees shown in Fig. 6.2 from 0.25 trees ha^{-1} before harvest, to a density of 0.04 trees ha^{-1} after harvest. This six-fold reduction in density resulted in an average spacing between trees of approximately 527 m in the harvested stand. Whether or not the pollinators – presumed to be a small species of bee or moth (Styles & Khosla 1976) – can maintain pollen flow at these distances is unknown.

It is technically possible to overcome these barriers and produce sustained yields of mahogany. Harvest rates can be set at or below sustainable levels, and deficiencies in natural regeneration can be compensated for with silviculture. This is done most successfully by line planting, which involves planting seeds or seedlings in lines throughout secondary forest, with sufficient clearing or thinning of the canopy to encourage good growth (Weaver 1987; Ramos & del Amo 1992). A good replanting programme will ensure that genetic variation is maintained, that seeds originate from outcrossed trees of good form, and that follow-up silvicultural treatments are conducted to ensure the recruitment of planted seedlings. Foresters in the first half of this century had an apparently viable silvicultural system developed for mahogany in Mexico and Central American (Lamb 1966), and successful results from enrichment planting of natural forests with mahogany have been achieved in many neotropical countries (Weaver 1987). As with many other natural resources, the problem is not whether sustained yields are technically feasible, but whether they are likely given the current socio-economic and political environment.

One of the biggest obstacles to implementing sustainable management of mahogany is the lack of financial incentives for producers to limit harvests to sustainable levels and to invest in regeneration. To maximise profits, trees should be felled at the size where their rate of increase in value drops below the prevailing interest rate (Pearce 1990). Delaying harvest beyond this point incurs an opportunity cost, because profits from logging could be invested at a higher rate elsewhere. The rate at which a tree increases in value is the sum of its growth and any real increases in value. In Bolivian forests, the minimum size at which trees are cut is about 40 cm diameter, well below the legal minimum diameter (R.E. Gullison, pers. obs.). At this size, trees are increasing in volume at a rate of about 4% per year (Fig. 6.4). Real price increases of mahogany between 1987 and 1994 have only averaged about 1% (Hardwood Review 1995), meaning that a 40 cm diameter mahogany tree increases in value at about 5% per year (the sum of the growth and the real increase in price), slowing down as the tree gets larger. On the other hand, real interest rates in Bolivia have averaged 17% for the last 5 years (Banco Central de Bolivia 1994, 1995). The disparity between the increase in value of the unharvested trees and local interest rates is large. Thus, there is a strong economic incentive to cut all trees with any value, including trees below the minimum cutting limit and seed trees, and invest the proceeds elsewhere. This finding is independent of ownership of the resource – in this case, tenure of concessions – but tenure insecurity can produce additional pressures for over-harvesting.

To determine the full costs of sustainable management to mahogany loggers, Howard *et al.* (1996) calculated the net present value of different harvesting regimes in the Chimanes Forest, Bolivia. Simulated harvests ranged from the unsustainable selective logging of mahogany (similar to current cutting practices) to a more sustainable harvest that included guidelines for leaving seed trees, harvesting a broader mix of species, and in one case even thinning out non-commercial species to increase the

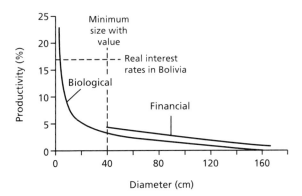

Fig. 6.4 The productivity of mahogany trees as a function of size. Financial productivity is the biological productivity combined with an annual 1% real increase in price. The minimum size at which trees have commercial value is about 40 cm diameter. At this size they increase at about 5% per year in value. To maximise their profits, loggers should fell trees at a size when the financial productivity of the tree drops below the prevailing interest rates. Because of scarce capital and high risk, real interest rates can be very high in developing countries. In Bolivia, the mean interest rate for the period 1989–94 was 17%. The disparity between interest rates and productivity of the trees provides a very strong incentive for over-harvest.

productivity of the remaining stand. The unsustainable logging of mahogany was five to eight times more profitable than the more sustainable alternatives, although the sustainable logging prescriptions did generate profits competitive with those available from other sectors of the Bolivian economy.

The tremendous value of the standing timber means that for sustained yields to be achieved, there must be a strong commitment and capacity to regulate harvest (FAO 1995b). This is almost completely absent from developing countries. Forest personnel are poorly paid and trained, and there is a lack of political commitment to good forest management (Poore 1989). For example, in Bolivia the forest service budget is approximately 7 cents per ha (J. Nittler, pers. comm., 1995). In the USA the USDA Forest Service has US$44.15 to manage the same area (House of Representatives 1996), and incidentally, thousands of violations by timber companies are still recorded each year (Rice 1989; Losos et al. 1993). There are virtually no forests being managed for sustained yields of mahogany from natural forest, with the possible exception of the Plan Piloto Forestal Project in the Yucatan. Even in this case, yields have been maintained in the region by reducing the minimum cutting diameter limits, and there are problems achieving regeneration (Snook 1991).

Furthermore, the establishment of plantations has not been on a scale that could compensate for the over-harvest of natural populations. A specialised herbivore, the mahogany shootborer (*Hypsipyla grandella*), is a severe problem when mahogany is grown in plantations (Newton et al. 1993). The shootborer eats the apical meristem of young trees, causing multiple shoots to form after the death of the leader. The resulting tree has poor commercial form and consequently is of little value. There has been considerable attention devoted to controlling the shootborer, but cost-effective techniques have yet to be developed. As a result, there are very few examples of mahogany plantations within its native range. Some plantations have been established in other regions, Indonesia and Fiji being the best examples. Even here, most of the plantations are very young, and will not be ready to supply significant quantities of wood to the international market for many years (Rodan et al. 1992).

In addition to the concern over the direct effects of harvest on mahogany populations, there is also worry about the indirect effects of mahogany exploitation. Much illegal logging has occurred on Indian lands, disrupting traditional life and introducing diseases (Watson 1996). In some cases, armed confrontations have resulted in the murder of Indians and loggers alike. A further problem is that settlers may use logging roads to invade previously inaccessible areas of forest for slash and burn agriculture (Lamb 1966). This is particularly true for mahogany logging because its low density and high value means that it is both necessary and profitable to construct very long roads for its extraction. For example, Verissimo et al.

(1995) report that Brazilian mahogany loggers have created a network of more than 3000 km of roads in only 30 years in South Para.

6.4 Conservation initiatives and enforcement issues

The obvious unsustainability of current logging practices and almost complete absence of management has caused serious concern about the conservation status of mahogany, paralleling an increasing public concern about the fate of tropical forests in general. This has resulted in a variety of conservation initiatives that range from the species to the ecosystem as the focus for action.

The most extreme approach, proposed by environmental groups such as Rainforest Action Network and Friends of the Earth, consists of implementing a total ban on the import of mahogany. The rationale is that the reduction or elimination of international demand will stop the harvest of mahogany. This logic runs directly counter to the 'use it or lose it' philosophy that is commonly used to justify logging of tropical forests. With a few exceptions (e.g. some municipal governments), bans on the import and use of mahogany have not been implemented. Unless all countries ban imports, exports will simply be directed to less discriminating buyers.

A second approach has been to attempt to list bigleaf mahogany on the Convention on International Trade in Endangered Species of Wild Fauna and Flora (CITES), the body responsible for regulating the international trade in species of conservation concern (Rodan & Campbell 1996). The actions that CITES takes depend on the degree of threat. Species listed on Appendix I are those species that are deemed to be threatened with extinction by international trade; an Appendix I listing bans commercial trade in these species entirely.

Species listed on Appendix II are 'all species which although not necessarily now threatened with extinction may become so unless trade in specimens of such species is subject to strict regulation in order to avoid utilisation incompatible with their survival' (Rodan & Campbell 1996). An Appendix II listing does not ban international trade, but requires that exporting countries certify that the species was obtained legally, and that the harvest will not be detrimental to the survival of the species. Importers must enforce the presentation of certificates. All import and export data is presented to CITES, which can use it to assess the conservation status of the species in question. An additional benefit of a CITES listing is that if it was enforced, it would eliminate the need for the timber from sustainable timber producers to compete with timber that can be produced more cheaply from unsustainable sources. This could increase the financial viability of sustainable forestry.

Both of the bigleaf mahogany's congeners, *S. mahogani* and *S. humilis*, are listed on Appendix II, but there is little or no trade in either. *S. macrophylla*, the focus of this chapter, was proposed for Appendix II listing in 1992, 1994 and 1997. The 1992 proposal was withdrawn at the last moment, and the 1994 and 1997 proposals failed to get sufficient support by a narrow margin. Opposition to an Appendix II listing has been strongest from some of the major range states, and from American hardwood importers (Rodan & Campbell 1996). Several countries have voluntarily put their own populations on CITES Appendix III, which requires that all CITES member countries provide certificates stating country of origin, but need not include information about the state of management or conservation status of the population.

The effectiveness of CITES in conserving commercial tropical tree species has yet to be demonstrated. In the case of mahogany, there may be enough range states that have a vested interest in maintaining trade that they continue to block any listing. Furthermore, CITES only addresses international trade. In countries where there is a sizeable domestic market for mahogany, even stopping international trade completely would not stop its harvest. Finally, a CITES listing would do nothing to address the underlying bio-economic incentives for over-harvest. However, CITES is one of only a few international legal mechanisms available to influence species harvest, and as such, has a potentially vital role to play in the conservation of timber species.

For those who hope to use mahogany as a means to achieve the more general goal of tropical forest conservation, a further problem is that the sustainable production of mahogany would not in itself ensure the conservation of high quality tropical forests. Mahogany is not dependent upon old growth (in contrast to the spotted owl in Pacific Northwest forests, where efforts to save it have resulted in the preservation of considerable tracts of old growth forest). In fact the opposite is true. Mahogany requires large disturbances to regenerate, not climax forest, and therefore managing forests for mahogany production could be to the detriment of primary forest if it was adopted on a large scale.

Sustainable forestry is an alternative approach to mahogany conservation which focuses on the forest ecosystem as a whole, rather than on individual species. There are many recently created initiatives to identify and certify 'sustainably managed' forests (summarised in IIED 1995). The guidelines for certification are comprehensive, and include principles and criteria ranging from socio-economic equity to biodiversity conservation. At present, the current guidelines define the issues that must be addressed in management plans, but do not generally provide quantitative targets for individual objectives, or frameworks for achieving compromise between conflicting objectives (Gullison & Cannon, in press).

In theory, sustainably managing a forest should ensure sustained yields of each of the commercial species, but this is not necessarily the case.

Because of the high species diversity of tropical forests, some silvicultural systems manage groups of species, and the demographic status of individual species are not specific management objectives. For a high profile species like mahogany however, it seems that a timber producer would in fact need to achieve sustained production at the species level to be certified.

Whether or not certification is implemented on a widespread basis will depend on the incentives that exist for its adoption. Certification can lead to two possible benefits to the timber producer (Bass 1996). First, it can be used to gain entry to markets if importers insist that wood products come from sustainably managed forests. For this to be a significant incentive, timber producers must not have access to alternative markets where they can sell uncertified products at similar prices. This is currently not the case for mahogany, with demand for unsustainably logged timber still outpacing supply and causing continued slow increases in price. A further problem is that mandatory certification would be a violation of the General Agreement on Tariffs and Trade, unless certification was required of all timbers, regardless of origin (Amilien 1994).

Second, certification can be used to command a higher price for certified products – referred to as the 'green' premium – currently estimated to be at most of the order of 6–10% (Conrad 1995). However, this is only a one-off increase, and as such cannot eliminate the financial incentive for over-harvest (annually compounding returns from conversion of profits from logging to higher yielding investments). For example, to eliminate the financial incentives for over-harvest in Bolivia, price premiums for timber from sustainably managed forests would need to increase at least 12% per year (the prevailing real interest rate minus the financial productivity of commercial sized trees), meaning that in 50 years mahogany would cost the green consumer approximately US$600 per board foot! This is clearly beyond what consumers will pay.

Moreover, price premiums do not provide much incentive for investing in regeneration. The high discount rates characteristic of developing countries mean that the net present value (NPV) of future production is essentially zero, with or without price premiums (Fig. 6.5). Growth data for mahogany in Bolivia suggest that it will take about 100 years to produce a 60 cm diameter mahogany tree in natural forest. Assuming this tree has a commercial height of 10 m, it will yield 2.83 m^3 of wood. R.E. Rice and A.F. Howard (unpublished) found that 1 m^3 of mahogany is worth approximately US$140 in profits and residual value accrued to logging companies. Accordingly, the NPV of a mahogany seedling under the prevailing interest rate of 17% is approximately one hundredth of a cent. Browder *et al.* (1996) estimate that it costs about 10 cents to establish a mahogany seedling, 1000 times more than the NPV of the commercial tree that will result.

Perhaps a more effective means to encourage companies to invest in regeneration is to make low-cost capital available to finance replanting activities. At interest rates below 8.7% it becomes profitable for companies to plant mahogany (Fig. 6.5). Note that it would also require that companies themselves are able to benefit from their investments through longer and more secure concession tenure. In the absence of cheap capital, there is no financial reason for loggers to invest in regeneration, and they will do so only if forced.

Only a small area of timber production forest has been certified to date (6 million ha out of 800 million ha total, and most of these are temperate forests; Bass 1996), and there is little hope that this will change in the short term. While the pursuit of sustainable forestry is an admirable goal, and may offer significant benefits over the status quo where implemented, the goal of this chapter is to present the general status of mahogany across the whole of its range. As such, certified timber producers are of little consequence. Even if sustainable forestry is ultimately implemented over larger areas of mahogany's range, the rate and scale over which mahogany logging is occurring means that these populations will largely have already been logged out.

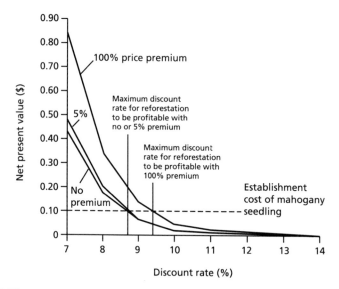

Fig. 6.5 The net present value (NPV) of a future mahogany tree planted now as a seedling, as a function of discount rate and with different price premiums. This calculation assumes (i) that it will take 101 years for a seedling to reach 60 cm in diameter growing in natural forest, yielding 2.83 m³ of wood (growth data from Gullison 1995), and (ii) that the wood will have a net worth of approximately US$140 per m³ (Rice & Howard 1995), giving a total value of the tree at maturity of US$396. At discount rates above 12%, the NPV is essentially zero, with price premiums having little effect. Discount rates must be below 8.7% in the case of no or small price premiums, or below 9.4% in the case of a 100% price premium, for the NPV of future production to exceed planting costs.

In summary, the high value and low productivity of mahogany trees, and the poor capacity for forest regulation in range countries means that sustainable use is not a viable strategy for conserving mahogany at present. A strong regulatory capacity is necessary to overcome the current large financial incentives for over-harvest and the lack of incentives for investments in regeneration. This will require significant political commitment, elimination of most or all corruption, and the allocation of resources and trained personnel (Fearnside 1989; Palmer 1989; Buschbacher 1990). The absence of these factors in many developing countries is the biggest barrier to enacting good forest management. At present, a more prudent mahogany conservation strategy would be to ensure that there were sufficient populations in protected areas so that the conservation status of the species would not depend on logged populations. Sadly, even mahogany populations in protected areas are being illegally logged throughout their range (Rodan et al. 1992; Verissimo et al. 1995; Snook 1996; R.E. Gullison, pers. obs.).

6.5 Concluding Remarks

Is mahogany an exception among neotropical tree species because of its low productivity or unusual regeneration requirements, or can we expect similar fates for other species as they become the focus of logging? Low productivity would appear to be a characteristic of natural forests in general (Fearnside 1989). Silvicultural systems for the management of natural forests in Queensland and Surinam, for example, achieve yields of $0.5–2.2 \, m^3 \, yr^{-1}$ (Vanclay & Preston 1989; Vanclay 1991), or in the range of 1–5% of standing forest biomass. Real annual price increases for other tropical timbers have averaged about 2% per year for the period 1950–92 (Varangis 1992). Thus, the financial productivity of other species in natural forest is similar to that of mahogany, and there are strong financial incentives to over-harvest these species as well.

While other species have the same incentives for over-harvest, some trees with less stringent regeneration requirements may be more robust to over-harvesting (Martini et al. 1994). It is still doubtful whether this would enable them to survive repeated unregulated logging. Until the economic climates of developing countries become more conducive to sustainable development, or until a serious regulatory commitment is made to restricting harvests, not only is logging not a viable conservation tool but it would appear to be a direct threat to the conservation of commercial timber species. It is not obvious how or even if these obstacles to good forest management will be overcome in the tropics.

Chapter 7: Cosigüina, Nicaragua: A Case Study in Community-Based Management of Wildlife

Vivienne Solis Rivera and Stephen R. Edwards

7.1 Introduction

This chapter examines a community-based management programme of two species of lizards (*Ctenosaura similis* and *Iguana iguana*) in Cosigüina, Nicaragua. The lizards are harvested for food and sale. This case study illustrates the problems and successes of a rural community participatory approach to the management of resources for sustainable use. Not surprisingly, the data are not complete. Villagers started management procedures but did not maintain them according to the schedule 'outsiders' believed to be important because they had other priorities such as planting and harvesting their crops and feeding their families.

The people are poor, but, from their perspective, they are better off today than they were in the past. Now they have land, they survived a war and women have a voice in the community decisions. Nevertheless, their principal concern is still having enough food for their families. Over time, their natural environment has become severely depleted. Agricultural land is marginally productive. The only assets they have are their labour, the meagre crops they produce and the wild resources with which they live.

In this chapter we describe the area where the village is located and review the history of their management programme from 1991 to 1996. An assessment of the sustainability of the management programme is provided, drawing on information developed through IUCN's Sustainable Use Initiative (Anon. 1996). The assessment is based on the hypothesis that sustainability is not determinant. Instead it is a goal, where sustainability is enhanced by balancing a number of social and biological factors, any one of which could influence the sustainability of the use. Our conclusions are contextual. That is, they are based on our current understanding of the conditions under which the Cosigüina management programme is being implemented. Those conditions are constantly changing and therefore the conclusions reached in our assessment of sustainability are also likely to change.

Our intent in this chapter is to document the dynamic nature of community-based management programmes. Elements of the management programme described have changed several times over the life of the project, reflecting the social dynamic of the community and shifts in interest of different sectors in the community. This is the norm when working with rural communities and typical of the conditions in which field conservationists work.

7.2 Profile of the region

Cosigüina Peninsula is approximately 430 km², forming the north-west corner of Nicaragua (Fig. 7.1). The Peninsula is bounded to the west by the Pacific Ocean and to the north and east by the Golfo de Fonseca. El Salvador and Honduras are easily reached by boat across the Golfo de Fonseca. The Peninsula is a flat plain dominated by the dormant Cosigüina volcano (859 m), from which the area gets its name. Ecologically, the area is divided into a tropical dry forest in the lowlands and sub-tropical wet forest over 400 m above sea level (Holdridge 1982). Only 22 000 ha of the original forest remain, which is about 19% of the area of the Peninsula. Most of the forest that remains is located around the volcano where the land is neither easily accessible nor suitable for agriculture.

The weather on the Peninsula is characterised as 'tropical savannah'. There is a dry season from November to April and a rainy season from May to October. Average annual precipitation ranges from 1800 to 2000 mm. The average maximum temperature is 30°C and occurs in April; the

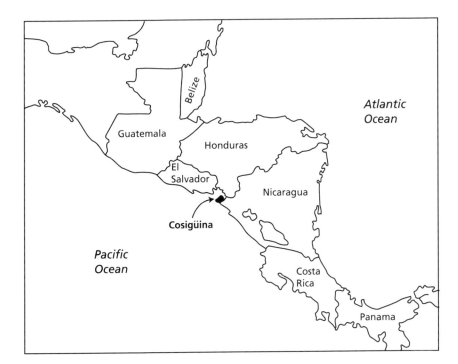

Fig. 7.1 Map of Cosigüina Peninsula.

average minimum temperature is 27°C and occurs in January. There are no permanent rivers on the plain. The only permanent water that is accessible occurs in small subterranean pockets around the base of the volcano in the area where the Cosigüina village we described is located.

For more than four decades agricultural production (e.g. cotton, sugar, bananas and sesame) for export has dominated the economy of the region. During this time farming was controlled by wealthy families who owned estates and relied on peasants to provide necessary labour. Inappropriate crop management technologies resulted in serious environmental degradation (IRENA & IUCN 1992).

In 1958 much of the land on the Cosigüina Peninsula was declared a wildlife refuge. The government's inability to enforce laws governing use of, and access to, the wildlife refuge led to widespread exploitation of the wildlife in the refuge with concomitant declines in wildlife populations. In 1983 the status of the area was changed to a nature reserve. Several animal species considered endangered by the government of Nicaragua are found in the area: guacamaya (*Ara macao*), caiman (*Caiman crocodilus*), peccary (*Tayassu tajacu*), white-tailed deer (*Odocoileus virginianus*), paca (*Agouti paca*), black iguana (*Ctenosaura similis*), green iguana (*Iguana iguana*) and nine-banded armadillo (*Dasypus novemcinctus*).

Between 1977 and 1979 the country was involved in civil war, during which many citizens fled the cities for the countryside. The strong dictatorial government that had ruled the country for several decades was succeeded by a Marxist-socialist government in 1980. In 1987 and 1988 parcels of land on the Cosigüina Peninsula were given to local peasants, irrespective of strong resistance against government policies from the land owners. This led to the formation of several co-operatives that have continued to operate since a democratic government was installed in 1991. The status of the nature reserve did not change through the succession of governments.

Today, the principal crops grown by the co-operatives are corn, beans and sesame seeds. Other crops include sorghum, pumpkin, yucca and watermelons. In the last few years agricultural production has been seriously affected by drought and nematode parasites. At the same time the shift to a free-market economy has resulted in increased costs for agricultural supplies and lower prices for products. The United Nations considers the region to be one of the very poorest in the country – 87.3% of the people in the region are considered to be living in 'extreme poverty' (UNICEF 1995). The social profile of the local population is typical of countries that have undergone protracted periods of civil conflict. Most of the resident population immigrated to the area from other parts of the country and from El Salvador, so they are not necessarily familiar with the ecosystem in which they now find themselves. The education levels are very low and most of the rural people in the region are illiterate.

The Omar Baca Co-operative, the principal focus of this case study, is representative of the people and conditions in the region. It was formed in 1987 by the heads of 39 households (36 men and three women), who built their homes in a cluster near their fields. Since the Co-operative was formed other families have immigrated into the area. Each household comprises six to eight people, including children, and the total population of the Co-operative is about 320 people. Thirty of the 39 families associated with the Omar Baca Co-operative have houses, most of which are made of palm thatch, mud and sticks and lack electricity. Furthermore, none of the houses have a toilet. Water must be drawn from wells and is often unsuitable for drinking (Davila et al. 1993; Lovo 1994). Malaria and other sicknesses are common in the community. A small school was established in the village in 1995. Twelve men in the village have completed the first grade and one has completed the third grade.

The Co-operative manages 384 ha, of which 70 ha are heavily degraded dry forest (Davila et al. 1993). The land is divided evenly among the 39 households who belong to the Omar Baca Co-operative, but individuals do not have legal title to the land they farm, which limits their ability to obtain credit. The average monthly income per household, as reported by the villagers, is Cordobas 800 (US$95) which is less than half the minimum income judged to be the poverty level (i.e. Cordobas 1500; US$178 per month for a family of three to four people). Taking account of the fact that the average family size in Cosigüina is twice that cited by the government it is clear that the villagers are very poor. Individuals from the village harvest fish and shellfish from the Pacific coast and Golfo de Fonseca and wood from mangroves bordering the Golfo de Fonseca. However, neither fishing nor wood harvests are organised activities of the Co-operative.

7.3 Cosigüina management programme

The subject of this management programme is two species of lizards: *Iguana iguana* and *Ctenosaura similis*. The former is the green iguana, the latter is called the black iguana or 'garrobo' by the local people. The two species are sympatric in the wild but, irrespective of their common names, are not closely related. Green iguanas are arboreal herbivores. Captive females lay between 22 and 28 eggs per year which are deposited in nests they dig in sandy soil, and it is estimated that in the wild females lay between 18 and 32 eggs per year. Several females use the same area to lay their eggs. The black iguana is a terrestrial omnivore, and can be very aggressive. Captive females lay between 36 and 40 eggs per year in nests they dig in sandy soil.

Management needs of green iguanas have been studied extensively (PNUMA 1985; Werner & Rey 1985; Werner 1986; Werner et al. 1987). Little research has been conducted on black iguanas and therefore management procedures have been developed through trial and error.

Nicaragua has protected green and black iguanas since 1980 (*La Gaceta No 240*; 13 October 1980). The law establishes minimum sizes for wild harvests (36 cm for green iguana and 26 cm for black iguana) and prohibits harvests between 1 January and 31 April of each year (i.e. during the period when the species are breeding and laying eggs). It also prohibits export of live or dead specimens without prior authorisation from government authorities located in the capital. Animals in the wild are considered the property of the government by the local people, but the animals managed in fenced enclosures are considered the property of the people who manage the enclosure.

In 1990 the villagers started managing iguanas. Their objective was to sustain subsistence harvests for food and local sales. Villagers have a tradition of eating black iguanas, especially on festival days like Holy Week, and at that time it was estimated that each family ate between 1.5 and 2 iguanas per week. Black iguanas were sold in the local market, and both black and green iguanas were sold in El Salvador. Their first step to manage the iguanas was the collection of about 5700 animals (both black and green iguanas) from nearby nesting sites, which the villagers released in a forested area owned by the Co-operative.

In 1991, IUCN, the World Conservation Union, with funding from the Norwegian Agency for Development, began providing technical assistance to manage black iguanas as a development alternative to traditional agriculture. The project was implemented under a partnership involving the Co-operative, the Natural Resource Ministry of Nicaragua, the National Autonomous University of León (UNAN) and IUCN's Regional Wildlife Programme. The principal goal of the project was to provide incentives for the local people to conserve black and green iguanas, and their associated habitat, in a manner that would contribute to the peoples' sustainable development. The project was conceived as a pilot, or demonstration, where the lessons learned would be communicated to other communities in the country and region.

Shortly after IUCN began assisting the Omar Baca Co-operative, villagers declared 80 ha of dry, secondary forest adjacent to the village as a 'protected area' and agreed not to harvest wood or iguanas from that area. Villagers mapped the area, prepared a management plan and constructed a fenced enclosure to manage captive animals. They agreed that harvests of adult animals would be limited to an area adjacent to the protected area. By the end of 1992 villagers had 1800 animals of both species in the fenced management area. At this point the villagers' management plan included three steps and involved three areas as illustrated in Fig. 7.2.

Management of the black iguanas was designed to address two practical objectives: to increase the number of animals available for harvest and to

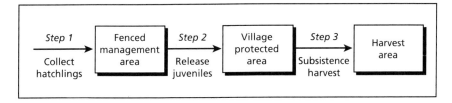

Fig. 7.2 Steps in the village iguana management plan.

establish a captive breeding stock. Management of the green iguanas did not have a particular objective. The fenced management area was used to protect hatchlings until they were able to avoid predation. Juveniles (8–9 months old) were released into the village protected area. Villagers believed that as the density of the population in the protected area increased, a percentage would expand into the harvest area. Also, by caring for the hatchlings and releasing them into the wild, the villagers did not have to bear the cost of maintaining the animals in captivity until they reached the size at which they could be legally harvested, which is about 4 years. The captive stock that was not released was maintained so that they would reproduce, thus reducing the need for collecting more hatchlings from the wild to maintain the stock in their fenced management area.

Members of the Co-operative surveyed the black iguana population in the protected area in May 1992 using Bailey's transect method (Esquivel 1992). Five transects were established in the protected area, each about 0.5 km long. Two groups, comprising six persons each, performed the survey. Each group counted animals along each transect. The process was repeated twice. A total of 108 animals were counted along the transects, leading the villagers to estimate the density of the population at 21.2 individuals per km (10.8 males and 10.4 females per km). Thirty-five animals were marked and recaptured. Based on the mark and recapture rate, the population size was estimated at 341 ± 192 (95 $\pm$ 52 males; 162 ± 111 females) in the 80 ha protected area. In 1992 villagers collected another 5381 hatchlings from the wild of which 4000 were released into the village protected area. The remaining 1381 animals were placed in the fenced management area. At the end of 9 months villagers released approximately 80% of the juvenile lizards into the village protected area.

By 1993 women began to take more interest in iguana management. Greater attention was given to managing green iguanas when villagers learned that they could be sold. Villagers contacted a local trader who bought green iguana hatchlings to export for the pet trade. He told them that he would pay between US$1 and US$2 for green iguana juveniles depending on their size and condition.

A revolving fund was established in 1993 with money provided by the project. The revolving fund was to be used to finance agricultural activities.

Crops included corn, sesame seeds, soybeans, beans, pumpkin and watermelon. The Co-operative agreed that 30% of the agricultural production would be used to feed the captive iguanas, 50% would be divided among the families in the Co-operative and the remaining 20% would be divided among the pregnant women and children in the village.

In 1993 the first conflicts between families involved in iguana management occurred. One family had dominated the management programme through its control of the Board of Directors of the Co-operative. A commission was formed by the members of the Co-operative to resolve the conflict. The commission included two representatives from the iguana project, one from a gender project and two from an agricultural project.

In 1994 villagers recorded data on the survivorship of black and green iguanas in the fenced management area based on the number of eggs incubated (Table 7.1). Black iguanas lay their eggs in February and March; green iguanas in March and April. Another 315 black iguanas were collected and placed in the fenced management area in 1994. Women collected 125 black iguana hatchlings and started a small-scale management programme in a fenced area that they had prepared.

During this year, several problems relating to the management of the animals in the fenced areas were identified during Co-operative meetings: how to control predation of hatchlings and juveniles by other animals; how to protect the green iguana nests from predation by black iguanas; how to prevent animals from escaping. They also discussed how the villagers should be organised to manage the animals in the enclosures, including individual family responsibilities.

The first formal evaluation of the programme was undertaken in 1995 with the full participation of the members of the Co-operative. The principal conclusions of the evaluation were:

• The villagers were losing interest in the management programme because they had not realised any substantial benefits. The programme was not achieving the objective of being an alternative to agriculture.
• Women were viewed as having a support role in relation to the

Table 7.1 Survivorship of black and green iguanas in the enclosed management facility.

Status	Black iguana		Green iguana	
	Number	% of incubated eggs	Number	% of incubated eggs
Incubated eggs	1524	—	1620	—
Hatchlings	1000	65.6	600	37.0
Survivorship at 5 months	530	34.8	206	12.7
Survivorship at 6 months	515	33.8	90	5.6

management programme and therefore were not considered eligible to receive benefits (Fig. 7.3).

• The deteriorating economic situation in rural Nicaragua exacerbated villagers' negative feelings about the management programme and limited the number of options they could pursue in solving problems.

• The fact that the people using the land have no legal title to the land (and the wild resources) was a factor inhibiting villagers' development.

• The villagers wanted more technical training to manage a broader suite of resources in the forests.

• The villagers wanted an organisation that allowed all of them to be more active in problem solving.

In 1995 the Board of Directors of the Omar Baca Co-operative was publicly criticised about their management of the iguana project in village meetings. Questions were raised in the village about the Board's management of project-related resources and the fairness of their decisions. As a result, a new Board was formed and officially recognised by the national authorities responsible for regulating co-operatives. A second, smaller fenced enclosure was constructed and stocked with 600 hatchlings of both species. The family that had dominated the Co-operative's Board through 1995, along with five other families, retained control of the larger enclosure with a stock of 1700 hatchlings of both green and black iguanas. Thirteen other families joined together to manage the second enclosure.

Steps were taken to acquire government approval to sell the hatchlings for export. They applied for export permits. Before the government approved the sale, the families managing the larger enclosure started to sell some of their stock, which was reported to government authorities. In consultation with the community, the government closed the first enclosure and transferred the stock to the second enclosure. No legal sale of iguana hatchlings occurred in 1995; however, villagers were offered US$0.5 per hatchling for as many as 1000 animals.

In May 1996 the families undertook a second survey of the wild

Fig. 7.3 Ensuring the full involvement of women in the project was an important issue that was discussed in the first evaluation of the project. (Photo by Loida Pretiz.)

population of the black iguanas in the village protected area using the same methodology they had used in the 1992 survey (Esquivel 1996). Ninety animals were counted along the five transects (in comparison to the 108 that had been counted in 1992) leading to an estimated 18.2 animals per km (7.8 males and 10.4 females per km). Twenty-eight animals were marked and released, and during the survey these marked animals were recaptured 32 times. Based on the mark and recapture rate, villagers estimated the total population to be 308 ± 158 animals (192 ± 141 males; 96 ± 53 females).

Total counts, population density and size estimates from the 1992 and 1996 surveys are provided in Table 7.2. The results of the survey do not show a significant change in the overall population size; however, there is a substantial shift in the relative numbers of males and females in the population between 1992 and 1996. This shift in the sex ratio may be the result of selective hunting of females with eggs. However, because this trend is based on data from two surveys that were conducted 4 years apart, and given the wide confidence limits, there may not be cause for concern. What is important is that the surveys should be conducted more often, and if this trend is verified, the harvest levels of females should be reduced.

Management of the green iguana appears to have been more a consequence of the collecting technique (i.e. hatchlings near the nesting sites of both species) rather than a conscious effort in the first few years of the project because the villagers did not see much value in them. The relatively lower importance of the green iguanas, from the villagers' perspective, may be linked to the low survivorship of captive animals of this species (see Table 7.1). No mechanisms appear to be in place to change the management of the captive green iguanas based on the information on survivorship that has been compiled by the villagers. The cash benefit from the recent sale of green iguanas may serve as an incentive for the villagers to take more interest in managing them. While the villagers have some understanding of the relationship between the captive population and the status of the species in the wild (i.e. harvest of animals for food is limited to the wild population) there is no defined activity that links the captive management activities to the harvest of wild animals.

Results	1992	1996
Total animals counted	108	90
Estimated number of animals per kilometre	21.2	18.2
Estimated population size:		
Total	341 ± 192	308 ± 158
Males	95 ± 52	192 ± 141
Females	162 ± 111	96 ± 53

Table 7.2 Results of black iguana surveys in 1992 and 1996.

A systematic participatory review of the project was undertaken in 1996 by the villagers with facilitation provided by IUCN staff and the other partners. Villagers analysed the programme from three perspectives: *personnel/organisational* issues, *technical/biological* factors and *economic/production* considerations. Achievements and problems were identified in relation to each perspective. Village discussions were facilitated with the assistance of an artist/communicator, who used a variety of techniques (e.g. drawings, games, charts, theatre plays) to illustrate different aspects of the management programme (Fig. 7.4). This approach helped villagers to communicate their common understanding about the management programme because most of them are illiterate. The drawings and written material are being reproduced for facilitators to use in other villages in the region.

Villagers concluded that considerable change had taken place in regard to the personnel/organisational aspects of the project. Of particular note were the attitudinal changes of the people involved in the management programme and the shifts in power between the families related to community leadership that had occurred because of the programme. From a technical/biological perspective the villagers recognised that they had gained considerable knowledge, skill and expertise in managing the iguanas. Their use of the black iguana was limited to subsistence harvests of animals in the wild, which they felt had made a substantial contribution

Fig. 7.4 Art at Cosigüina: part of the participatory review of the project. (Photo by Loida Pretiz.)

Enclosure	Size	Number	Sale price (US$)	Total (US$)
I	Hatchlings	100	1.50	150.00
II	Hatchlings	315	1.50	472.50
II	Juveniles	9	2.00	18.00
II	Juveniles	180	2.25	405.00
Total	–	604	–	1045.50

Table 7.3 Sales of green iguanas for the pet trade in 1996. (Data prepared by F. Esquivel, PRODEMUJER.)

to their diets. In fact, wild-harvested iguanas provide the primary source of protein for the villagers during the dry season which lasts for several months each year. All of the experiences were viewed as contributing to villagers' empowerment to manage the resource.

The first organised sale of green iguanas for the pet trade occurred in 1996. According to information provided by the Omar Baca Co-operative they earned a total of US$1045.50 from sales from both fenced management areas (Table 7.3). Villagers deposited US$800 in their 'revolving fund' and the balance was used for sesame seed and watermelon production.

Villagers reported that each household eats about three black iguanas per week, which represents 90% of the animals they harvest; 10% are sold in the local market for Cordobas 15 (US$1.79) each. Based on these reported harvest levels, the 39 households harvested about 6700 animals from the wild each year. Total income from sales is estimated at Cordobas 10 140 (US$1207) per year. Assuming each household receives an equal share of the income, iguana sales contributed about Cordobas 260 (US$31) per year to each household.

Table 7.4 lists the 1996 values (in US$) of both green and black iguanas in 'local' and 'non-local' markets. The term 'local market' refers to sales to other villagers in Cosigüina or in nearby villages. The term 'non-local

Species/age class	Local market value	Non-local market value
Green iguana		
Hatchlings	0.18	1.50
Juveniles	0.71	2.00–2.25
2–3 years old	1.19	2.98
4–6 years old	2.38	4.21
Black iguana		
Hatchlings	0.12	0.00
Juveniles	0.71	1.43
2–3 years old	1.19	2.41
4–6 years old	1.79	4.17

Table 7.4 Values in US dollars of different age classes of green and black iguanas in local and non-local markets in Nicaragua. (Data prepared by F. Esquivel, PRODEMUJER.)

market' refers to sales destined for use, consumption or resale outside of the immediate area. Sales of hatchling and juvenile green and black iguanas in the local market are for stocking local captive management programmes. Most lizards sold in the local market are adult female black iguanas with eggs, because the highest demand is for the eggs which the local people eat. Sales of green iguanas in the non-local market are destined for the pet trade.

A comparative analysis of the economic and social impact of the iguana management project on the Omar Baca Co-operative and a second Co-operative in the same area, which is harvesting iguanas without a management programme, was undertaken in 1996 (Gutierrez-Montes 1996). It was found that there were no significant differences in income between the two co-operatives. However, the study did document substantial differences in the perceptions and attitudes of the families involved in the two co-operatives. Families of both co-operatives had migrated into the area seeking freedom, survival and protection from the civil conflict. Today, families in the Omar Baca Co-operative in Cosigüina are seeking affection, recreation, leisure time and protection while the families in the other Co-operative still desire survival, knowledge and protection. The fact that the families in the Omar Baca Co-operative are placing higher importance on personal endeavours underscores their enhanced personal and collective security.

Seven people from a community-based management project in Panamá visited Cosigüina for 4 days in mid-1996. Subsequently a delegation from Cosigüina was hosted by the villagers in Panamá. During these exchange visits the villagers shared their experiences, problems and the lessons they have learned in their respective management projects.

7.4 Assessment of sustainability

Natural resource management programmes, like the one in Cosigüina, are complex systems that vary according to the social context in which the management is being pursued and the biological characteristics of the resource that is being used. Consequently, there are a multitude of configurations of biological and social conditions under which sustainability can be achieved. To assess the sustainability of the Cosigüina iguana management programme we have structured our analysis around four key *components* which are characteristic of organised wild resource management systems:

1   the *resource* that is being used;
2   the *management plan* or actions followed in using the resource;
3   the *incentives* that motivate the managers to continue the management;
4   the *institutional structures* within which the management is taking place.

For each of these components we identify the two *factors* that we consider

Table 7.5 Framework for assessing the sustainability of the managed use of iguanas in Cosigüina.

Components	Factors	Indicators
Resource	Biological status of the species	Status of wild populations has been established Habitat has been assessed Species requirements are known
	Access rights to the resource	Legal rights are established Managers are accountable
Management plan	Monitoring procedures	Wild populations are monitored Social and economic influences are monitored
	Risk management	Management procedures are flexible
Incentives	Economic incentives	Profit has been achieved Income is re-invested in management
	Non-financial incentives	Managers are secure Traditional uses of the resource are sustained
Institutional structures	Community-based institutions	Co-operative leadership is stable
	Co-management agreements	Government authorises management Partners continue their commitment

to be most important in determining the sustainability of the managed use in Cosigüina. For each factor we define two or more *indicators* against which we can assess progress. Table 7.5 lists the components, factors and indicators used to assess the sustainability of the Cosigüina management programme.

When considering this assessment there are several points that should be kept in mind. First, this assessment should have been conducted by the local resource managers in a participatory process involving other local stakeholders, i.e. the Nicaragua Natural Resource Ministry, IUCN Regional Wildlife Programme and the National Autonomous University of León. If they had undertaken this assessment it is highly likely that they would have identified other factors and indicators that they consider more important than the ones we have chosen. Second, the relative importance of individual factors will vary according to the resource being used, and the social, cultural, political and economic context in which the management is taking place. The importance of individual factors will also vary over time as the management programme is being implemented.

The resource

Biological status of the species

Neither sophisticated population surveys nor detailed studies of the habitat are required to manage a wild species. However, it is necessary to have sufficient information about the species and its habitat when the management programme is started to provide a baseline against which changes in the wild population or habitat can be measured over time. It is also wise to know if the target species has any unique requirements (e.g. behavioural, food, reproductive) that need to be addressed for successful management. To assess the status of this factor in Cosigüina we examine three indicators:

- *Status of the wild population has been established.* The status of the black iguana in the wild was established by the population surveys that the villagers conducted in 1992 and 1996 (see Table 7.2). The status of the green iguana has not been established. The results of the black iguana surveys document no significant change in the overall size of the wild population between the two surveys. However, as noted above, the data indicate that there has been a substantial shift in the ratio of males and females in the population.
- *Habitat has been assessed.* Villagers have declared an area of 80 ha as a protected area in which they do not harvest wood or lizards. The area has been mapped, but the villagers have not compiled any data on the status of the habitat. Consideration of the habitat in the protected area appears to be passive.
- *Species requirements are known.* The managers know a lot about the biology of the two species (e.g. dietary requirements, when they reproduce, where they lay their eggs, how many eggs are laid per female). The reproductive capacity of the species is high: captive black iguana females produce between 36 and 40 eggs per year; green iguana females between 22 and 28 eggs per year.

Access rights to the resource

Clearly defined legal rights to use the resource and means for holding the managers accountable for their actions are crucial to the sustainability of management programmes. Two indicators are examined:

- *Legal rights are established.* Legal rights of access to the resource are not clearly established. Wild animals are the property of the government. Legal rights to hunt green and black iguanas are conveyed to individuals by the Natural Resources Ministry under a national hunting law that establishes the minimum size that can be hunted, provides a closed season (during the reproductive period of the species) and establishes fines for violators. Animals that are harvested in the wild are the property of the individual hunters. They either eat the lizards or sell them and the income from the sale goes to the hunters' households. Another office in the Ministry (i.e. the CITES Management Authority) authorises the sale of live animals for export. A third government agency, which registers co-operatives, has formally recognized the Omar Baca Co-operative and their captive management of the lizards. The Co-operative determines how benefits from the sale of the lizards from the fenced management areas are allocated to individuals and the Co-operative account.

• *Managers are accountable*. The villagers have demonstrated their capacity to hold individuals accountable for their actions. The family that attempted to sell animals before the government authorised sales was reported to government officials. The consequence of this action was rather severe: government suspended sales to non-local markets in 1995, confiscated the captive stock from the family who had violated the law and reallocated it to other members of the Co-operative.

The management plan

Monitoring procedures

Procedures to assess the impact of the harvest or use on the target population and the social and economic factors that can influence the management plan are crucial to sustaining the use of a wild resource. Our assessment examines two factors:

• *Wild populations are monitored*. The two surveys of the black iguana do not constitute a monitoring programme. The interval between the surveys is too long to provide any information that could guide management decisions. There is no monitoring of the wild green iguana population.

• *Social and economic influences are monitored*. There have been two substantial reviews of the project in which the villagers participated. These reviews have led to adjustments in the management programme. The socio-economic study documented substantial changes in the attitudes of the people involved in the Cosigüina management programme.

Risk management

The sustainability of management programmes is influenced by events under the direct control of the managers (i.e. internal) and events beyond their control (i.e. external). Because the social and biological conditions under which resource use is taking place change over time, due to both internal and external events, the management systems must be flexible and have the inherent capacity to adjust to changes in the conditions in which the use is taking place. Adaptive management is the most cost-effective and expedient method of dealing with risk and uncertainty associated with these changes. That is, information compiled in the monitoring programme should guide decisions governing subsequent uses of the resource. To assess this factor we examine one indicator:

• *Management procedures are flexible*. Because there is no active management of the wild populations, there are no mechanisms to adjust activities to respond to changes in the status of the wild populations. The shift in the sex ratio of the black iguanas documented in the 1992 and 1996 surveys should prompt adjustments in the hunting levels. The Co-operative provides a mechanism to discuss problems and issues, and to adjust activities to accommodate new information. Thus far this appears to be limited to issues related to the governance of the Co-operative and management of the captive populations.

Incentives

Economic incentives

For economic incentives to influence the sustainability of the use, the income obtained from the use must cover the full cost of managing the resource (e.g. re-investment) plus a predictable profit. In relation to economic incentives, two indicators are examined:

• *Profit has been achieved*. To date cash income from the sale of the lizards has been minimal. Only 10% of the wild harvested black iguanas are sold. Acccording to information provided by the villagers this yielded an estimated US$31 income per year per household. Sale of the green iguanas in 1996 generated a little over US$1000, which represented a substantial income to the Co-operative. The principal economic benefit appears to be the alternative food source from the wild-harvested animals. Ninety per cent of the wild harvested animals are consumed by households. Also, the average number of lizards consumed per household increased from 1.5–2 in 1990 to 3 in 1996.

• *Income is re-invested in management*. The villagers put a substantial amount of the money they received from the sale of the green iguanas into their revolving fund. This fund is used to finance agricultural activities (e.g. purchase seed, fertilisers, etc.) from which they anticipate a profit that will benefit the Co-operative. A percentage of the agricultural production is allocated to feeding the captive lizards. Thus far, there is no obvious link between the income or other benefits received from the sale of the lizards and the management of the resource in the wild.

Non-financial incentives

Non-financial incentives may also influence the sustainability of natural resource management systems. While there is no doubt that the project funding has had considerable influence on the sustainability of the management programme so far, we believe that certain non-financial incentives

have also been important, and may have considerable influence in sustaining the management programme in the future. The two indicators that we believe are important in assessing the sustainability of the Cosigüina management programme in relation to non-financial incentives are:
- *Managers are secure*. Security relates to the villagers' capacity to plan for their future and their commitment to contributing to the programme. The socio-economic study (Gutierrez-Montes 1996) documented substantial differences in the perceptions and attitudes of the families involved in the Omar Baca Co-operative. Families had migrated into the area seeking freedom, survival and protection from the civil conflict. Today, they place a higher priority on affection, recreation and leisure time, which underscores their enhanced personal and collective security.
- *Traditional uses of the resource are sustained*. The villagers have sustained a long-standing tradition of eating black iguanas during festivals or holidays. That tradition had been sustained, at least in part, because they release wild caught and captive-raised animals in the village protected area.

Institutional structures

Community-based institutions

The organisation and structure of the institutions under which the people manage wild resources, and derive the benefits from those uses, are crucial to enhancing the sustainability of the use. Different modes of use require different institutional structures to favour sustainability. In Cosigüina the Omar Baca Co-operative is the principal institutional structure for community involvement in the management programme. The Co-operative provides a forum for the members to exchange information, monitor progress and contribute to the development and implementation of the management programme. The key indicator to assess sustainability with regard to the community-based institution is:
- *Co-operative leadership is stable*. The Co-operative has evolved since 1990 when the villagers began their management of the lizards. The composition of the Co-operative has not changed substantially during that time, but the leadership has changed. The second fenced management area that was established by another group of households also reflects a division in the leadership of the Co-operative. Nevertheless, the authority of the Co-operative has been sustained. Villagers are able to express themselves freely in meetings and decisions taken at these meetings appear to be implemented.

Co-management agreements

Co-management agreements formalise relationships between the rural people who are engaged in the day-to-day management of the resource and the government and/or other institutions, such as conservation organisations. In Cosigüina the relationship between the local managers and the government and other partners are important indicators:

• *Government authorises management*. Different government offices have been very supportive of the Cosigüina management programme; the Omar Baca Co-operative is recognised. Sale of the green iguanas for export was authorised in 1996. However, to our knowledge, the government has not yet formally licensed the management programme. In the future, licensing of the management programme in relation to agreed criteria will be the only mechanism that government will be able to use to ensure that the villagers are accountable for their actions.

• *Partners continue their commitment*. IUCN-Central America, the National Autonomous University of León, PRODEMUJER and others have assisted the villagers to develop and implement their management programme. Their continued support of the programme is dependent on the availability of external funding.

7.5 Progress and Prospects

The status of the 15 indicators of sustainability that we examined is summarised in Table 7.6. Seven of the indicators have been achieved. The status of three indicators is inconclusive. The remaining five indicators have not been achieved.

Based on this assessment, we are obliged to conclude that the Cosigüina management programme is not sustainable as presently configured. However, as noted above, sustainability is not an 'either–or' situation and therefore, if certain management needs are addressed, the sustainability could be enhanced substantially in the future. Of greatest importance is the need to implement a monitoring programme to assess the impact of the use on the wild populations. At the same time the managers need to adopt mechanisms to adjust their activities based on information from the monitoring programme. Sustainability of the use will also be enhanced if the government formalises the villagers' rights to manage and harvest the animals from the wild under the authority of the Co-operative. Only when such access rights are clearly defined will the government have the ability to hold the managers accountable for their actions. At this point the villagers' principal incentive to continue their management of the lizards is probably the funding provided under the project. Formal recognition of the management programme by government, through the granting of an authority to the Co-operative, could serve as a powerful incentive for the villagers in Cosigüina to sustain their management programme.

In the long term, the sustainability of the use (and conservation) of the species will depend on the profit villagers receive from their management

Table 7.6 Status of the indicators examined to assess the sustainability of the Cosigüina management programme.

Factors	Indicators	Achieved
Biological status of the species	Status of wild populations has been established	Yes
	Habitat has been assessed	No
	Species requirements are known	Yes
Access rights to the resource	Legal rights are established	Not clear
	Managers are accountable	Yes
Monitoring procedures	Wild populations are monitored	No
	Social and economic influences are monitored	Yes
Risk management	Management procedures are flexible	Yes
Economic incentives	Profit has been achieved	No
	Income is re-invested in management	Some
Non-financial incentives	Managers are secure	Yes
	Traditional uses of the resource are sustained	Yes
Community-based institutions	Co-operative leadership is stable	No
Co-management agreements	Government authorises management	No
	Partners continue their commitment	Maybe

of the resource. Therefore, villagers' knowledge of the markets will be very important. The villagers know the values of the animals in the local and non-local markets (see Table 7.4). They also know there is a market for hatchlings and young green iguanas for pets in the USA and Europe, but they do not know what these animals sell for in the market countries. In order for the Co-operative to negotiate a fair price for their product (and enhance their profit) there is need for the partners to provide information on the markets, including prices and demand.

Finally, irrespective of our assessment of the sustainability of the use, from a conservation perspective there are several positive outcomes from the programme. There is evidence that the population of black iguanas is being conserved. The population in the wild has been stable for 4 years even though villagers continue to harvest wild black iguanas at relatively high levels (estimated to be about 6700 per year). The villagers have designated the protected area, which is ensuring that this forested land is being kept in a relatively natural state and may represent a substantial asset to the community in the years to come. The attitudes of the villagers have changed, and they are more aware of the need to conserve their natural resources today than when the project started.

Chapter 8: Sustainability of the Falkland Islands *Loligo* Squid Fishery

Sophie des Clers

8.1 Introduction

According to the latest FAO *Review of the State of World Fishery Resources* (FAO 1995a), many of the world's largest fisheries have either collapsed from over-exploitation or appear imminently at risk (Cook *et al.* 1997). On the eve of a new millennium, the sustainability of marine commercial fisheries is strongly questioned (Loftas 1996), and exploited fish species are for the first time being considered for the IUCN Red Lists of threatened species (Hudson & Mace 1996). The demise of the Canadian northern cod fishery in 1992 has greatly shaken the confidence of all involved, fishermen, scientists, managers and environmentalists, who are calling for new approaches. In order to tackle our incomplete understanding of complexity and natural variability, a strong case is now being made for widespread use of marine protected areas (Walters & Maguire 1996), and for an ecosystem-based (Christensen & Pauly 1995; Larkin 1996) and generally more cautious and responsible approach to fisheries management (FAO 1997).

By the end of 1996, the fishery for the loligonid Patagonian or common squid, *Loligo gahi*, in Falkland Islands waters had been under regulation for 10 years. The fishery has always been exploited by a fleet of large factory trawlers, which have been blamed elsewhere as the main factor underlying fisheries collapse, for example, in North America (Stump & Batker 1996).

Costanza and Patten (1995) argue that sustainability cannot be predicted, but only observed after the fact. The purpose of this case study is to illustrate how cautious and straightforward management can lead to sustainable fisheries exploitation, even when many aspects of the fishery are poorly known to start with.

8.2 Development of the Falkland Islands fisheries

The Falkland Islands lie in the southern Atlantic Ocean, off the tip of South America on the edge of the Patagonian shelf at 60° W and 52° S (Fig. 8.1). In the austral summer, the ocean climate west and north of the Islands is influenced by warmer waters originating from the coastal branch of the Brazilian Current. South and east of the Islands, the colder Falklands

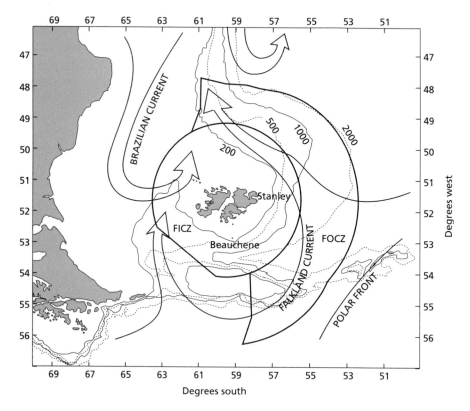

Fig. 8.1 Falkland Islands Conservation Zone (FICZ), Outer Conservation Zone (FOCZ) and the main currents. Depth contours (200–2000 m) are also shown.

Current flows eastward from Cape Horn, around the edge of the continental shelf and north, with a secondary branch flowing north from Patagonia and to the west of the Islands in the winter. In the south, sea surface temperatures vary between 5 and 11 °C on the main *Loligo* fishing grounds around Beauchene Island. In the north, the amplitude of seasonal variations is higher, with a stronger cold influence in winter and warmer temperatures in summer (Fig. 8.2).

The general climatic and weather conditions for mariners in the region are described as 'an almost unbroken series of depressions and troughs that move East across or close South of the area (i.e. southern tip of South America); the weather in all seasons predominantly wet and stormy' (Hydrographer of the Navy 1993). These difficult sailing conditions, together with the Islands' remote location and dangerous coasts probably explain a late development of commercial fisheries.

History

Commercial fisheries on the outer edge of the southern Patagonian shelf

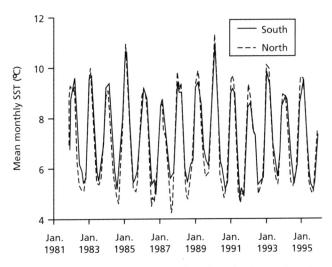

Fig. 8.2 Sea surface temperature (SST) at two locations, in the north and south of the *Loligo* fishing grounds. (From Reynolds *et al.* 1996.)

began in the late 1960s, initially targeting hake (*Merluccius* spp.) and shortfin squid (*Illex argentinus*), followed by southern blue whiting (*Micromesistius australis*) and other finfish (FAO 1983). Until the early 1980s *Loligo* hardly had a mention in the South Atlantic fisheries statistics, with recorded catches below 1000 tonnes per year (FAO 1983); although it is clear that commercial fishing for *Loligo* had started in 1982, first by USSR vessels followed in 1983 by Spanish, Japanese and East German vessels, and in 1984 by Bulgarian vessels. In 1985, the annual reported catch had increased to 50 500 tonnes (Csirke 1987), but estimations from trans-shipment activities in Falkland waters indicated more than 100 000 tonnes.

A rapid increase in catches observed in all squid and finfish fisheries around the Falklands reflected two important trends in the world's fishing fleets throughout the 1980s. First, in the late 1970s there was a re-distribution of distant fleets to new grounds as access to traditional grounds was restricted following many coastal states' declaration of a 200-mile Exclusive Economic Zone (EEZ). Second, from the mid-1980s there was a rapid expansion of the world's commercial fishing fleet capacity, notably of factory freezer vessels (see des Clers, 1998 for changes in the European fleet of large factory trawlers).

By 1985, it had become clear that the uncontrolled expansion of fishing effort on *Illex*, the most abundant squid species to the north of the Islands, threatened the fishery with collapse. After several unsuccessful initiatives to obtain some multilateral co-operation on stock conservation, the UK government asserted the Islands' fishery limits up to 200 nautical miles (FIGO 1989). A conservation zone was established October 1986 and a management regime put in place for all fisheries from 1 February 1987.

The Falkland Islands fisheries conservation zones

The Falkland Islands Interim Conservation and Management Zone (FICZ; Fig. 8.1) represents an area of 150 nautical miles radius from a point at the centre of the Islands. It covers the edge of the Patagonian shelf around the Islands, and most of the slope between the 200 and 1000 m contours. The Falkland Islands Outer Conservation Zone (FOCZ) extends to 200 nautical miles from the baseline, and was opened to fishing in 1994, notably to an experimental deep sea longline fishery for Patagonian toothfish (*Dissostichus eleginoides*, see des Clers *et al.* 1996) on the edge of Burdwood Bank in the south.

Currently six fisheries are managed as separate entities. Two squid fisheries target (i) shortfin squid (*Illex argentinus*), and (ii) common squid (*Loligo gahi*; Fig. 8.3); the original finfish fishery is split into three with (i) finfish including hake (*Merluccius hubbsi* and *M. polypepis/australis*), (ii) finfish without hake (i.e. mostly southern blue whiting (*Micromesistius australis*) and hoki (*Macruronus magellanicus*)), and (iii) rays (*Raya* sp., *Bathyraya* sp.); and finally a deep water longline fishery which targets Patagonian toothfish. Average yearly catches were 220 000 tonnes of squid and 83 000 tonnes of finfish per year between 1987 and 1995 (Fig. 8.4).

The management regime

In setting up a fisheries management regime in 1987, the Falkland Islands government had three objectives: to conserve the resource, to maintain the economic viability of the fisheries and to enable the Islands to enjoy greater benefits from the resource (FIGO 1989).

At that time, it was already clear at an international level that catch limitation alone could not protect fish stocks from over-exploitation by rapidly expanding fleet and fishing power (Pearse 1980). Fishing licence schemes to control fishing effort directly (Beddington & Rettig 1984) were

Fig. 8.3 The Patagonian common squid, *Loligo gahi*. (Photo by S. des Clers.)

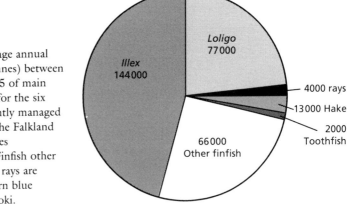

Fig. 8.4 Average annual catches (in tonnes) between 1987 and 1995 of main target species for the six fisheries currently managed separately by the Falkland Islands Fisheries Department. Finfish other than hake and rays are mostly southern blue whiting and hoki.

being implemented in many fisheries, often in addition to a quota or catch limitation system (FAO 1984). For coastal states developing their own fleet, restrictive licensing was also a flexible and natural way to manage access of foreign fleets (Huppert 1982; Beddington & Clark 1984).

Since its inception in 1987, management of the Falkland Islands fisheries has been based on effort control complemented by the possibility of closure before the end of the fishing season, and for each fishery by a specific set of technical conservation measures including gear restrictions, closed areas and closed seasons. Restrictive licensing, to the exclusion of catch limitation, was chosen on the basis of the particular circumstances of the Falkland Islands. First, the Islands' subsistence economy, based on sheep farming for wool production, could not provide the financial resources needed for fisheries surveillance, monitoring, management or research. It was therefore obvious from the start that the full cost of these had to be borne out of fisheries revenue. Fee-paying fishing licences offer a simple way to generate such revenue. Furthermore, the 2000 strong population of the Islands had until then been little involved with commercial fishing activities, which were conducted offshore and by foreign fleets. In 1986, for an estimated 19 000 factory vessel-days or an average of around 100 vessels fishing throughout the year and around 30 crew per vessel minimum, there were more people fishing offshore than the entire population of the Islands. Second, the provision of data, past and present, could easily be made a condition of licence granting, while quota systems were notorious in encouraging misreporting of catches. As catches were not landed, there was also little scope for shore-based catch control, or enforcement of catch limitation measures. Third, effort limitation was also particularly appropriate for the special case of short-lived squid species (Caddy 1983) whose abundance was expected to vary greatly from year to year (Beddington *et al.* 1990).

Fisheries sector development

Trans-shipment fees have been charged for vessels anchored in territorial waters (12 miles from baseline) since 1984, and licence fees charged from 1 February 1987. From the start, revenues from licence fees and the return on investments from government-backed fishing companies have provided for the development of a local fisheries sector. Joint venture companies have been encouraged, as well as the creation of businesses providing services to the fleet such as fuel, or agency services (FIGO 1989). The Fisheries Department has been kept small considering the size of the fisheries it manages (Fig. 8.5), and has developed over the last 10 years, progressively assuming tasks that were initially contracted out. At present, the Director of Fisheries has a staff of 20, but much of the scientific research, fisheries assessment and scientific management advice is still generated through contracts overseas.

8.3 The *Loligo* fishery

Loligo gahi are caught by trawlers during two separate seasons of 4 months (February to May) and 3 months (August to October). From 1990, access to the *Loligo* grounds (the '*Loligo* Box') has been limited throughout most of the year to licensed vessels only, excluding demersal trawlers licensed in other fisheries. There are no minimum size or mesh size restrictions in the *Loligo* fishery.

Life cycle and ecology

Generally, life cycles of cephalopods are lesser known than for finfish (see Boyle 1983, 1987), partly due to the relatively recent development of commercial fisheries, but also because of difficulties in determining age (Jackson 1994). *Loligo gahi* is no exception (Hatfield & Rodhouse 1991). The Falkland Islands government has commissioned research and research cruises on *Loligo* since 1987. Results so far suggest that the stocks targeted during the two fishing seasons are likely to be separate (Patterson 1988), although the lack of genetic differences (Carvalho & Loney 1989; Carvalho & Pitcher 1989) is probably explained by some interbreeding resulting from a combination of migration and extended spawning periods for males. Most commonly there are two broods in the first season, and often two in the second season, with some possible overlap of the first season second brood into the second season (Hatfield 1996). *Loligo gahi* spawn inshore (Hatfield & Rodhouse 1994a), migrate down the continental slope as they grow and mature, and return inshore at the end of their life to spawn (Hatfield *et al.* 1990). Growth is fast, with daily checks within the statoliths, and an estimated lifespan of around 1 year for both sexes

Fig. 8.5 The Falkland Islands Fisheries Department. (Photo by S. des Clers.)

(Hatfield 1991). The analysis of growth is complicated by the continuous migration of successive broods. During the fishing season growth is linear, between 0.3 and 0.7 mm per day, and faster for males than females (Patterson 1988; Hatfield & Rodhouse 1994b).

Sea temperature (along with food availability, photoperiod and lunar cycle) plays a key role in the life cycle of cephalopods, determining growth, development and sexual maturation and therefore the pattern of migrations between nursery grounds inshore, feeding grounds offshore where the fishery operates, and back to inshore egg laying grounds. A model after Forsythe (1993) for the seasonal growth of squid has recently been proposed, to explain the size of *Loligo* caught in the fishery as a function of sea temperature experienced by juvenile squid in coastal waters during the previous 6 months (Grist & des Clers 1997).

Management scientific advice

The number of vessels allowed to fish is fixed before the season starts on the basis of an assessment of the same season a year previously. Traditionally, *Loligo* vessels have been licensed for the whole duration of the season; thus the only possibility to reduce fishing effort during the season is to close the season early.

From mid-season onwards, or whenever catch rates drop to low levels mid-season, stock assessments are attempted on a weekly basis to estimate current population size in the presence and absence of fishing. A ratio of the two gives the proportional escapement. In the past a threshold of 40% or 30% has been used as a conservation target, below which the season was either closed early (second season 1989) or effort was significantly reduced the following year (first season 1988).

Estimates of vessel catchability coefficients are also obtained from the assessment, and are used to translate the escapement target into allowable vessel-units for the following year (see Beddington *et al.* 1990). The Director of Fisheries uses the recommended total allowable fishing effort (TAFE) in vessel-units and estimated vessel-units for each vessel size category, to operate the effort limitation scheme for the following year. In the *Loligo* fishery effort has most often been greatly over-subscribed, and vessels have been selected according to criteria published in the Fisheries Policy.

Stock assessment

Assessments are used at the end of each season to set maximum effort targets for the following year, and more rarely during the season to assess the need for early closure.

The *Loligo gahi* fishery is assessed using a modified Leslie–DeLury model developed for the Falkland Islands *Illex argentinus* fishery (Rosenberg *et al.* 1990; Basson *et al.* 1996). Few assessment methods have been developed specifically for short-lived squid species (Pierce & Guerra 1994). The Leslie–DeLury method has also been used for *Loligo vulgaris reynaudii* in South Africa (Augustyn *et al.* 1993), and *Loligo pealei* on the west coast of North America (Brodziak & Rosenberg 1993). Also called the depletion method, the Leslie–DeLury model has been modified to include natural mortality, and to provide estimates of the initial population size and catchability coefficients from different sub-fleets operating at the same time. For each season separately, the fishing fleet is split up, according to flags and vessel sizes, into homogeneous sub-groups as far as catch rates are concerned. Catches are analysed in numbers per week. Thus, individual vessel daily catch reports (in metric tonnes) are converted using length distributions collected daily at sea, and weekly weight–length functions produced on land by scientific observers throughout the season.

There are two main assumptions behind this model. The first one requires that, apart from natural and fishing mortality, the population is demographically and geographically closed over the period assessed, i.e. that there are no other sources of mortality, no recruitment and no migrations during that time. The second assumption is that natural mortality is known and constant over the period. Given these conditions, it follows that if removals from fishing are large enough compared to the initial population size, then abundance decreases as the population is depleted by exploitation.

The assumption of population closure is the most critical for *Loligo*. It often breaks down, notably in the second shorter season, when two broods are present and depletion is not usually observed for more than 4 to 6 weeks. In some of these cases, only part of the data has been used, and an *ad hoc* correction for immigration has been estimated (Tingley *et al.* 1990; Tingley *et al.* 1992). In others (Tingley *et al.* 1994), averaged estimates for catchability coefficients from previous seasons have been used to determine the following year's effort limitation.

Fisheries performance

At an average of 77 000 tonnes per year, total *Loligo* catches between 1987 and 1995 have been relatively stable (Table 8.1; Fig. 8.9). In 1987, the first year of management, very few regulations were introduced except for a split of the FICZ into two zones, north and south of latitude 51°20′ S, which was aimed at controlling the number of trawlers present in the *Illex* fishery in the first season and in the *Loligo* fishery in both seasons. In that first year of management, most trawlers operated as they did in previous years when access was open, exploiting a mixture of demersal resources, squid and finfish, depending on local abundance and time of year. Thus, vessels specifically targeting *Loligo*, i.e. for which *Loligo* represented at least 50% of the daily catch, were also catching some hake and other finfish, and some *Illex* in the first season (Table 8.1). In 1988, *Loligo* catches were poor

Table 8.1 Total annual catch (metric tonnes).*

Year	Total catch (t)	*Loligo* licences (t)	Other species caught in *Loligo* fishery (t)			*Loligo* by-catch by finfish boats (t)	% of total *Loligo* catch
			Illex	Hake	Other finfish		
1987	83 000		14 700	8500	3000		
1988	54 200		5700	12 300	4300		
1989	118 600	117 900	200	2000	1100	700	0.59
1990	83 300	82 900	100	2000	1200	400	0.48
1991	53 800	53 600	2300	200	400	200	0.37
1992	83 000	82 100	700	300	600	900	1.08
1993	52 300	51 600	100	300	2300	700	1.34
1994	65 700	64 900	–	300	1100	800	1.22
1995	98 400	97 900	–	–	800	500	0.51
Average	77 000		2600	2900	1600		0.80

*All figures rounded to nearest 100 tonnes.

(54 200 tonnes, Table 8.1), and *Loligo* vessels increased their pressure on hake (12 300 tonnes). This led to the creation of a specific finfish licence, and to the delimitation of the *Loligo* Box the following year (Fig. 8.6). From 1989, the *Loligo* Box has served two purposes. It has kept *Loligo* vessels from fishing elsewhere unless they have a finfish licence and use a minimum mesh size of 90 mm, but above all, it has protected *Loligo* from being caught incidentally in other fisheries. From 1989, when access to finfish and *Loligo* was separated through different licences, the average reported by-catch of *Loligo* in other fisheries has been less than 1% of the total catch of licensed vessels.

The efficiency of the *Loligo* Box as a conservation measure is greatly increased by a stable geographical distribution of *Loligo gahi*'s feeding migrations across the 200 m depth contour on the edge of the Falklands

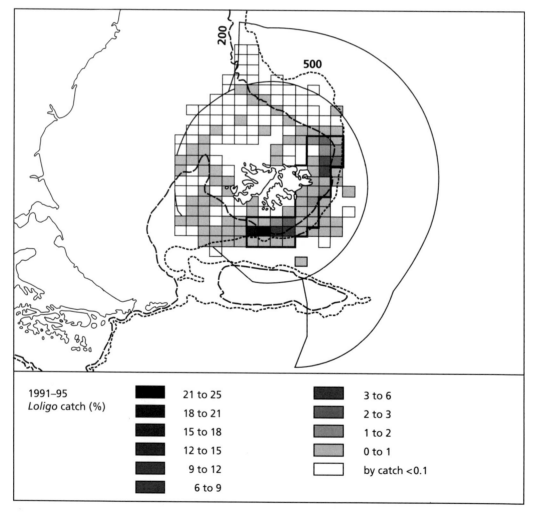

Fig. 8.6 Map of *Loligo gahi* catches in the *Loligo* Box restricted to licence holders, and outside as by-catch in other demersal fisheries. (The *Loligo* Box is denoted by a thick black line.)

Current (Fig. 8.1). However, when necessary the Fisheries Department has also extended the *Loligo* Box slightly to the north or to the south-west for the duration of a season in order to accommodate changes in migration patterns.

Another important feature of the *Loligo* fishery is that once caught, every squid is kept, washed, graded and frozen, including broken pieces. Thus, the disastrous statistics of 20% of the world fisheries catches being discarded and unaccounted for (Alverson *et al.* 1994) do not apply. Furthermore, very few finfish are caught on *Loligo* grounds relative to the rest of the fishing zone (between 200 and 1500 tonnes or less than 3% weight per season).

Effort

For most years between 1987 and 1995, the first season produced more squid than the second. Until 1991, it also had a larger effort. A first assessment of the *Loligo* fishery at the end of the first season in 1987 showed alarming signs of excess capacity and an estimated proportional escapement of less than 20% of the unfished biomass equivalent, when the target had been set to 40% (FIGO 1989). Consequently, the number of vessels licensed to catch *Loligo* in the first season of 1988 was reduced by a third (Fig. 8.7a). The 1988 season was poor, vessels concentrated on *Loligo*

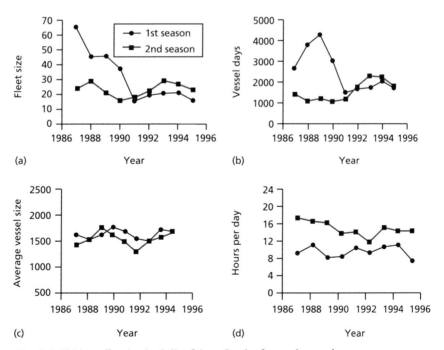

Fig. 8.7 Fishing effort in the *Loligo* fishery for the first and second seasons. (a) Numbers of different vessels. (b) Days fishing. (c) Average vessel size (gross registered tonnage). (d) Hours fished per day.

Fig. 8.8 One of the larger *Loligo* trawlers (2150 gross registered tonnage). (Photo by S. des Clers.)

and fished a lot more days in the season than in 1987. Another increase in days fished per vessel was observed in 1989, and the number of vessels was further decreased in the first season of 1990 (Fig. 8.7a,b). Over the same period, the average size of *Loligo* vessels and the number of hours fished per day have changed relatively little (Fig. 8.7c,d; Fig. 8.8).

The years 1989 and 1990 were very good and led to production exceeding the capacity of the European market. Prices tumbled, and in 1991 nearly half of the fleet didn't come back for the first season because of financial difficulties. From 1992, *Loligo* fishing effort has been similar for both seasons, apart from the numbers of hours fished per day (Fig. 8.7d) which is always larger in the second season. Monthly catches (Fig. 8.9c) have also stabilised, and are more evenly distributed than before regulation and until 1990, but this could also be due to changes in ocean climate.

Catches in the two seasons are becoming comparable, possibly also as a result of effort limitation in the first season and a gradual increase in effort for the second season (see Fig. 8.7).

8.4 Long-term sustainability

After 13 years of full commercial exploitation and 10 years of management through fishing effort limitation, the *Loligo* squid fishery in Falklands waters is showing good signs of sustainability. Ten years may seem too short a period from which to judge, but *Loligo gahi* is an annual species, and 10 non-overlapping generations is thus a reasonable time scale. It is widely recognised that the collapse or threatened state of northern cod has been caused by a combination of poor management and environmental change. Environmental change created the conditions for weaker recruitment, lower survival or deceptively fast growth, while poor management allowed fishing fleets to expand well beyond levels that could be sustained by the resource.

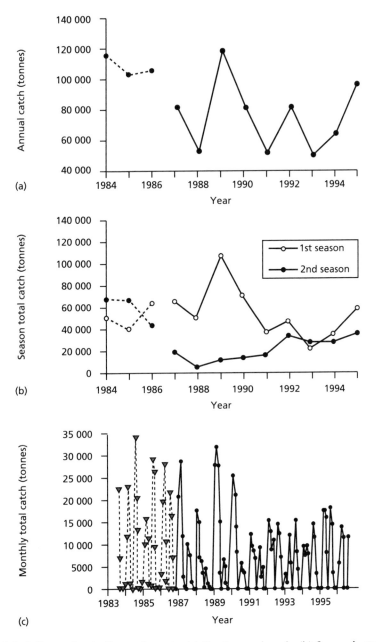

Fig. 8.9 *Loligo* catches by licensed vessels. (a) Total annual catch. (b) Seasonal catch. (c) Monthly catches. Catches between 1984 and 1986 (dotted lines) have been estimated from trans-shipment records and are thus not directly comparable.

Environmental variability

Future research is needed to describe and to better understand the ocean climate around the Islands and its effect on fisheries yield. *Loligo gahi* is a

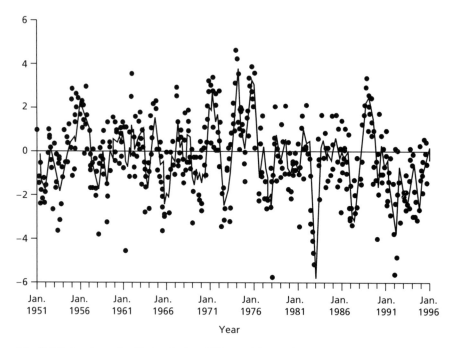

Fig. 8.10 Southern Oscillation Index. (Climate Diagnostic Bulletin, Climate Analysis Centre website, January 1996.)

colder water species than *Illex argentinus* in Falklands waters, which is similar to *Illex illecerebrosus* and *Loligo pealei* squid species off the east coast of North America (Dawe *et al.* 1990). However, 10 years of data are too short a time series to analyse long-term variability of *Loligo* catches in relation to sea temperature. There are already some indications that annual *Loligo* catches in the northern half of the *Loligo* Box may vary inversely with annual *Illex* catches, suggesting a dynamic equilibrium between water masses to the north and east of the Islands. It is also likely that El Niño–Southern Oscillation events influence the relative strengths of the cold north-flowing Falklands Current, and the warm south-flowing Brazilian Current, which in turn would have a direct effect on annual *Loligo* catches. It is interesting to note, for example, that 1989 was a record year for *Loligo* and for *Illex* in the Falklands, as well as for the South African chokka squid (*Loligo vulgaris reynaudii*) fishery (Roberts & Sauer 1994). A year of strong Southern Oscillation Index (Fig. 8.10) may have meant a year of strong upwelling, and ample food for squid.

Fisheries policies

It is currently argued by most fishermen in Europe that the biggest threat to the perenniality of fish stocks they exploit is the European Common

Fisheries Policy (CFP). The CFP is made up of four parts: the Structural Policy, the Market Policy, the Conservation Policy and International Fisheries Relations. Holden (1996) identifies four major shortcomings of the CFP: first, a major contradiction between the Structural Policy and the Conservation Policy; second, the absence of specific policy objectives; third, the absence of economics as a basis for the policy; and finally the importance of political compromise, often involving other sectors such as agriculture or industry, in the decision-making process.

The Falkland Islands *Loligo* Fisheries Policy has been immune from such problems for the simple reason that the Islands have not had their own fishing fleet until recently. Therefore there was no need for a Structural Policy to restructure, renew or re-adjust its fleet to be more efficient, more in line with resource abundance, or to find new resources. Furthermore, conservation imperatives have always been put before any other aspects of fisheries development, management or rent extraction. However, fisheries socio-economic aspects have been largely ignored so far, as there have been very few fishermen and vessel owners. Fisheries policy matters remained relatively simple as long as there was no domestic fleet.

This is not the case any longer. *Loligo* vessels have gradually been switching to the Falkland Islands Shipping Register since 1992. Since 1995 a renewed European interest in the Falklands Register has occurred as a direct consequence of the CFP Structural Policy, which generously subsidises vessels to leave the crowded European Register in order to meet fleet reduction targets. At the end of 1995 six vessels, or 25% of the *Loligo* fleet in vessel number and tonnage, had made the permanent move out of the European Register into joint ventures with Falklanders (des Clers 1998). Pressure from Europe will probably continue to encourage vessel owners to apply for registration on the Falklands Register. Thus, the Falkland Islands are entering a new phase in the development of their fisheries sector, when Islanders are becoming joint or sole owners of fishing vessels. As a result, there will be an increased need to integrate fisheries economic and social aspects in the Fisheries Policy in order to avoid the pitfalls of most systems in European waters (Holden 1996; Crean & Symes 1996) and the North Atlantic (Hannesson 1996).

Fishing power

'Considerable excess of fishing capacity' is leading to over-exploitation of most of the world's marine fish resource (FAO 1995a). Except for the year 1991, when European *Loligo* prices collapsed pushing many vessel owners to bankruptcy, access to the *Loligo* fishery has always been greatly over-subscribed. However, vessel numbers have been cautiously controlled by the Falkland Islands Fisheries Department.

A chronic problem of effort limitation regimes is the continuous increase in fishing mortality per unit of effort linked to the technological upgrading of fish-finding equipment, fishing vessels and processing equipment. Although few vessel characteristics were precisely known for the first years of the *Loligo* fishery, on-going research is suggesting that daily freezing capacity is directly limiting daily catches, and that it has been continuously upgraded by most vessels over the years. Increased efficiency does not threaten the fishery if vessel numbers are reduced accordingly. It also has a potential to extract higher rents from more efficient vessels, and thus does not threaten licence revenue. Still, an ever smaller fleet brings its own management problems of reduced flexibility and negotiating power for the fisheries manager.

Fisheries assessment

Many crucial aspects of *Loligo gahi*'s fisheries biology, ecology and dynamics are not known. Multiple broods, extended spawning periods and separate migrations of the two sexes combine to produce a complex picture during each fishing season and between the north and south ends of the *Loligo* Box. The ocean climate on *Loligo* spawning grounds in coastal waters around the Islands, for example, is not well known. The different scales of environmental variability need to be described, in order to identify those important to *Loligo* recruitment, growth and maturation. Pre-season research cruises would be needed to investigate the two main dynamic determinants of a good fishing season, namely recruitment and migrations. Similarly, food availability and predation remain to be investigated as there is currently little insight on population regulation, either from within (cannibalism), from other species (predation, competition, food limitation), or from the fishery itself (see Murphy *et al.* 1994). The early parts of *Loligo gahi*'s life cycle are still poorly known, and laboratory experiments are needed to determine the influence of environmental parameters on growth, maturation and lifespan. There is a great need to understand *Loligo*'s ecology and ecosystem on a more global scale in order to understand its population dynamic processes more fully, and potential limits to its sustainability (Southwood 1995).

However, catch and effort data are reported daily, and the current management regime protects the *Loligo* fishery from systematically inaccurate data, the most common ailment plaguing stock assessment exercises. There are no discards, little by-catch or illegal catches. Although the current absence of research cruises to survey abundance before the season starts means that assessments cannot be fine tuned, they still provide a basis for rational management (Augustyn *et al.* 1992).

Economics

Hannesson (1996) identifies the roots of poor management in terms of governments putting political considerations before economic sense in fisheries matters. His analyses of the Norwegian, Faroese, Icelandic and Newfoundland cod fisheries show clearly that a heavy dependence of national or provincial economies on fishing is no guarantee of good husbandry. He also argues that perverse effects arise when fisheries' development is subsidised as a means of employment and economic activity in remote coastal areas. This is undoubtedly the case in the Islands, although on a very small scale due to a small population and abundant resources. Moreover, from the onset of management, strong precautionary principles have put the health of the resource before economic or social gain. Clearly, as a local fisheries sector is now developing in earnest, research on fisheries economics aspects will become crucial, notably on *Loligo* markets, prices, and production costs in order to develop basic bio-economic models.

Governance

As many fisheries resources have been threatened by excessive fishing pressure, they have become increasingly regulated. In this respect, the Falkland Islands Fisheries Management regime is much simpler than any faced by European fishermen. The rules and regulations relevant to the *Loligo* fishery have been limited to essential measures, simple to enforce and with obvious conservation benefit. So far, there has been little local demand for property rights. The European principle of relative stability, which guarantees access according to historical involvement, and gives fishermen a false assurance of permanent access with every encouragement to race for fish, does not apply (Garrod, Preface to Holden 1996). The fisheries resource has been treated as a public good which the owner, the Falkland Islands government, does not exploit but has a keen interest to conserve. Until now most tenants pledging access to the resource have been foreigners with their own agenda. This has undoubtedly made the task of fisheries management easier in years when the number of licensed vessels had to be reduced.

For the Islands, an important challenge in the near future will be to identify a sustainable scale for the economic development of the local fishing industry relative to its ecological life-support system. Although sustainability cannot be predicted (Costanza & Patten 1995), it has been observed in the *Loligo* squid fishery after 10 years of an undoubtedly cautious management regime put in place by the Falkland Islands government.

Chapter 9: Recreational Use of Coral Reefs in the Maldives and Caribbean

Andrew R.G. Price, Callum M. Roberts and
Julie P. Hawkins

9.1 Introduction

Tourism and travel are among the world's most lucrative enterprises, with turnover now rivalling that of the global oil industry. Ecotourists alone spend around US$12 billion annually on their enjoyment of the natural world (Primack 1993). Since the 1970s coral reefs have increasingly become the focus for tourism in many developing countries, providing the basis of recreational activities such as diving, glass-bottom boat tours and fishing. Their natural beauty, associated sandy beaches and warm climate attract thousands of holiday-makers annually to islands in the Caribbean, Maldives and other tropical destinations.

Coral reefs are the most diverse ecosystems in the sea, supporting thousands of species in dense and spectacular concentrations around limestone structures, themselves the culmination of thousands of years' growth by innumerable coral colonies. Atoll islands, such as the Maldives, owe their physical existence to coral reefs. The islands consist of an unstable and dynamic base, composed of coral debris overlying reef platforms built upon the submerged slopes of extinct underwater volcanoes. They exist through a balance of reef growth and erosion and so a healthy marine environment is fundamental to the economy of such countries. Even in continental and high island states coral reefs provide an important coastal protection function, sheltering coasts from the fury of tropical storms which would otherwise render large tracts of land unsafe. Coral reefs can also be critical for food and finance for those coastal states heavily dependent on marine tourism and fisheries, such as Australia and many Caribbean islands.

The geological solidity of reefs belies a biological fragility which renders them remarkably vulnerable to the activities of people. The rapidly growing coastal populations of developing countries derive much of their income and food from reefs, in many areas resulting in serious over-exploitation (Polunin & Roberts 1996). Land development is leading to increasing levels of reef degradation by sediment and nutrient pollution (Roberts 1993; Ginsburg 1994). Even tourism, an apparently benign activity bringing many benefits to rich and poor countries alike, can damage the very environment tourists come to enjoy (Hawkins & Roberts

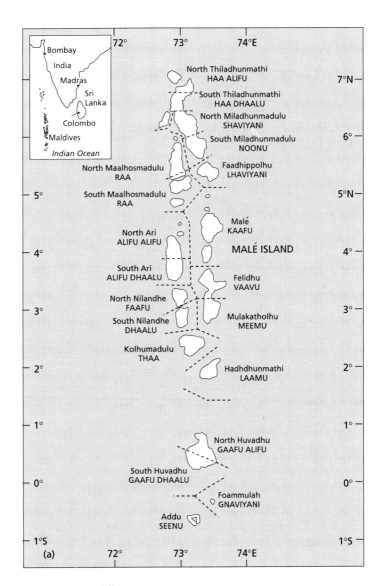

Fig. 9.1 Map of the Maldives.

1994). In large numbers tourists cause significant wear and tear to reefs, damaging their biological integrity.

This chapter assesses the sustainability of reef-based tourism in the Maldives and the Caribbean island of Saba (Figs 9.1, 9.2). We examine growth and environmental impact of the industry on the ecology of the two regions, conservation initiatives, and future prospects for tourism. These regions are geographically widely separated tourist destinations: one in close proximity to North America (Caribbean), and the other increasingly accessible to European and Japanese travellers (Maldives). Other

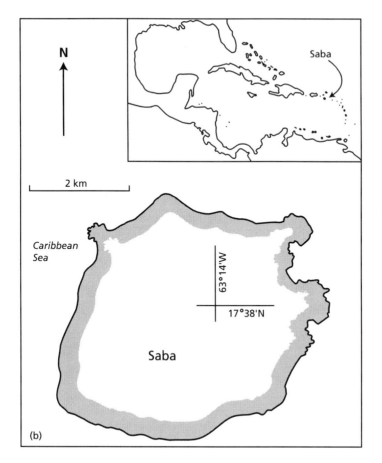

Fig. 9.2 Map of Saba in the Caribbean.

differences are evident, both ecological and socio-cultural, yet they also share a number of common features. Comparison between the two regions provides useful insights not only into the immediate and indirect environmental implications of tourism, but also into wider issues surrounding human resource use and decision-making.

One other important reason for choosing Saba and the Maldives is that both are still early in the process of tourism development. They remain in a position to make crucial choices about the direction that development may take. This contrasts with long-established resort areas such as the Florida Keys or Hawaii where reefs are now suffering from widespread overuse (Wells & Hanna 1992). In Saba there is a great awareness of these dangers to reefs, and a conscious effort is being made to avoid exceeding the island's tourism capacity. Environmental awareness is also appreciable in the Maldives. Unlike Saba, however, this country is an archipelagic state, having the luxury of other islands for future expansion of tourism, and for use if the industry on existing resort islands becomes unsustainable.

9.2 History of use of the system

The first tourist resorts in the Maldives opened in 1972. They were developed close to the airport and the capital Malé. Cumulative bedspace increased exponentially from 1972 to 1995 (Fig. 9.3a). Until 1981 more than 91% of available bedspace was in resorts on North and South Malé atolls. Thereafter new resorts on other atolls, especially North and South Ari atolls, accounted for most of the increase in bedspace. The steady development of new and upgraded resorts in the Maldives was mirrored by an equally dramatic growth in tourist arrivals (Fig. 9.3b). The number of visitors peaked at 300 000 in 1995, spread over 74 resorts. Clearly, demand has more than kept pace with supply. By the year 2000 tourist arrivals may be 400 000, and some models predict a demand of more than 1 million by the year 2001 (Hameed 1993).

Fisheries remain the cornerstone of Maldivian society, although generating less income than tourism. Exports in 1992 amounted to US$31.6 million (MPHRE 1994). Pelagic tuna make up the bulk of catches, in particular skipjack (*Katsuwonus pelamis*) which represented 78% of catches in 1990. There is also a reef-dependent fishery (providing bait for tuna fishing) which will become increasingly important as a source of food on resort islands. Use of coral for building material has been extensive in the

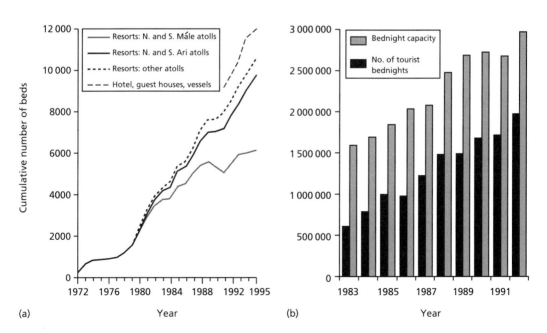

Fig. 9.3 (a) Cumulative bedspace in the Maldives, 1972–95. (From Nethconsult/ Transtec 1995.) (b) Increase in tourist bednights and bednight capacity in the Maldives, 1983–92. (From Nethconsult/Transtec 1995.)

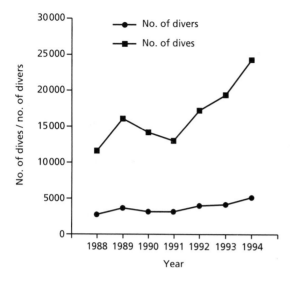

Fig. 9.4 Trends in diving tourism in Saba. Figures show the number of divers passing through commercial diving centres and the total number of dives they made each year. These data do not include dives made from visiting yachts.

past. However, coral mining has a major impact on reefs (Brown & Dunne 1988; Brown *et al.* 1990; Dawson Shepherd *et al.* 1992), and is now permitted only at specified sites. The practice continues, but due to the recent reduction of duty on imported construction materials (to 10%), it is on a smaller scale.

On the tiny island of Saba tourist development has been a more recent phenomenon but has rapidly assumed central importance to the economy. Environmental tourism, based around the island's reefs and forest, has been strongly promoted overseas since the mid-1980s. This triggered a move from the traditional fishing industry into scuba diving and hotel construction and there are now only a handful of commercial fishermen active in Saba. Growth in scuba diving tourist arrivals has been rapid and is reflected by increases in the number of dives made annually (Fig. 9.4). Nevertheless, a tourism-based economy is subject to external influences and the decrease in diving in 1991 reflected a fall-off in international tourism caused by the Gulf War. Although not shown in the figure, more recently, tourist arrivals have dipped once again due to severe hurricane damage to nearby islands in 1995. Despite these setbacks tourism is set to continue expanding well into the next century.

9.3   Ecology of the reefs

Coral reefs have enormous intrinsic appeal. Many of their direct values to tourism and recreation are a reflection of their high biodiversity. However, the coastal protective function of reefs (from waves and storms) is also of major practical importance to coastal developments such as tourist resorts and often entire islands throughout the Maldives and Caribbean.

Maldives

The Maldives form the central and largest part of the Lakshadweep (Laccadive)–Chagos ridge in the central Indian Ocean (Fig. 9.1). It comprises 26 atolls and some 1200 islands, of which only about 200 are permanently inhabited and 74 used for tourist resorts. Despite extensive territorial waters ($67\,000\,km^2$) and a vast Exclusive Economic Zone ($c.$ $840\,000\,km^2$), total land area is only $186\,km^2$ (H. Hansen, pers. comm.). Further, 80% of Maldivian land area is less than 1 m above mean sea level and so threatened by impending climate change and sea level rise. The islands consist of coral rubble and sediments, providing a dynamic and unstable environment shaped by the changing monsoons, storms and other episodic events.

Coral reefs have been well studied in the Maldives. Conservative estimates indicate 161 species of reef corals (Sheppard 1987; Sheppard & Wells 1988), with a broad trend of increasing coral diversity from north to south (Sheppard 1981). Coral fish are also diverse, with more than 1200 species known. Several physical features of the atolls, reefs and islands of the Maldives vary geographically (Table 9.1). These features are a major determinant of natural resource availability and many have implications for tourism. Those relating to atoll robustness and stability are particularly important. In southern atolls there is a more continuous atoll rim than in the north, so providing better natural protection from wave action.

Marine ecological studies have been undertaken on reefs in several atolls, although few resort island reefs have been studied directly (MPE 1992, cited in Price & Firaq 1996). However, reefs of 43 of the 74 resort islands were recently assessed, based on analysis of questionnaire data (Price & Firaq 1996). Reef attributes were assessed in terms of perceived present state and perceived magnitude of change in recent years (Table 9.2). Perceived current overall reef condition was correlated most strongly with number of reef fish species, and also significantly with extent of live coral

Table 9.1 Gradients in physical conditions in the Maldives. (From Anderson 1992.)

Climatic factors	N–S pattern
Annual rainfall	Increases to the south
Strength of monsoon reversal	Increases to the north
Incidence of severe tropical storms	Increases to the north
Atoll features:	
Depth of atoll lagoon	Increases to the south
Continuity of atoll rim	Increases to the south
Proportion of atoll rim with islands	Increases to the south
Occurrence of faroes (ring reefs)	Increases to the north
Island height	Increases to the north

Table 9.2 Median (Mn) values for coral reef attributes and change in coral reef attributes for Maldivian resort islands using a ranked scale of 0 to 5 for current conditions (0 lowest, 5 highest), and −4 to +4 for change. Also shown is the significance of change based on two different tests (NS = not significant). (From Price & Firaq 1996.)

Coral reef attribute	Mn value of current state	Mn value of change in state	Significance of change in state	
			Wilcoxon's test	Mann–Whitney U-test
Number of divers	3.0	2.0	$P < 0.01$	$P < 0.01$
Overall reef condition	4.0	−1.0	$P < 0.01$	$P < 0.01$
Live coral cover	3.5	0	NS	NS
Coral species	3.0	0	NS	$P < 0.01$
Algal abundance	2.0	1.0	$P < 0.01$	$P < 0.01$
Reef fish abundance	4.0	0	$P < 0.05$	NS
Reef fish species	4.0	0	NS	NS
Pelagic fish abundance	3.0	0	NS	NS
Urchin abundance	2.0	0	NS	NS
Coral bleaching	(48% of sites)			

cover and number of coral species. Coral bleaching (loss of symbiotic algae, a sign of stress) was reported for 48% of the resort islands, and (not surprisingly) is associated with significantly reduced coral cover. Other analyses reveal a significant increase in algal abundance, increase in the number of divers and decrease in overall reef condition.

Caribbean and Saba

Caribbean coral reefs are less diverse than those of the Indian Ocean but still support a profusion of species with, for example, around 150 species of fish readily seen by divers. Reefs also differ from those of the Maldives in that coral growth has typically been less vigorous throughout the Holocene, especially around the volcanic islands which make up most of the Lesser and Greater Antilles. Consequently, in most places reefs have not developed the extensive very shallow flats which typify much of the Indo-Pacific.

In recent years the ecology and structure of Caribbean reefs have been transformed by outbreaks of coral and sea urchin diseases which occurred in the mid-1980s. Prior to 1983 the needle-spined urchin, *Diadema antillarum*, was exceptionally abundant and had been perhaps for centuries (Jackson 1997), but a little-known pathogen then swept through the region killing more than 99% of urchins (Lessios 1988). This greatly reduced grazing pressure by herbivores leading to vigorous algal growth at the expense of slower growing corals (Hughes 1994). About the same time disease wiped out populations of staghorn and elkhorn coral (*Acropora*

cervicornis and *A. palmata*), important reef-building species which dominated shallow water (Gladfelter 1982). Recent figures suggest that disease
killed more than 95% of these corals over large areas (Bythell & Sheppard
1993).

The origins of these diseases can only be guessed at but their consequences are clear: Caribbean reefs are currently undergoing serious declines in coral cover and diversity (Ginsburg 1994). The stressed condition
of coral reefs has compounded the effects of natural (storms, hurricanes)
and human (e.g. pollution) impacts (Rogers 1990; Hughes 1994; Sladek
Nowlis *et al.*, in press) with the result that recovery of coral reefs from
recent disturbances has generally been slow or non-existent (Connell
1997). Saba's coral reefs are in better condition than many. Although
suffering the coral and urchin mortalities described above, living corals still
cover between 25 and 40% of the surface of deeper reefs and algal cover is
kept in check by healthy populations of herbivorous fishes (Roberts 1995a;
Roberts & Hawkins 1995).

9.4 Tourist use and its environmental consequences

Current reef uses for tourism

Reef-based tourism in the Maldives is now the principal source of national
revenue, generating 17% of GDP, 90% of foreign currency earnings and
40% of government revenue (Price & Firaq 1996). The island chain
appeals to specialist groups, in particular divers, ecotourists and travellers
seeking remoteness. About 300 000 dives occur annually in the Maldives
(Nethconsult/Transtec 1995). Cruise ships and yachts also visit the
islands. Some sport fishing is undertaken and both reef fishing and game
fishing have potential to expand.

Throughout the Caribbean tourism is rapidly superseding the importance of traditional agriculture and fisheries. Tourism revenues in the
region were US$7.2 billion in 1988 (Dixon *et al.* 1993). Like the Maldives
many islands attract specialist holiday-makers such as scuba divers. For
example, in Antigua 65% of visitors go diving or snorkelling (Bunce *et al.*,
in press), in Saba many of the 23 000 or more annual visitors to the island
dive (Fernandes 1995) and reefs generate an estimated 20% of the island's
revenue (Framheim 1995). Similarly, in Bonaire over 70% of GDP is
generated by diving tourism (Dixon *et al.* 1993).

Impact of tourism infrastructure

Tourism expansion carries environmental and social costs. Environmental
impacts accrue both directly from use of reef resources and indirectly from
construction and pollutant discharge (Table 9.3). Effects of tourist infra-

Table 9.3 Environmental impacts and issues associated with resort construction and their estimated severity (1: minor, 2: moderate, 3: high) in the Maldives (M) and Saba (S). (See also Nethconsult/Transtec 1995; Roberts *et al.* 1995; Price & Firaq 1996.)

Impact or issue	Severity M	Severity S
Limited understanding of environmental carrying capacity of an island and reefs prior to resort development	2	3
Construction of solid jetties, piers, groynes and breakwaters, which restrict seawater circulation and can increase erosion, resulting in the need for costly beach replenishment. Locally severe disturbance to some fringing reefs has also occurred	2–3	2
Creation of access routes by blasting of reefs for channels and vegetation clearance, which can undermine the reef's protective capacity and increase soil erosion	2	2
Clearance of peripheral vegetation and seagrass, without the role of the latter being assessed, and disturbance of nesting turtles and seabirds	2	2
Reclamation of more land for additional rooms and/or resort infrastructures	1	1
Contamination of aquifer by pathogens during resort construction and from chemicals to clear/control vegetation; contaminants may also reach the marine environment	1	1
Discharge of sediment onto reefs during construction	2	3
Improper disposal of construction materials	1	1
Non-sustainable use of coral and timber as construction materials	1	1

structures are multiple and difficult to predict, since reefs and islands are naturally highly dynamic (see Pernetta & Sestini 1989; Pernetta 1993). Detection of change is compounded by limited baseline data, particularly in the Maldives (Price & Firaq 1996). Resort siting, construction and operation are critical factors, because they affect the sustainability of tourism even before users get in the water. The principal problems arise from lack of understanding of the fragility of tropical coastal environments and many are preventable. For example, sediment discharge is a serious form of stress to reefs (Rogers 1990) but can be controlled through careful construction practices. The dual problem of erosion and sedimentation has been particularly acute in Seenu atoll in the Maldives, due to earlier projects interlinking several of the islands. This had adverse effects on the reefs and reef-dependent fisheries. Opening up the causeway between Gan and Fehydhoo to re-establish water flow during the 1990s increased the abundance of bait fish in the lagoon. Environmental issues of moderate concern include the removal of island vegetation and seagrasses. Seagrasses are irritating to tourists in the Maldives, as they dirty lagoon bottoms and their leaves litter beaches. Much effort is spent clearing this vegetation, without its ecological role being assessed.

Following construction, environmental impacts from tourist operations continue (Hawkins & Roberts 1994). The main environmental concerns stem from discharges of waste, including solid waste, sewage, pesticides and oil. Small islands by their very nature have only a limited capacity to accommodate tourist waste. Tourists produce more solid waste per capita than is generated by locals. Waste production by resorts in the Maldives is typically $40–200 \, t \, yr^{-1}$ (Hameed 1993). Some of this ends up on the reefs (e.g. plastics, fishing lines and building materials) but much increases pressure on local landfills or is improperly disposed of. In many Caribbean islands, rubbish is simply dumped into steep ravines causing pollution on land and sea. Sewage disposal has both health implications (e.g. water contamination), and environmental consequences (e.g. eutrophication) for shallow lagoons and the reef edge.

In areas with porous rocks and thin soils, nutrients from septic leach fields can be rapidly washed into the sea (Roberts *et al.* 1995). Where piped sewerage is employed, sewage is often discharged without treatment, although clear effects of discharge onto reefs are typically apparent only very close to outfalls (e.g. MPE 1992; Grigg & Dollar 1995). Background nutrient enrichment is recognised as an important cause of reef degradation worldwide (Bell *et al.* 1989; Ginsburg 1994). Nutrient enrichment encourages algal overgrowth of corals and can help transform actively growing reefs into eroding ones (Bell *et al.* 1989). Risk *et al.* (1994) report lower coral cover and diversity and higher number of associates (commensals, epizoans, epiphytes and internal bioeroders) in sites at which sewage is the likely source of stress. Other differences at affected sites included higher coral growth rates and higher values of ^{15}N (used a sewage tracer). Proliferation of seagrass (and perhaps also algae) in lagoons is increasingly reported as a problem to tourists on resort islands. Restricted water circulation due to sea walls, piers and other solid structures adds to the problem of nutrient enrichment from sewage effluents.

Declines in coral cover and number of species, and increase in algal abundance in the Maldives (Table 9.2) may be partly linked to increasing nutrient levels. In the Caribbean, sewage stress exacerbates the problem of herbivore loss due to sea urchin death and over-fishing (Roberts 1995b). In places like Dominica, which are severely over-fished, the problem of nutrient enrichment by sewage (both from locals and hotels) has become very severe with rampant growth of algae (including toxic blue-greens) displacing corals and sponges (Fig. 9.5). By contrast, there have been no harmful effects in Saba which relies entirely on septic tanks and discharges very little sewage into the sea.

(a)

(b)

Fig. 9.5 (a) A healthy Caribbean coral reef from the island of Bonaire where the reefs are managed by an excellent marine park. Such reefs typically have a high cover of stony, or reef-building corals and low cover by algae. (b) A reef degrading through algal overgrowth on corals in Dominica. Loss of herbivores due to sea urchin mass mortality and over-fishing has led to greatly increased cover of algae on Caribbean reefs. These algae outcompete corals for space. The problem can be compounded by nutrient discharges, such as sewage, acting as a fertiliser and stimulating algal growth further. The Saba Marine Park includes a zone which is completely protected from fishing, allowing the build-up of herbivorous fish populations which help to control algal growth. (Photos by Callum Roberts.)

Impacts of diving and snorkelling

Reefs are living ecosystems easily damaged by scuba divers, snorkellers and bathers. The greater the use the more wear and tear there is (Hawkins & Roberts 1997). Direct impacts have been detected on reefs throughout the word (e.g. Red Sea, Hawkins & Roberts 1992a,b, 1993; Bonaire, Dixon *et al.* 1993; Great Barrier Reef, Rouphael & Inglis 1995). A study at Kurumba in the Maldives assessed the effects of snorkellers on the reef flat/crest at depths down to about 1.5 m (Allison 1995). Results indicate that 18% of all *Acropora* corals are broken each month, so most or all coral colonies of this genus are damaged each year. In the Red Sea, 10–15% of coral colonies show physical damage at heavily used sites (Riegl & Velimirov 1991; Hawkins & Roberts 1992a, 1993). By comparison, Saba's coral reefs are still only lightly used for scuba diving and only 2–3% of corals show evidence of recent breakage (Hawkins & Roberts 1997). Snorkelling, trampling and scuba diving may also have local effects on reef fish since many species shelter in branching coral colonies (Brown *et al.* 1990).

In the Maldives, corals and 'tortoise-shell' from hawksbill turtles are among species collected from reefs and sold as souvenirs to tourists (despite leaflets on turtle conservation in resort rooms). Collection of such species (by licensed Maldivians) is apparently being phased out, although items on sale in local shops did not appear to decline between June 1993 and June 1995. Maldives is currently (1998) considering becoming a party to CITES. All curio collection is prohibited in Saba, but in some other regions of the Caribbean shells and corals are collected for sale to tourists. In some places, such as Puerto Rico and the US Virgin Islands, local prohibition of reef curio collection has led to large-scale imports from other regions, such as South-East Asia.

Boating, fishing and other effects of tourism

In the Maldives cruises and boating are likely to expand, in line with the national trend towards tourism diversification. Some charter and cruising vessels are not allowed in lagoon and reef areas of certain resorts, in order to prevent anchor damage (H. H. Maniku, pers. comm.). Discharging sewage in lagoons and near resorts is unacceptable in most areas. A holding tank is required, with the contents pumped out only in deep water away from inhabited islands.

Speedboats and jetskis pose a hazard to bathers, divers and marine life such as turtles, and are also noisy. Environmental problems, including erosion from propwash, groundings and anchor use on seagrass beds, are most pronounced in enclosed areas. Sport fishing is undertaken from several resort islands; favoured species are generally large pelagics not associated with the reef fishery. Tourists also undertake limited fishing for groupers, snappers and barracudas, which are important fishery species. Brown *et al.* (1990) consider that reef fish landings by tourists probably make up no more than 2–3% of the total reef fishery.

Use of pleasure boats in Saba is currently severely restricted by lack of harbour facilities. Nevertheless, a growing number of local inhabitants are purchasing boats following increased prosperity from tourism, and fishing for pelagic species is becoming popular.

Problems in the Maldives which have their origins at least partly in tourism include migration and urban pressures, particularly in Malé. Wealth from tourism leads to secondary businesses, which may impinge upon the environment. More directly, fishing increases to meet tourist food demands. On the other hand, fishermen are increasingly lured to better job opportunities at resorts, leading to reduced fishing (C. Anderson, pers. comm.).

9.5 Sustainability and future prospects

Up to now we have described many of the actual or potential impacts of tourism on coral reefs. In some regions of the world the prospects for sustainability look bleak, as development initiatives transform previously remote destinations into venues for mass tourism, for example, on the Red Sea coasts of Egypt and Israel (Hawkins & Roberts 1994). However, if carefully handled, tourism can have many positive effects. For example, divers in the Maldives like to see giant clams and protested when fishing began; removal of clams is also destructive to the reef. Divers' protests were accepted, and the time from the start to the end of the fishery was only 1 year (C. Anderson, pers. comm.). Another example is the reef shark fishery. Sharks are attractive to divers but are caught at night by fishermen. It is estimated that a single shark is worth US$3000 generated annually from divers, compared with US$30 per shark dead for export (C. Anderson, pers. comm.). In a similar vein tourism has afforded protection to reefs from coral mining, which is prohibited on resort reefs in the Maldives. Hence, tourist islands function as 'protected areas', safeguarding reefs from an activity that can be much more destructive than tourism pressures, provided that resorts are not constructed from coral mined from other islands (Price & Firaq 1996). While it encourages economic growth of many kinds, coastal tourism can also deter more damaging forms of development, such as industrial and port facilities.

Tourism can also force the pace of conservation initiatives in developing countries. In Saba, a marine park was established in 1987 which surrounds the entire island. The development of the park grew out of an awareness of the need for conservation measures to be implemented alongside tourist development. Recently the park became the first self-funding marine park in the world on the basis of user fees levied on scuba divers and yachts (Framheim 1995).

Saba is rather unusual in promoting conservation at such an early stage in the development process. In general, conservation initiatives respond to problems reactively and lag behind damage inflicted rather than preventing it. Conservation measures can be implemented to reduce environmental damage by tourism and enhance sustainability. In the following sections we describe some of the initiatives taken in Saba and the Maldives.

Conservation initiatives

Many conservation measures and environmental legislation are in place in the Maldives and Saba, some broadscale, others targeted more specifically at tourism-related issues. In the Maldives, environmental impact assessments (EIA) are now required for major developments. Environmental policy options identified for the 1995 Maldivian Tourism Master Plan are:

- integrating tourism with coastal zone management;
- multidisciplinary EIA;
- incorporation of new technologies and conservation measures (e.g. for waste disposal);
- further development of protected areas and a zoning system;
- development of contingency plans for oil spills and other environmental hazards;
- environmental public awareness;
- combining tourism with marine research (Nethconsult/Transtec 1995).

In Saba, the marine park is closely involved with issues relating to the immediate coastline but as yet has little say in regulating activities on land which may be damaging to marine resources. Even so the park works in close collaboration with the Saban government and another NGO, the Saba Conservation Foundation, whose remit is more terrestrial in focus.

In Saba, resort waste is dumped in the island's landfill site. In the Maldives islands are highly dispersed and each must make their own arrangements. Dumping at sea is the most common approach, and practised by 78% of 32 resorts (MPE 1992). Incinerators and compactors have become mandatory at all resorts since November 1987 (Pernetta 1993). Beach clean-up is also given high priority, costs for which average US$300–500 per month (Hameed 1993). New and innovative conservation efforts include those of certain airlines. For example, a German airline issues passengers with large plastic bags for storing all their refuse while on a resort island. On leaving the bags are 'checked in' and taken back to Germany for disposal or recycling. Measures in the Maldives to protect the environment in cases of direct sewage discharge include the siting of outfall pipes 100 m from the island and 30 m below mean sea level (Hameed 1993). Some new resorts are using the latest technology in sewage treatment, for example a plant at one involves UV irradiation and can treat sewage to produce virtually pure water. Such technologies will do much to limit environmental problems on remote resort islands, particularly as resorts will soon have to comply with newly proposed environmental standards for discharge (MPE 1992).

In the Maldives, conservation measures include the selection of 15 special dive sites for protection in five northern atolls. Although detailed management plans have yet to be developed, these represent a step towards coastal zoning and a protected areas system. A high conservation ethic prevails among dive bases and most tourists, although in itself this will not ensure protection. Practical measures are also necessary. The extent of damage by tourists depends a great deal on the care taken by resort operators to educate tourists as to the problems and the degree of investment in infrastructure to control damage. Anchoring by boats is the single greatest threat to reefs from divers but can easily be controlled by installation of moorings such as has occurred in many parts of the Caribbean. In Saba the marine park has installed moorings at all dive sites including those for yachts so eliminating anchor damage completely. Records are kept of the number of dives made at each site. Since 1993 diver impacts have been measured in an annual monitoring programme involving a subset of sites and the findings will help determine if sites are being overused (Roberts & Hawkins 1995). Such monitoring is central to the objective of remaining within the capacity of the environment to cope. Recent research in the Red Sea suggests that per-capita damage rates by divers can be reduced by 60% with a single briefing at the start of the holiday on how to behave on a coral reef (Medio et al. 1997) The effect of diver education by the marine park and by diving centres is also clear in Saba where despite a 42% increase in diver pressure damage levels have actually decreased recently (Hawkins & Roberts 1997).

Integrating reef-based tourism with coastal zone management

Economics, equity and ethics

In the Maldives, national economic benefits from tourism are considerable. These include employment for over 5000 Maldivians whose wages amount to US$9 million, and US$7 million profits to Maldivians (cf. at least US$7 million flowing out to foreign interests). Tourism accounted for 45% of government revenues in 1993 (Nethconsult/Transtec 1995). In addition, there are indirect income effects from tourism (e.g. income to construction workers and fishermen). Economic benefits from tourism have spread beyond the tourism industry, raising overall Maldivian living standards. However, national figures can mask fluctuations from the average; many outlying islands have only a modest or even meagre standard of living. Socio-cultural impacts from tourism include separation of resort staff from their families, essentially 'voluntary displacement', and problems from migration to the capital Malé.

Equity and ethics are also embedded in the concepts of 'benefit' and 'sustainability', becoming manifest through religious or other social codes of behaviour. The Maldives is an Islamic country; Saba and most of the

Caribbean is Christian. While Islamic principles provide the right and privilege of people to use nature's resources, each successive generation is also entitled to benefit from them. The law is flexible and adaptive to local conditions or changing needs (see Child & Grainger 1990). Nevertheless a commitment not to misuse or over-exploit natural resources is acknowledged or implied. The relevance of this clearly extends to tourism and recreation. Hence, in Islam, and currently in Christianity, people are seen as custodians rather than owners of the environment.

In Saba tourism is the primary industry, and the very considerable economic benefits it generates reach almost all parts of the small population in one form or other (a fortunate position shared by few other tourist-based economies in developing countries). Nevertheless, while the marine park is making strenuous efforts to ensure that tourism does not damage the coastal environment, social impacts are becoming evident. Until a few years ago the island received few visitors; this contrasts dramatically with today's position where more than 10 times the population size visit annually. While the reefs may still be some way from reaching their biological carrying capacity for tourism, the island may soon reach a social carrying capacity beyond which the benefits of tourism are bought at an escalating social cost.

Cross-sectoral planning

Many environmental problems and issues are not unique to the tourist industry (e.g. waste disposal, erosion), but common to several development sectors. Hence, there is an increasing need to plan and manage tourism not in isolation from, but as an integral part of economic development in general. Experience elsewhere points to the benefits of integrated management plans (Pernetta & Elder 1993; Price *et al.* 1993), to help ensure convergence between conservation and development. Under these regimes protected areas are seen as an integral component.

Immediate socio-economic factors largely determine which resort atolls and islands are to be developed in the Maldives, as space is currently not a constraint and the environment is still mostly pristine. However, the number of islands developed may gradually increase without full awareness of possible consequences, either to tourism or other forms of economic development. Impact from resorts relates not only to the nature and size of each resort, but also to the total number of islands developed. A step towards optimising coastal use in the Maldives might be zoning of the entire Maldivian environment and associated human uses. Further evolution of the protected area system already in operation is also desirable. In any zoning plan, at both the micro and macro (i.e. national) scale, environmental, social and economic concerns will need to be well integrated.

Effective integration of management across institutional barriers has often been very difficult to achieve, especially with regard to the coastal zone where land and sea are generally the responsibility of different government departments. Small island developing states, such as Saba and the Maldives, have a much greater chance of attaining fully integrated management because land and sea are more intimately connected and government responsibility less divided (Roberts *et al.* 1995).

Enforcement, politics and institutional factors

In the Maldives, environmental work is undertaken principally by the Ministry of Planning, Human Resources and the Environment, and the Marine Research Section of the Ministry of Fisheries and Agriculture. For issues relating to tourism (e.g. resorts) there is collaboration with the Ministry of Tourism. However, shortage of physical and human resources is a problem nationally, for example in applying environmental standards and undertaking EIAs, especially underwater biological components. Increasing use carries with it growing demands for resources to police and enforce regulations.

Without adequate enforcement or regulations marine protected areas can be virtually useless. Unfortunately this is the case for many of the more than 100 designated marine protected areas in the Caribbean. However, tourism revenue can help support protection measures. Saba represents a regional model for how marine parks can be made to function more effectively. Fees from recreational users go directly towards supporting the park and there are daily boat patrols to collect fees, inform visitors of park regulations and prevent violations. Surveys of willingness to pay by scuba divers for marine park management have shown that the amount visitors are willing to pay is substantially greater if the fees go directly to the park rather than being routed through government (A. Smith, pers. comm.). This model of management supported by fees from users is now being taken up by many protected areas in the Caribbean.

Carrying capacity and sustainability

Overall, reefs in the Maldives and Saba are in good condition (Table 9.4). In the Maldives the impact of tourism includes some decline in reef quality and conflicts between tourism and other sectors (e.g. fisheries); in Saba the impacts are predominantly social. Clearly, it is important that the effects of tourism do not foreclose other coral reef uses, and vice versa. Such issues are critical to the sustainability of all human activities dependent on reefs. To ensure the long-term sustainability of tourism based around coral reefs a much greater attention must be focused on carrying capacity. Estimates have been made of carrying capacity based on the direct effects of tourists on reefs, and place sustainable use at between 4000 and 6000 scuba dives

Table 9.4 Summary comparison of the Maldives and Caribbean (Saba) concerning the coral reef environment and tourism.

Feature or issue	Maldives	Caribbean/Saba
Reef environment	26 geographic atolls and *c.* 1200 coral islands of 186 km² land area	One island of 13 km² (Saba); 27 countries, >1000 islands
Diversity of reefs*	High	Low
Environmental status of reefs	Good nationally, but reef state declining towards Malé	Moderate to heavy deterioration in Caribbean from natural and human impacts; human impacts on Saba's reefs still very low
Nature and extent of tourism	Diving, snorkelling, boating, fishing; 74 islands used for tourism	Saba – low density environmental tourism. Caribbean tourism varies widely from mass to specialist interests
Extent of conservation	National and international measures considerable, but constrained by physical and human resources	Saba is considered an excellent model for sustainable tourism. Caribbean has many marine protected areas but over 65% poorly managed
Short-term prospects for tourism	Favourable	Favourable
Medium- and long-term prospects for tourism	Moderate to good, but increasing need to implement environmental measures in tourism planning and operations; the consequences of climate change and sea level rise of much concern	Good but environmental limits to tourism surpassed in some areas and rapidly being approached in others

*Coral reefs are high diversity systems. These figures are relative and while Caribbean reefs are low diversity for this system they represent the highest diversity shallow water ecosystem in the Atlantic Ocean region.

per site per year (Dixon *et al.* 1993; Hawkins & Roberts 1997); estimates of limits to development have been made based on these (e.g. Hawkins & Roberts 1994). These are rough and ready estimates and carrying capacity figures will not be fixed but will depend on local reef characteristics, environmental awareness and diving experience of visitors (Hameed 1993; Hawkins & Roberts 1997). The cumulative effects of infrastructure development on the reef environment have scarcely been addressed to date and require concerted research effort. Healthy reefs are likely to better withstand adverse effects of global warming and sea level rise than heavily degraded reefs. This may well become an increasingly important factor influencing the sustainability of reef uses on low-lying islands such as the Maldives.

Mass tourism is incompatible with conserving the diversity, productivity and beauty of coral reefs. The Maldives and Saba have so far avoided the perils of mass tourism. To ensure that they remain this way it is vital that issues of carrying capacity and sustainability are tackled now and cautious long-term limits to growth determined. In this way both reefs and people will continue to coexist well into the future.

Chapter 10: A Century of Change in the Central Luangwa Valley of Zambia

Joel Freehling and Stuart A. Marks

The environment has suffered more neglect at the hands of social scientists than any other comparable subject . . . [I]t has fallen to natural scientists to understand environmental change without recourse to the methods and analytical tools of social science [and as a result], . . . thinking about the environment has become divorced from social and economic theory. (Michael Redclift 1987)

10.1 Introduction

The Valley Bisa of Zambia and their relationships to and uses of wildlife suggest important lessons for conservation policy and theory. Since descending into the agriculturally marginal Central Luangwa Valley several hundred years ago, Bisa livelihoods have depended upon wildlife. In more recent times these biotic resources have become of interest to others, who in pursuit of their goals have increasingly placed restrictions on residents' uses, even putting the Bisa at risk. Outsider tactics have included: trade in slaves and ivory, restrictions on traditional methods of taking wildlife, alienation of large parcels of formerly occupied land and its resources, increasing restrictions and surveillance of local wildlife offtakes, including progressive costs to individuals, households, and lineages if apprehended, together with the loss of autonomy and self-determination in the allocating of manpower and resources.

Using primary data on Bisa life and livelihood collected during the past 30 years complemented with impressions of early explorers, hunters, and colonial administrators, this study examines how governmental directives have affected Bisa wildlife use, social groupings, and cultural practices during the past century. These data suggest that regulations instituted through centralised management policies have largely failed to preserve wildlife. Instead, more recent schemes have contributed to progressive declines in most wildlife populations and in the standard of living for rural people. We argue that these declines in rural human welfare and wildlife numbers are causally linked.

10.2 The Nabwalya study area

The Valley Bisa are a scattered population of thousands (now estimated between 7000 and 8000 people) currently residing in the Munyamadzi Game Management Area (GMA) in eastern Zambia. Here they live along the banks of the perennial rivers flowing into the Luangwa River. This 3100 km² corridor is located between the Luangwa River and the Muchinga Escarpment on its east and west boundaries respectively, and between the North and South Luangwa National Parks (Fig. 10.1). Under both colonial and independent governments, the Valley Bisa lost land and hunting rights through the gazetting of these vast Luangwa game reserves in the late 1930s (which in 1973 became national parks), and through the classification of their current living estate as a GMA in the 1950s. Yet historically, wildlife was and continues to be an important component of household livelihood. Wildlife is also a critical source of animal protein since the presence of tsetse fly prohibits livestock production.

Within many Bisa lineages (the basic unit of social organisation), 'hunter' and 'hunting' are culturally important roles and activities for some men. However, both these roles and activities have undergone major transformations this century. Today, legal hunting is limited to those

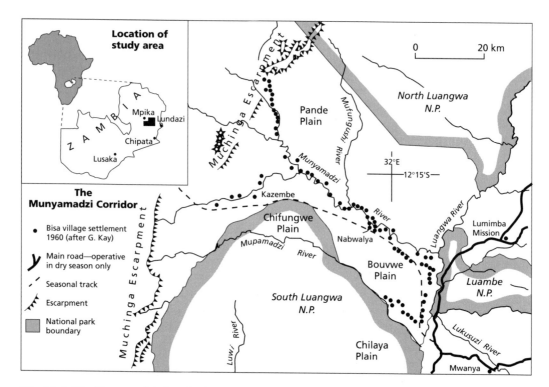

Fig. 10.1 The Munyamadzi Corridor in the Luangwa Valley. (From Marks 1984.)

individuals holding area access permits and specific hunting licences. Pro-hibitively expensive for most valley residents, these permits are more readily purchased by affluent outsiders who can afford the national taxes levied on hunting. Even where concessions were made to local residents, these permits were limited in number and also applicable to civil servants and town residents who were often more strategically situated to acquire them.

In 1987, the Munyamadzi GMA became part of ADMADE (Adminis-trative Management Design), Zambia's 'community-based' wildlife man-agement programme. Under this initiative, the Sub-district Management Authority (a committee composed of the local chief, wildlife unit leader, civil servants, and a few selected residents) is given a portion of the revenues generated from safaris and from access fees. Provided these funds are accessible, they may be expended on 'approved' development projects. Officials expect that in making these funds available locals will refrain from taking wildlife and report all game violations to authorities. However, a project evaluation in 1992 showed that very little money (less than 2%) left governmental control (DeGeorges 1992). Other external reviews contend that the Bisa seem unwilling to forego their dependence on wildlife in exchange for development projects such as schools, dispensaries, or grinding mills (Gibson & Marks 1995). Many residents note that private enterprise and government agencies provided these 'goods' previ-ously without attaching such prohibitions on their activities.

At varying intervals since 1966, the Valley Bisa and their interactions with wildlife have been studied in a 60 km^2 area surrounding the residence of the Chief of Nabwalya (Marks 1976, 1984, 1994). Located near the centre of the Munyamadzi GMA, this tract, designated as the Nabwalya Study Area, contains an admixture of major habitats and wildlife species found throughout the GMA and valley. As the residence of the government-recognised chief, the Nabwalya Study Area is the main arena for transactions between the resident community and the outside world. During the 19th century, Nabwalya was located along a major trade route into the interior of Central Africa. This site later served as a colonial administrative post (1899–1908), and as a station for trypanosomiasis research until it was abandoned for a more favourable location on the adjacent plateau. Consequently, written records on both the Bisa and their surrounding environment date from the late 19th century (Vail 1977). In the 1980s, Nabwalya became once again a major target for development programmes involving the Zambian government and international donor agencies. Today, a large contingent of government civil servants, including school teachers, clinic staff, court clerks, agricultural assistants, wildlife officers, and Peace Corps personnel maintain residence there.

The research methods utilised over the past 30 years have altered to accommodate the changing circumstances of the Valley Bisa and the stricter enforcement of game laws under ADMADE. In the 1960s and

1970s, studies relied upon normative anthropological techniques such as participant observation, interviews, and surveys. Methodology shifted in the 1980s and 1990s as several Bisa became actively involved in data collection, in conducting interviews, and in reflecting upon and interpreting their behaviours and activities. Contemporaneously, enhanced local enforcement of game laws in the late 1980s under ADMADE seriously constrained residents' willingness to share information about their uses of wildlife. Local Bisa also began to take game more surreptitiously. As a consequence, recent records on local wildlife utilisation are more conservative and less representative than were previous accounts.

10.3 Trends in wildlife numbers

Archival accounts of early British administrators, the published manuscripts of sports hunters and European travellers, together with the recollections of Bisa residents are the basis for profiling the population trends for the major wildlife species in the Central Luangwa Valley (Fig. 10.2). These data suggest that the relative abundances of most mammals have undergone oscillations during this 100 year time frame. A significant downturn in population numbers for most species occurred with the rinderpest epizootic of the 1890s. Wildebeest and buffalo were

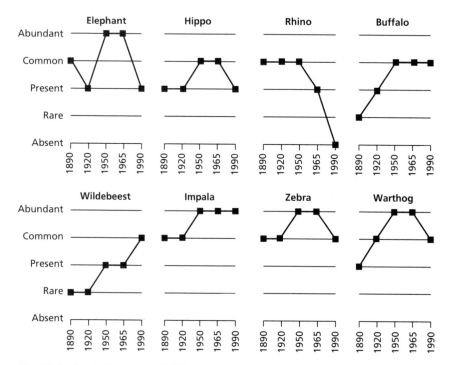

Fig. 10.2 Trends for major wildlife species in Central Luangwa Valley, 1890–1990. (From various manuscripts including FAO 1973; Marks 1976, 1984, 1996.)

particularly affected and suffered dramatic declines in numbers (Hall 1910; Melland 1938). Although not as affected by this disease as some other species, elephant declined due to the flourishing ivory trade carried out by foreigners and locals alike during the late 19th century. The ivory trade was linked to the traffic in slaves, the latter transporting the ivory to the coast. For their protection during this time, human residents lived within stockaded villages strategically located along the trade routes, or led a shadowy existence, moving in small bands and hiding in the thickets and in the caves of the escarpment.

Most wildlife species have subsequently increased from the low population levels reached at the turn of this century, with their numbers peaking on the study area at different times. A victim of the massive commercial 'poaching' during the 1980s, the black rhinoceros is the only species to become extinct within the Central Luangwa Valley. In contrast, buffalo were slow to recover from the epizootic but have remained plentiful for the past 40 years. Also, wildebeest, a victim of rinderpest and rare for most of this century, have become abundant during the past decade. Elephants have suffered a decline similar to that of rhino but one less permanent. Their populations increased gradually for most of this century and reached their zenith in the 1960s and 1970s, then their numbers fell precipitously due to the large-scale commercial and illegal slaughter during the late 1970s and throughout the 1980s (Leader-Williams *et al.* 1990). While elephants still occur at Nabwalya, residents contend that the population is primarily composed of young bulls, cows and calves. They also note that the proportion of tuskless elephants has risen (Jachmann *et al.* 1995). The increase in elephant numbers anticipated under ADMADE protection will bring renewed conflicts with local farmers who will not readily accept crop losses to these large mammals.

Similar to the trend for elephants, most wildlife species were at their highest levels in the Nabwalya area during the 1960s. While correlations remain speculative, several socio-economic changes in the Bisa community occurred contemporaneously with this observed abundance. For instance, for some years prior to these increases in wildlife, local hunters were comparatively few in number. Most adult Bisa men (up to 70% in many years) were employed as migrant labourers outside of the valley. Consequently, human predation on local wildlife was at relatively low levels. Furthermore, residents used muzzle-loading guns rather than traditional weapons, such as snares, pitfalls, a variety of traps and poisoned arrows, to kill mammals. Muzzle-loading guns had recently become available to the Bisa and were a major objective which men sought to acquire from their wages and employment elsewhere. Once brought back to rural villages, these weapons became lineage heirlooms and potent symbols of masculine power and prestige for their 'owners'. Utilising these guns, hunters predominantly took male prey and their overall offtake of wildlife was limited by the weapon's inefficiencies. Also, during the 1950s to 1970s, control of marauding elephants in fields was a priority for wildlife officials who distributed the meat from these elephant kills widely among residents. This meat provided a substantial subsidy for most households at Nabwalya. In 1966–67, the amount of fresh meat consumed per adult was estimated at 92 kg. Circumstances were soon to change.

In 1964, the completion of a vehicle track down the escarpment and into Nabwalya greatly facilitated contact with the District offices at Mpika. This road, while providing closer markets for game meat, also extended the presence of government within the Central Luangwa Valley. Completion of this road into the Munyamadzi corridor can be linked to the subsequent decreases in wildlife. Likely causes for this reduction in wildlife numbers are the following agents in various combinations: increasing human presence and changing land uses; a deteriorating national economy; wildlife commercialisation by all parties and the facilitation of transactions possible with motorised transport; changes in government policies and practices; intermittent and prolonged droughts; and the increasing differentiation and marginalisation of some residents locally and regionally.

Wildlife counts since the 1960s illustrate the cumulative effects of these influences. These numbers also reveal significant aggregate and specific declines for wildlife species adjacent to Bisa settlements. A comparison of counts taken in 1966 and 1988, using the same methodology, suggests a 50% decline in wildlife abundance (Fig. 10.3). These wildlife counts, taken every decade by or in conjunction with local hunters, are based upon species observed per unit of search time, in minutes (Fig. 10.4). Search is defined as that time during which an individual is actively looking for

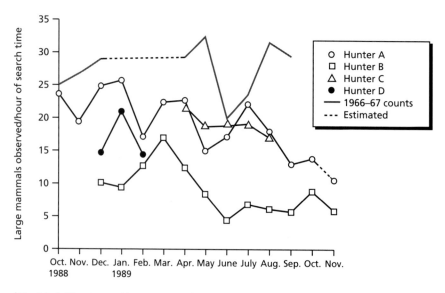

Fig. 10.3 Numbers of large mammals observed per hour of search time 1988–89 by four local hunters. (From Marks 1994.)

Fig. 10.4 Two local hunters recording observational data on wildlife while making a foray around their villages, 1993. The hunters are carrying their muzzle-loading guns for protection. (Photo by Stuart Marks.)

wildlife. Search begins when the individual leaves his village and continues until his return. This time is separate from other activities which an individual may engage in while afield, such as hiding, stalking or butchery. The high numbers of wildlife observed during the 1960s came from transects made very close to Bisa villages, in the same area assessed by hunter B during 1989. The higher counts for both hunters A and C are based on their searches further afield (Marks 1994).

Similarly, a comparative study of the time between a hunter's departure from the village and his first encounter with wildlife also suggests an absolute decline in numbers around villages. Indicative of a widening space separating wild mammals from Bisa households, these studies denote a four-fold increase in time to first encounter over these four decades (Fig. 10.5). While the absolute numbers of wildlife within this system

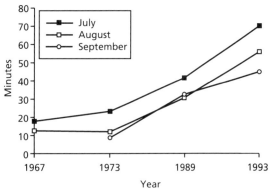

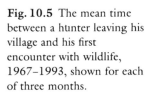

Fig. 10.5 The mean time between a hunter leaving his village and his first encounter with wildlife, 1967–1993, shown for each of three months.

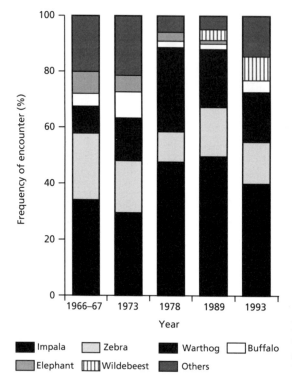

Fig. 10.6 The species composition of wildlife encounters by Bisa hunters, 1966–1993.

Impala   Zebra   Warthog   Buffalo
Elephant   Wildebeest   Others

have diminished, its composition remains largely unchanged. For this time frame, small prey such as impala, warthog, wildebeest, and perhaps zebra, have shown modest increases. There have been corresponding decreases for the larger species such as elephant and hippo (Fig. 10.6).

10.4 History of Bisa hunting

An examination of life histories for Bisa hunters suggests some aspects of their traditions continue as before (Fig. 10.7), yet this apparent succession masks significant discontinuities and transformations in hunting processes which have occurred. The story begins with the break-up of the traditional hunting guilds, colonial economic policies, and labour demands for European enterprise at the close of the 19th century. The chronicle ends with droughts and a faltering national economy in more recent years.

A major transformation began when the tax and labour requirements of the colonial regime forced men to seek employment outside of their homeland (Chipungu 1992). Conscription became necessary to field the teams of carriers ferrying supplies and messages between far flung government outposts. Local labour was also essential for new European initiatives in agriculture and in business. Expanded underground mining in the Northern Rhodesian copper belt after the 1920s required continuous and

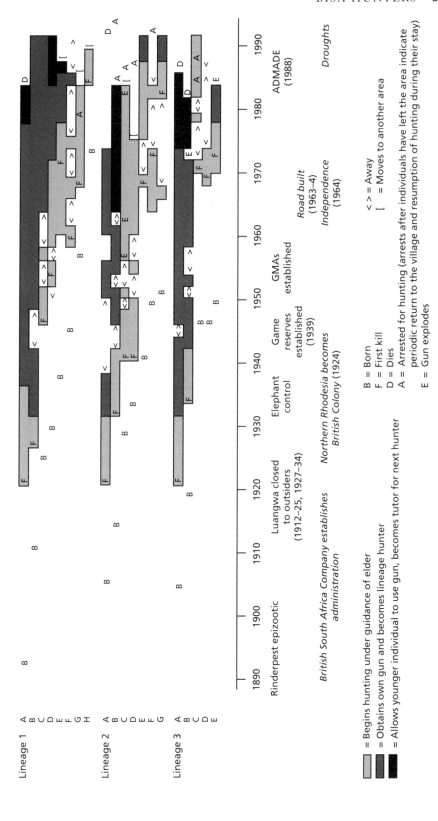

Fig. 10.7 Significant events in the hunting careers of individuals in three Bisa lineages, 1890–1993.

large numbers of unskilled workers. To meet these requirements, colonial officials promulgated policies forcing men to take employment in these growing ventures. Labour migration for African men became an expedient to meet these new European demands.

Through their participation in migrant labour, some men were able to acquire the capital needed to purchase a muzzle-loading gun, a new weapon that caught on with explosive force. These weapons were readily adopted into village life and lineage customs, and rapidly replaced the more easily acquired and locally produced technologies, such as bows, poisoned arrows, game pits, and a wide assortment of traps and pitfalls. Although purchased by and licensed to individuals, these guns became lineage property and important heirlooms for men.

As more egalitarian methods gave way to centrally controlled techniques, guns and their control became a potent symbol of the changing social order. As the main hunting weapon, muzzle-loading guns symbolised masculine prestige, privilege, and patronage. Only a few men were allowed use of these weapons and only after tutelage under an elder. Presiding over lineage weapons and access to wildlife, some elder men consolidated their control over others, especially women and younger men. Youths were recruited into professional hunting cults by having their dreams interpreted by elders and by serving as an apprentice under an adept. During this apprenticeship, young men learned the techniques, lore, and mores of lineage traditions. They used their instructor's guns during forays and participated in the distribution of meat that granted hunters a visible and privileged position in society. Given time and experience, initiates acquired prescriptions and trophies attesting to their bush and survival skills. Furthermore, they could look forward to inheriting custodianship of the lineage's weapon and to developing followers and clients of their own (Marks 1979). The recruitment of successive generations of hunters within three lineages indicates the continued strength of these traditions (Fig. 10.7).

Changing circumstances were to undermine the social cohesion fostered by these small-scale kinship ties. Participation in urban employment brought exposure to new skills and relationships. The monetarisation and market orientation of the national and local economies, together with increasing literacy, contributed to progressive factionalism within villages. The diverse experiences gained by individuals living and working in other environments eventuated into differing expectations for young and old, between the haves and have-nots, between men and women, and to increasing hostility, crime and distrust among all community residents. Throughout the 1960s and 1970s, the integrity of lineages as the main residential and social unit was implicit in the well-defined village boundaries. In the mid-1980s, the death of a chief who had reigned for more than 50 years, and the prolonged struggles over his replacement signalled an end to this well-defined social order. Exacerbated by the decline in the local economy, the scattering of smaller nucleated settlements across the landscape reflected growing austerity and social isolation. Isolated from neighbours and relatives, individuals and households have become increasingly competitive for scarce resources. Today each residence remains shielded from the scrutiny of its neighbours by stretches of bush and fields.

Game laws furthered these processes. Previously, hunting rights were part and parcel of residence within the community and came with lineage membership, but hunting restrictions eroded the control of lineages over wildlife and usurped local use rights. By making possession of wild game illegal for those not holding permits, game laws drove local hunting processes underground. The expanded ranks of wildlife scouts have intensified households' use of snares, a traditional technique which kills prey effectively and leaves little evidence. However, unlike guns, snares indiscriminately take game of any age and sex which greatly reduces the ability of the species to reproduce. The loss of earlier controls on local hunting processes also opens these activities to all residents, particularly the young seeking ways to circumvent the authority of their elders. When hunting was more locally managed, requiring tutelage, access to lineage weapons, and sponsorship by elder men, only 25% of men in residence actively pursued wildlife (Marks 1984). As these restrictions no longer apply, snaring and secrecy abound (Firey 1960).

Further, prey captured privately by snares no longer provides public mechanisms for demonstrating kinship relationships, which were at the heart of earlier survival strategies in this harsh environment. In a previous era, each ritually important kill required a public celebration in remembrance of lineage ancestors. This openness offered an occasion to express and to commemorate lineage cohesion.

Social dislocation began during the colonial period and continued in more modern times, thus altering the nature of Bisa wildlife use. With the loss of lineage authority, hunting techniques and targeted species have changed, and these changes have been destructive to wildlife and human livelihood alike. Such transformations reinforce Michael Painter's (1995) conclusions that the solution to environmental destruction lies not in reorienting the relationship between 'man and beast' but between the different segments of human society competing over diminishing resources. In the next section, we address this issue.

10.5 The significance of wildlife to users

Cumulatively, the social processes outlined above have led to progressive inequalities in wealth and in nutritional status among Bisa households. In the 1960s and 1970s, game meat was shared more evenly within the villages and was more widely available to individuals (Fig. 10.8). Larger

Fig. 10.8 Bisa men salvaging meat from a recently killed zebra carcass after chasing away a pride of lions, 1973. The younger man butchers while the elders instruct. (Photo by Stuart Marks.)

mammals, such as buffalo, were differentially targeted by local hunters, comprising 38% of their kills. Moreover, culling of marauding elephants contributed significantly to household larders, providing nearly 40% of the game meat consumed in 1966–67 (Marks 1976). Lineage membership and kinship ties afforded most individuals rights to the meat from these offtakes.

These processes crumbled in the late 1980s with the imposition of more strenuously enforced game laws under ADMADE and the faltering of the national and local economies. Under increased vigilance from scouts, hunters were forced to concentrate on smaller mammals and to take game more discreetly (Gibson & Marks 1995). Smaller mammals, such as impala and warthog, are hoarded and apportioned among extended family rather than dispersed widely as are larger mammals such as buffalo. Furthermore, an increasing human population and fewer employment opportunities elsewhere mean greater competition for the diminishing resources in the valley.

The effects of these changes are apparent in studies on the percentage of game meat consumed in Bisa meals. Bisa repasts consist of a flavoured stew, described here as the relish dish, combined with boiled maize or sorghum meal which forms a pudding-like consistency. In 1966–67, a nutritional study showed that game meat comprised 38% of daily relish dishes consumed in 19 households from three villages (Marks 1976). Later studies conducted between 1973 and 1978 found that the percentages of wild meat as the basic component of relish dishes were 32 and 41% respectively (Marks 1984). More recent studies conducted in 1988–89 showed that the proportion of game meat in these dishes had decreased to 29%. In addition, this latest survey indicated considerable disparities in animal protein intake between various segments of the local population. Observations on household relish dishes showed that meat and fish as a proportion of total annual dishes consumed among various households ranged from 61% to a low of 14%. At the lower end of this spectrum were households headed by women and wives in polygamous marriages.

Adding to this impoverishment is the loss of subsidies from wildlife culls carried out by officials. Today, the meat from carcasses of wildlife killed because of their depredations on cultivated lands (particularly elephants) is dried and carried elsewhere as state property. Consequently, many Bisa are nostalgic about the past and disillusioned with these actions by state authorities. As noted by one Bisa elder: 'Government a long time ago was good . . . [It] gave the elephant control guard a gun and bullets and it was easy for him to kill elephants. People then shared the meat. When an elephant is shot now the meat is taken [outside] and not locally shared.' Currently, there is no compensation for local loss of life, limb, or livelihood.

The economy has also worsened appreciably in these intervening decades. Declining copper revenues (which provide 80% of Zambia's foreign exchange) during the 1970s and increasing costs of imports in the 1980s have adversely affected the vast majority of Zambians. Falling standards of living and rising death rates are apparent in national statistics. The International Fund for Agricultural Development (IFAD) labelled Zambia in 1992 as having one of the highest rates of rural poverty in the world.

Although declines in household livelihood are evident for most Bisa families, it is not true that fewer animals are being killed in the corridor. Visible wildlife is an important source of revenue for national conservation and development schemes. Deteriorating economic conditions place considerable pressures on managers to transform this 'wealth on the hoof' into monetary income. These concerns are evident in the minutes of the 1991 Munyamadzi Management Authority: 'As a measure of strengthening revenue generation base for the Munyamadzi and as a measure of good wildlife management by reducing excess wildlife species to bring them in line with habitat potential, [a] culling programme will take place in Munyamadzi this year . . . [with the] following species cropped: Bushbuck 2, puku 2, waterbuck 2, warthog 8, buffalo 9, wildebeest 10, hippo 18, impala 48.' Despite restrictions on Bisa offtakes and noticeable declines in most wildlife populations, management recommended a rather substantial culling (Marks 1996). In total, this offtake represents nearly 17 000 kg in carcass yield, roughly twice the amount allocated for local Bisa licences (Table 10.1). Premised upon providing inexpensive game meat for sale to residents in compensation for denying them access, this project has not lived up to its stated objectives. Residents note that most of the culled meat is sold to government civil servants and to outsiders. Among these groups, the meat commands a much higher price than among residents, who know that sloppy accounting procedures allow scouts and other governmental agents to pocket most of this money that should go instead to support community projects.

These statistics, however, ignore the large offtakes by licensed hunting, the primary revenue source for the government and for ADMADE. With its

Table 10.1 Culling and resident hunting quotas, Munyamadzi Game Management Area, 1991. (From the minutes of the Munyamadzi Management Authority.)

	Culling quotas		Resident hunting quotas	
	Number allocated	Carcass yield (kg)*	Number allocated	Carcass yield (kg)*
Buffalo	9	2871	8	2552
Bushbuck	2	47	6	140
Bushpig	—	—	10	452†
Hippopotamus	18	11 340	5	3154
Impala	48	1600	12	400
Puku	2	81	8	325
Waterbuck	2	225	4	449
Warthog	8	361	10	452
Wildebeest	10	406‡	—	—
Zebra	—	—	5	656
Total	99	16 931	68§	8580§

* Estimated by averaging carcass yields for both sexes given by Marks 1976.
† Estimated yield 45.15 kg per animal (value given for warthog).
‡ Estimated yield 40.61 kg per animal (value given for puku).
§ Not included are quotas for 10 baboons (which are not eaten by residents) and 10 unspecified 'birds'.

rich stocks of prized mammals the Munyamadzi corridor is one of Zambia's most important safari hunting areas. In 1994 alone, access fees and hunting licenses issued for the corridor generated more than US$145 000 (ZNPWS 1994). Theoretically most of these funds are destined for use in the corridor for community development and in wildlife management, but diverse needs and interests compete for these scarce resources. Furthermore, other official offtakes also occur in the valley. Whereas the offtakes from safari hunting are monitored and limited, a number of special licences are issued each year for individuals to hunt in the corridor. The animals killed under these general licences are not recorded.

Wildlife though is more than protein, calories, and revenue; it embodies symbols, values, and meanings that are culturally coded and which are also subject to contestation by the competing parties in this system. According to sociologist Michael Redclift (cited in Painter 1995): '[T]he environment, whatever its geographic location, is socially constructed. The environment used by ramblers in the English Peak District, or hunters and gatherers in the Amazon, is not merely *located* in different places, it means different things to those who use it' (emphasis in original). This conclusion is equally applicable for wildlife, whose meanings vary culturally. For instance, the panda, one of the most recognised symbols in conservation, is used by the World Wide Fund for Nature (WWF) to represent its activities throughout the world, yet Sally Zalewski (cited in Pearce 1991) offers a powerful anecdote about the cultural specificity of this symbol. She recounts an incident in which African villagers seeing the panda image during conservation seminars, inquisitively asked 'where it lived, did it exist in their country, and could it be eaten?'. Clearly the panda meant something different in this African community than for Western conservationists.

Similarly, wildlife signifies discordant cultural idioms of the various parties present in the Nabwalya area. For the state, wildlife represents flows of revenue and wealth maintained through its control of licences and access. This wealth can be utilised to provide roads, schools, clinics, foreign exchange, personal enrichment, and is symbolised by key concepts such as 'development' and 'progress'. For officials stationed in the corridor and a few upwardly mobile Bisa, wildlife also represents salaried employment, political patronage, and perhaps access to influential outsiders. However, self-proclaimed 'poachers', as many residents reflectively refer to themselves, have other uses for wildlife. It is not clear that they are willing to trade their ways for material progress defined by others (Gibson & Marks 1995; see below).

Wildlife for these Bisa is reconfigured in a whole range of resource categories, among which are capital, land, labour, information, time, and identity. Transference of items among these categories are essential for the continuation of Bisa livelihoods within this marginalised landscape where few resources are available for generating wealth. For example, a local hunter may invest some of his time and labour to discern mammal movements (environmental information) and village scout locations (social information). He can then use this knowledge together with additional time and labour to kill an animal (a product of the land). The hunter can convert this animal into social or monetary capital, either through distributing portions among his relatives or selling other shares to neighbours and outsiders. In the process, he may further clinch his local reputation as a patron (identity) upon whom others can depend for some of their needs, whether in cash or in meal forms. Furthermore, his activities, if illegal, may visibly heighten the boundaries between an insider's and outsider's construction of livelihood and appropriate behaviour, providing an identity for himself and for those with whom he consorts. Thus, wildlife is utilised not only as food, but to cement social bonds, to provide meaning and identity, and to maintain ties to ancestors and the past. All of these possibilities are culturally important things not accessible or accountable through monetary currencies.

In systems with contested resources such as the Munyamadzi GMA, the persistence of local wildlife use highlights the difficulties of operationalising Western notions of conservation in other areas of the world. Interests not only compete for the right to use these resources but also over how to use them and towards what ends. In the Nabwalya area, the government has launched a programme to discredit local uses and to force residents to accept modern management practices, yet the application of force and use of incentives have proven unsuccessful in eliminating 'illegal' activities. 'Poaching' continues in this and many other rural areas in Africa despite the risks of arrest, fines, and physical abuse.

10.6 Conservation initiatives and law enforcement

For more than a century local uses of African wildlife were derided as unsustainable by Westerners and the users portrayed as slaves to tradition and custom. In tandem with this portrayal, restrictive laws sought to separate locals from their adjacent resources (Marks 1984; Anderson & Grove 1987; MacKenzie 1988). This separation rendered both wildlife and people more manipulable for external purposes. As resources diminished, governmental interventions became more forceful and repressive, and its rhetoric increasingly combative and disparaging. African offtakes were characterised as destructive and demonised as 'poaching'. Furthermore, managers have sought to supplant local hunting with more profitable activities, such as culling programmes, yet these modern uses are frequently detrimental to the very local communities they are attempting to improve (Leach 1994). Today similar misconceptions are embedded in programmes such as ADMADE and other community-based wildlife programmes. While perhaps more enlightened than most projects, AD-MADE still assumes that rural communities are wasting resources which can be better managed by an outside agency.

A useful framework for testing the assumptions underlying 'community-based' wildlife management programmes against empirical data has been suggested for the Bisa case (Gibson & Marks 1995). Using economic decision modelling and game theory, hypotheses can be generated with reference to how rural residents should respond under increased threats of arrest and under more equitable sharing of the licence proceeds. This model suggests that as enforcement gets stronger, arrests should initially rise, but should then decline as hunters find it in their interest to refrain from illegal activities. Such decisions are affected by the material benefits gained from the re-distribution of revenues and from the associated development projects that occur with compliance.

Tables 10.2 and 10.3 document the numbers of arrests following the initiation of ADMADE in the community and the deployment of local scouts. Contrary to expectations, arrests for wildlife violations continued to rise. Moreover, hunting offtakes reported by three local hunters did not decrease between 1988 and 1993, despite the risks of arrest and the inducements of government-sponsored development projects (Fig. 10.9). These data indicate that 'while the benefits and increased enforcement have changed locals' tactics and prey, the programme has not convinced

Table 10.2 Law enforcement by wildlife scouts, Munyamadzi wildlife subauthority, 1988–93. (From Gibson & Marks 1995.)

Year	Arrests	Convictions	Pending	Dismissed	Escaped
1988	4	3	0	0	1
1989	5	5	0	0	0
1990	8	7	0	0	1
1991	11	9*	0	0	2
1992	21	10	0	9	2
1993	19	4†	12	1	2

* Includes two punishments handed down by the chief.
† Includes three punishments handed down by the chief.

members of the rural community to conserve animals; neither has it stopped their illegal hunting practices' (Gibson & Marks 1995). Such independent assessments of government projects are critical for evaluating the effectiveness of these programmes in the future.

Although not altering their behaviours as expected, ADMADE does exert a considerable effect upon the lives of most Bisa. The number of arrests, complaints of increased harassment, physical violence, and economic decline all attest to this effect. Residents contend that these policies are inappropriate because they target the innocent and limit historically sustainable practices. 'My uncle taught me both to hunt with a gun and with wire snares, just in case the situation changed as it has today with many game guards,' explained a Bisa elder. 'Their presence, which prevents us from pursuing game,' he continues, 'is contrary to the wishes of our [ancestors]. This is a bad situation. Our ancestors settled in this country of wild animals, killed, and fed upon game, yet the animals never decreased. That's why their sons and grandsons have got on with what their forefathers used to do . . . Outside people tell us "you have finished the animals". Now I ask you, how can a muzzle-loading gun finish animals when it fires only once in comparison with semi-automatic rifles? Rather those who come from [outside] for business are the ones who have diminished our animals.' Discursive conflicts over the causes of wildlife 'destruction' remain and will probably persist into the future. However, given the disparities in power between the local community and outside

Table 10.3 Law enforcement by the Zambian Wildlife Department, Luangwa command, 1985–90. (From Gibson & Marks 1995.)

Year	Arrests	Convictions	Acquittals	Pending
1985	112	97	8	5
1986	149	116	5	28
1987	130	119	4	7
1988	155	137	2	16
1989	165	136	8	21
1990	197	183	7	7

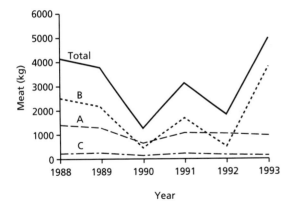

Fig. 10.9 Kilograms of meat taken by three local hunters. (From Gibson & Marks 1995.)

agencies, the local economy will probably continue to decline as will the wildlife numbers.

10.7 Conclusions

The Bisa and their reactions to state-imposed programmes have much to tell about sustainability and survival, but also about injustice and misguided projects. The Bisa are not ecologically 'noble savages', any more than they are despoilers of the garden of Eden (Århem 1984; Redford 1990). Rather, they are a community of men and women struggling to survive and to sustain traditions that provide them with comfort, a livelihood, and meaning. The debate about use versus preservation continues, and the same rooted causes tend to dominate the discussions. However, the voices of those people living with wildlife must be given standing, even when they contradict the simplistic notions which those living elsewhere hold about the world and about 'wilderness' (Nash 1978; Oelschlaeger 1991). Recently, scholars have described these Western concepts as narrative because they are myths and stereotypes often at odds with reality. Maybe successful implementation of biodiversity conservation at the local level will only be successful when local narratives are considered as compelling as others.

Chapter 11: The Economics of Wildlife Conservation Policy in Kenya

M. Norton-Griffiths

11.1 The problem

Although Kenya is some 580 000 km² in area, only 15% supports continuous agricultural production (Fig. 11.1a). About 80% of the population of 21 million are concentrated in this area (GOK 1994) along with over 90% of all livestock (Norton-Griffiths & Southey 1993). The agricultural and livestock industries are well organised and highly profitable, contributing (in 1994) around 35% of GDP and generating some 56% of all foreign exchange earnings (GOK 1996).

The remaining 500 000 km² of the country consists of rangelands. While some of these rangelands are of high agricultural potential most are arid drylands or semi-deserts. These rangelands support some 4 million pastoralists with their livestock, the majority still following a traditional, nomadic lifestyle.

Conservation policy in Kenya is based primarily on the network of protected areas (PAs), the national parks and the national reserves, most of which lie within these rangelands (Fig. 11.1b). The PAs are of international scientific and conservation interest. Each year they attract literally hundreds of thousands of overseas visitors, they generate vast tourism revenues (US$400–500 million a year) and they attract significant international aid. Nonetheless, the continuing conservation of wildlife in Kenya is beset by many problems and uncertainties.

The scale of the problem is shown by data recently released by the Department of Resource Surveys and Remote Sensing (DRSRS) which has been monitoring the size and distribution of wildlife and livestock populations throughout the rangelands since 1977 (GOK 1995a,b). Kenya has lost 44% of its wildlife over the last 18 years (Table 11.1), while over the same period livestock populations have been relatively stable. It was never the avowed policy of the government to lose half its wildlife, so this clearly indicates a major policy failure.

Closer inspection of these DRSRS figures reveals insights as to what has gone wrong. First, other DRSRS data (GOK 1995c) show that the majority of wildlife, well over 70%, live either permanently or seasonally on the rangelands outside the PAs rather than inside. Second, the PAs are at least partially effective, for losses within them over the last 18 years are 31%

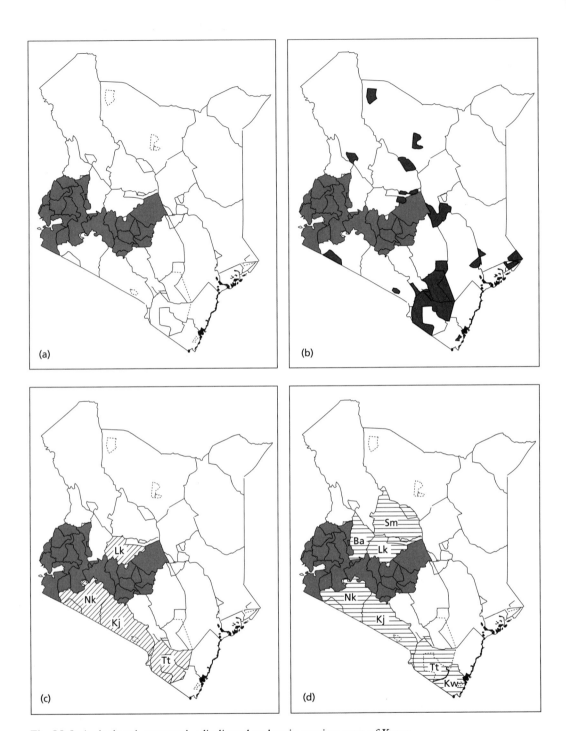

Fig. 11.1 Agricultural, protected, adjudicated and main tourism areas of Kenya.
(a) Agricultural areas (shaded) occupy only 15% of the country, the remainder being
semi-arid or arid rangelands. (b) Most of Kenya's protected areas (dark shading) lie in
the rangelands. (c) Land has been adjudicated in only four rangeland districts: Kajiado
(Kj), Laikipia (Lk), Narok (Nk), Taita Taveta (Tt). (d) The seven most important
districts for wildlife-based tourism: Baringo (Ba), Kajiado (Kj), Kwale (Kw), Laikipia
(Lk), Narok (Nk), Samburu (Sm), Taita Taveta (Tt).

Table 11.1 Population trends for livestock and wildlife on the 18 rangeland districts of Kenya, 1977–94.

		Rate of change per annum (%)	Percentage change in 18 years	*P* value of population trend
Livestock	All	+ 0.60	+ 11.0	0.398 NS
Wildlife	All	– 3.24	– 44.0	< 0.001
Wildlife	Inside PAs		– 31.0	
	Outside PAs		– 48.0	
Wildlife	Adjudicated land	– 2.04	– 30.0	< 0.001
	Unadjudicated land	– 3.36	– 50.0	< 0.001
Wildlife	Tourism districts	– 2.16	– 32.0	< 0.002
	Non-tourism districts	– 4.56	– 55.0	< 0.001

Census data from GOK (1995a), trend analysis by OLS using logged data and dummy variables for each district and each wildlife species.

compared with 48% from outside (Table 11.1). Clearly, the main conservation problem facing Kenya lies with wildlife on the rangelands outside the PAs, for 84% of the total wildlife lost during these 18 years was from outside and only 16% was from inside the PAs. This chapter therefore considers policy options for wildlife conservation on land outside the formal conservation areas of Kenya.

11.2 Property rights and land tenure

Property rights (Bromley 1991) and land tenure are central to the debate. Within the PAs, the government has retained all property rights but has transferred operational control over the national parks to the Kenya Wildlife Service (KWS) and over the national reserves to the appropriate local county councils (Bragdon 1990). Outside the PAs, where 4 million pastoralists live, land is either adjudicated or unadjudicated (Fig. 11.1c). On adjudicated land, the government has assigned property rights either to individual landowners, who accordingly have individual tenure to a single ranch or landholding, or to groups of landowners, who accordingly share among themselves the property right and tenure to a group ranch (Galaty 1980, 1992). These property rights are legally enforceable, so tenure is strong and landowners can, within reason, do what they like with their land (Norton-Griffiths 1996). In contrast, the property rights on unadjudicated land remain held in trust by the county councils on behalf of the landusers. They at best have usufruct rights, based on their traditional lifestyles, but tenure is weak, and with no formal or legally enforceable property rights there are continuous problems over land alienation.

The DRSRS data show how important property rights and land tenure are to wildlife conservation. Wildlife losses have been 30% over the last 18 years in the four districts where most land apart from the PAs is adjudicated (Fig. 11.1c), compared with 50% in the districts where land remains unadjudicated (Table 11.1). Clearly, conservation of the wildlife resource is favoured by secure title to land and enforceable property rights.

Unfortunately there is a downside to both private and group title in that land holdings tend to become split up and sub-divided into smaller and smaller units. There are strong socio-economic forces driving this trend (Galaty 1992), including rising populations, fragmentation of landholdings following inheritance, the need to raise capital, and the fear of being marginalised by stronger groups of landowners. The danger of this for wildlife conservation is shown clearly by data from 32 ranches in Laikipia District ranging in size from 1500 to over 40 000 ha. Wildlife numbers and diversity are significantly lower on smaller than on larger holdings, and wildlife is effectively absent from ranches of under 2000 ha (Fig. 11.2).

Land sub-division is politically highly sensitive in Kenya and deep policy conflicts are apparent. The government, for example, is encouraging the rapid adjudication of land in the rangelands and the transformation of the group ranches into (smaller) individually owned ranches. In contrast the KWS, which is also responsible for managing all wildlife outside the PAs, is trying to discourage land sub-division and instead encourage landowners and landusers to form wildlife associations to jointly manage their wildlife (KWS 1995a, 1996), much as neighbouring landowners in Namibia, Zimbabwe and Europe do for shooting and hunting.

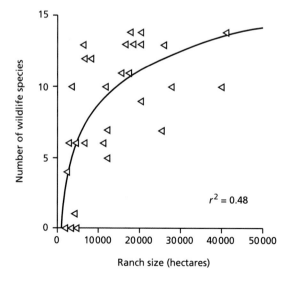

$r^2 = 0.48$

Fig. 11.2 Influence of ranch size on the number of wildlife species present on the ranch.

11.3 The influence of tourism

Tourism is a major bastion of conservation policy in Kenya, second in importance only to the network of PAs. Tourism is meant to provide the flow of benefits to support conservation activities and in general it has a beneficial influence. Wildlife losses in the seven districts which between them account for over 95% of all wildlife visits (Fig. 11.1d) are 32% over the last 18 years compared with 53% in the districts where tourists barely venture (Table 11.1). Clearly, however, even these vast revenues generated by tourism (some US$500 million annually) are not adequate in themselves to ensure the conservation of the wildlife resource on which the industry depends.

The PAs generate revenues from entry, bednight and concession fees. Revenues from the national parks accrue to the KWS while those from the national reserves accrue to the county councils, in each case ostensibly to meet both the direct costs of conservation and for social investment at national or district levels. Most of the remainder of the vast tourism revenues are captured by the tourism industry (the travel, transport and hotel operators) and only a tiny amount (as little as 1%) accrues to the landowners and landusers on whose land the majority of the wildlife resides (Douglas Hamilton 1988; Talbot & Olindo 1992; KWS 1995b, c; Norton-Griffiths 1995, 1996).

Table 11.2 ranks five of the major wildlife tourism districts in the order of the severity of their wildlife losses. It is clear that these losses are not related in any simple way to visitor numbers, or therefore to total revenues. They are, however, related to the distribution of revenues between central government, the tourism industry and landowners.

In Narok and Samburu Districts (Nk and Sm in Fig. 11.1d), tourism revenues are captured mainly by the tourism industry and by the county councils, and more than 50% of all wildlife has gone. Taita Taveta (Tt) is much the same (39% losses) with revenues going mainly to the tourism industry and to the KWS. In contrast, in Kajiado (Kj), where there has been a continuous programme over the last 21 years to share wildlife revenues and benefits with group ranch owners (Berger 1993) and where substantial wildlife development funds have been disbursed to landowners over the last few years, there are roughly the same numbers of wildlife today as there were 18 years ago. Laikipia (Lk) is the most interesting, for conservation in the district is carried out solely on private land by private landowners developing their own privately financed wildlife-based activities (Thouless 1993). Landowners have reaped these benefits directly, and wildlife numbers have increased.

This is strong evidence that from the perspective of landowners and landusers that the contemporary distribution of wildlife benefits is inequit-

Table 11.2 Wildlife losses and tourism benefits in five districts of Kenya.

District	Narok	Samburu	Taita Taveta	Kajiado	Laikipia
Loss of wildlife 1977–94	– 65%	– 33%	– 29%	+ 2%	+ 12%
Main conservation area	National reserve	National reserve	National park	National park	Private land
Tourist numbers (1994)[1]	138 000	90 000	238 000	160 000	$\simeq$ 50 000
Control of access and concession fees	Local county council	Local county council	Kenya Wildlife Service	Kenya Wildlife Service	Land owners and users
Years of community-based, wildlife extension work[2]	5	2	4	21	3
Disbursement since 1992 from the KWS Wildlife Development Fund[3]	US$0.1m	US$0.2m	US$0.2m	US$1.2m	US$0.2m
Revenue distribution[4]					
to central GOK revenue	**	**	***	***	*
to city council	***	***	*	*	*
to tourism industry	***	***	***	***	*
to landowners and users	*	*	*	**	***

[1] GOK (1996); [2] Berger (1993); [3] KWS (1996); [4] Minor * through to major ***.

able, and is a major contributor to the problem of wildlife losses. Direct benefits are clearly more important than are any indirect benefits through social investments (WCMC 1992; Goodwin 1996).

11.4 Other costs and benefits of wildlife to landowners

The KWS and the county councils enforce their property rights to the PAs by granting access only to tourists and by excluding neighbouring landowners and landusers. Furthermore, the game laws (Bragdon 1990) allow the KWS to enforce property rights to all wildlife outside the PAs on both adjudicated and unadjudicated land.

The enforcement of these property rights imposes significant external costs on landowners and landusers. First, important natural resources are alienated from them, for there is relatively more high potential land inside PAs than outside, and relatively more low potential land outside than inside. Second, wildlife significantly raises the costs of livestock and agricultural production. Wild animals compete for grazing, they spread disease, they kill and maim people and livestock, they damage property, and they raid crops. In response, owners and users of land must undertake all kinds of defensive activities, such as building wildlife-proof fences and stockades and even moving away from areas seasonally infested by wildlife.

Conditions are accordingly ripe for major conflicts between the economic interests of landowners and landusers and the social and scientific interests of the government and conservationists. Indeed, a recent study by the KWS has shown that the vast majority of landowners and users in pastoral Kenya would like to see all wildlife eradicated and the PAs opened for development (KWS 1995c).

Many observers, including the KWS itself, point out quite correctly that one root cause of these conflicts under current conservation policy is that wildlife *benefits* to landowners have been effectively zero, especially since the ban in 1977 on hunting and all other consumptive forms of wildlife utilisation. At that time, consumptive use generated annual revenues of some US$24 million (nearer US$80 million in today's money) of which some 10–15% went to landowners and users. Furthermore, compensation schemes for death or injury to persons, and for losses to crops or livestock, were suspended long ago because of corruption.

11.5 Economic appraisal of new policy options

In response to all these problems there has been a significant change in policy thinking since the KWS was established in 1989 (KWS 1995a, 1996). Some consumptive utilisation of wildlife is now allowed under special KWS permit (over 60 wildlife cropping, ranching and farming operations are now licensed) and sport hunting might one day be reintroduced. Plans are well under way to license wildlife associations made up from neighbouring landowners and landusers to whom KWS will grant wildlife user rights, given certain conditions. The KWS has also introduced variable entrance fees for parks to even up the distribution of tourist visits, and is trying to attract tourists to more districts.

Furthermore, the new Community Wildlife Service (CWS) programme of the KWS is providing tangible benefits to landowners and users in at least five districts through the disbursement of wildlife development funds which are themselves generated from tourism revenues (see Table 11.2). The CWS is also helping landowners and landusers to negotiate more advantageous concession fees, and set up their own privately financed tourist operations such as camp sites, tented camps and camel trekking.

The clear objectives of these new policy orientations is to ensure that the *benefits* of wildlife to *landowners* create *incentives* to invest in wildlife conservation so that landowners (and users) will become partners in conservation with the KWS rather than opponents. Policy objectives are to create an enabling environment within which the private sector (landowners and landusers) has incentives to support the public sector in achieving national conservation objectives (KWS 1995b, 1996; Kock 1995).

Nonetheless, the complexity of the linkages between wildlife conservation and development introduces seeds of doubt and uncertainty about the new KWS approach.

A wildlife production function

Let us start with a simple example of a landowner (everything from now on is about landowners and wildlife on people's land) who, following the new KWS policy initiatives, has decided to keep wildlife on his land. The net benefits of wildlife (NB_W) to him can be expressed very simply in terms of the direct benefits of wildlife ($DirB_W$), the management costs of wildlife ($MgmtC_W$), the compliance costs of wildlife ($CompC_W$) and the social benefits of wildlife (SB_W). Let:

$$NB_W = DirB_W - MgmtC_W - CompC_W + SB_W \qquad (11.1)$$

where NB_W is a function primarily of the difference between the direct benefits of wildlife (represented by the stream of benefits from a tented camp, access fees for game viewing, hunting or bird shooting, or from cropping) and the management costs of wildlife (represented by all the costs associated with creating and capturing those benefits). Clearly, net benefits will be positive so long as the direct benefits are larger than the management costs, under which conditions a landowner will look favourably on wildlife as a resource.

However, we must not overlook $CompC_W$, the costs of *compliance* with all the rules and regulations put in place by KWS or other agencies in order to use wildlife. If KWS insists on too many committees, utilisation plans, monitoring, regulations, complicated licensing arrangements and reports then costs will outweigh benefits. If:

$$DirB_W < MgmtC_W + CompC_W \qquad (11.2)$$

the landowner will give up in despair. KWS policy documents show serious signs of imposing crippling compliance costs onto landowners with too much unnecessary regulation. This may negate the very objectives of their new policies.

The final term SB_W represents all those intangible social benefits of having wildlife around. For some landowners these social benefits seem to outweigh all other costs and they gain great pleasure and satisfaction from conserving the resources on their land. For others, of course, these social benefits are strongly negative and they will never tolerate wildlife under any conditions.

If a landowner cannot capture benefits from wildlife, or if compliance costs are too high, then the decision whether or not to conserve wildlife depends solely on SB_W, which is risky to say the least (for most landowners and users this was indeed the situation following the ban on all consumptive use of wildlife in 1977). In principle, therefore, policy initiatives from KWS which allow landowners to both create and keep benefits from wildlife will in general be effective in creating incentives to conserve the resource.

A ranch production function

This would be as far as we had to go if it were not that wildlife also enters into the agricultural and livestock production function of the landowner. Let:

$$NB_P = DirB_P - MgmtC_P - CompC_P + SB_P - IDirC_W \qquad (11.3)$$

where the net benefits of production (NB_P), from either agriculture or livestock or both, is simply a function of the direct benefits of production ($DirB_P$) less the management costs ($MgmtC_P$). Compliance costs ($CompC_P$) are represented here by local taxes, veterinary or other regulations, movement restrictions, etc., while the social benefits of production (SB_P) cater for the landowner who, for example, keeps 2 ha of maize among the coffee because he likes home grown maize, or the landowner who wants to keep a small herd of livestock because he always has and always will.

The indirect costs of wildlife to the producer ($IDirC_W$), some of which were mentioned earlier, include competition for grazing and water resources and for space; and the costs of crop damage, predation and injury to livestock, death and injury to persons, disease, and damage to property. They also include the costs of defensive activities such as moving away from migratory herds, building strong bomas to keep lions out and children in, extra veterinary requirements, and electric fencing around fields.

Clearly, these $IDirC_W$ add to the *production costs* of a landowner and reduce both his profitability and his efficiency. One recent study (Norton-Griffiths 1996) showed that grazing competition alone reduced net benefits of livestock by some 35–40%, while another (Omondi 1994) highlighted the costs from predation and crop raiding. Equation 11.3 shows that, all things being equal, a ranch or farm with fewer wildlife around will be more efficient and profitable than will one with lots of wildlife.

Under the old conservation policy in Kenya, NB_W in Equation 11.1 was effectively zero (SB_W apart) so it was not possible for a producer to offset any of the $IDirC_W$. The consumptive use of wildlife by landowners had been banned, all compensation schemes had been closed, and tourism benefits were in the grip of a powerful tourism cartel. Conservation policy in fact created *disincentives* for landowners to conserve wildlife, as getting rid of it reduced production costs.

However, Equation 11.3 also shows that policies which simply allow landowners to make profits from wildlife (Equation 11.1) are no guarantee at all that it will be in their economic or financial interests to do so. The key relationship is between the net benefits of wildlife (NB_W) and the indirect costs of wildlife on production ($IDirC_W$), for if

$$NB_W < IDirC_W \tag{11.4}$$

then wildlife will remain a net cost to a landowner even if individual wildlife utilisation activities themselves yield net benefits. It is Equation 11.4 which explains why the group ranches in Narok have lost more than 50% of their wildlife over the last 17 years (Broten & Said 1995; Norton-Griffiths 1996) despite the truly massive tourism income generated on their lands, and why the ranches in Laikipia have kept theirs.

Development pressures on land

There persists the romantic notion that pastoralists coexist with wildlife in a harmonious relationship. The truth is, of course, quite different and what one observes and interprets as coexistence is in fact a shortage of capital and technology on the part of the pastoralists, restricting their ability to change the *status quo*. Perhaps in the past, when population densities were low, pastoralists could indeed afford to ignore wildlife, but today, population growth across the country in cities, towns and villages, and on farms and ranches, leads to a demand for increased production, while expanding markets at home and abroad, real increases in prices, growing personal expectations, and advances in agricultural technology all create overwhelming pressures to raise the productivity per unit area of land.

Pastoralists simply can no longer afford the extra costs of production associated with wildlife. Indeed, Table 11.1 shows the success with which they have eradicated wildlife from the rangelands over the last 20 years or so, despite gains from land adjudication and from tourism, and despite all the well-meaning efforts of conservationists.

The growth in rangeland production over the last 18 years in response to these economic, social and market forces has been astonishing. The sales and slaughter of livestock in the pastoral districts of Kenya have each grown at over 4% per annum (Table 11.3), incidentally demonstrating a fundamental change in pastoral production strategy, for it was achieved without any increase in the actual numbers of livestock (see also Scoones 1994). Pastoralists have also shown themselves to be extremely price sensitive in that numbers sold and slaughtered increase by 0.7% for every 1% increase in prices. Pastoralists are also investing more in agriculture, for planted hectares in the rangelands have been growing on average by 7% per annum (but by up to 18% per annum in districts with high agricultural potential, such as Narok) with a 0.6% increase in area for every 1% increase in producer prices.

Table 11.3 demonstrates clearly that landowners and users can distin-

Table 11.3 Trends in livestock slaughter and auctions, and cultivation, in 17 pastoral districts of Kenya, 1977–94. Annual district level data (1977–94) were obtained from district records and reports, and from internal records and reports in the Ministry of Health and in the Ministry of Agriculture and Livestock Development.

	Rate of change[1] (% per annum)	Price elasticity[2]
Livestock slaughter	4.84***[3]	0.38%** KShs kg^{-1} meat
Livestock auctions	4.12**	0.32%** KShs per carcass
Cultivated hectares	8.64***	0.55%*** District producer price for maize

[1] Trend analysis by OLS of logged data with dummy variables for each district, for each livestock species and for drought years. [2] Price elasticity shows the percentage increase in slaughter, auctions or cultivated hectares for each 1% increase in price. [3] Significance of trends: $P < 0.05*$, $< 0.01**$, $< 0.001***$.

guish between the relative benefits of development (i.e. livestock and agricultural production on their land) and conservation (the benefits from keeping land undeveloped for wildlife). Figure 11.3 portrays this pure development-conservation dynamic in terms of the marginal benefits of *development* (curve D-DD) and the marginal benefits of *conservation* (curve C-CC) with an initial equilibrium (Q^{***}) where the marginal benefits of one are matched more or less by the marginal benefits of the other (marginal benefit curves show the benefit arising from the 'next' unit as a function of the number of units already produced or in production).

This equilibrium will be displaced by policies and events which change the relative values of these marginal benefits. For example, should the marginal benefits of development increase (to D'-DD') following, perhaps, further growth in demand, markets and prices, then the equilibrium will shift away from Q^{***} towards a new equilibrium at Q^{**} characterised by more development and less conservation. Similarly, should the marginal benefits of conservation fall (to C'-CC') following, perhaps, increased competition from Tanzania and Southern Africa, or a downturn in the global market for tourism, then the equilibrium will shift even further away from Q^{***} towards a new equilibrium at Q^*.

Of equal relevance to policy is the concept of the rate of change through time of these marginal benefit curves. If the marginal benefits of conservation are increasing (an upturn in tourism), but those of development are increasing even faster (new and expanding overseas markets), then the equilibrium position will still shift towards more development at the expense of conservation.

Figure 11.3 contains clear policy implications for conservation. In general terms, population growth, expanding markets, improving agricultural technology and real gains in producer prices will all act to increase the marginal benefits of development relative to those of conservation, and increase them at a faster rate. In the Kenyan context, this makes it more difficult for wildlife to pay its way and it makes it more sensible for landowners to get rid of it.

The opportunity costs of conservation

The opportunity costs of conservation to a landowner are the forgone benefits from development, which can be quantified in terms of the expected net returns to land under contemporary levels of development and with contemporary technology and land uses (Norton-Griffiths & Southey 1995). Net returns to land are high in areas of high agricultural potential so the opportunity cost to a landowner of leaving such land undeveloped (for conservation) is also high. In contrast, net returns to land are much lower in arid areas, so the opportunity cost to a landowner of maintaining land for conservation will be less, and it will be easier for wildlife-generated benefits to meet these opportunity costs or even to surpass them.

In terms of wildlife conservation policy, it is the net opportunity cost to a landowner of keeping land relatively undeveloped *for the benefit of wildlife* which is important. This can be expressed in terms of the opportunity costs (given full development) and the current *net* returns from ranch production (NB_P from Equation 11.3) and wildlife conservation (NB_W from Equation 11.1).

$$Net\ OC = Expected\ net\ returns\ (Full\ Development) - NB_P - NB_W \qquad (11.5)$$

Quite simply, the greater these net opportunity costs are, the greater will be the economic incentives to the landowner to develop his land, and the harder it will be to protect conservation interests. A recent example is given by a study of the Maasai Group Ranches surrounding the Maasai Mara National Reserve in Kenya (Norton-Griffiths 1995, 1996) where there is much discussion of the impact on conservation values from the Maasai developing their land. If their land was fully developed (just like similar land elsewhere in Kenya but outside Maasailand), net revenues to the Maasai landowners would be some US$28.8 million each year, compared with contemporary net earnings of US$2.4 million from ranch production (NB_P) and US$0.4 million from wildlife tourism (NB_W). The net annual opportunity cost to these landowners is accordingly:

$$Net\ OC = \$28.8m - \$2.4m - \$0.4m = \$26m \qquad (11.6)$$

This equation has four important policy implications for conservation. First, these net opportunity costs (of US$26 million) represent an awesome financial incentive to develop the land. If net benefits from wildlife

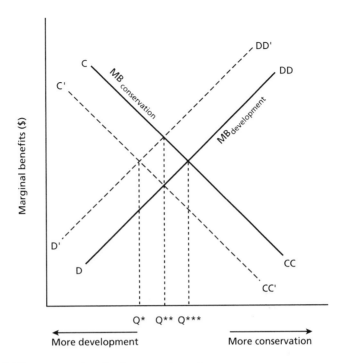

Fig. 11.3 The conservation-development dynamic.

cannot match these opportunity costs then it is inevitable that the land will become developed (either by the Maasai or by outsiders alienating the land from them) and wildlife conservation values will suffer greatly. Second, Equation 11.5 is also linked to Fig. 11.3, for opportunity costs will increase as market forces push up the marginal value of production from D-DD to D'-DD'. Third, if the Maasai were to freeze development on their land at contemporary levels to maintain conservation values then they would forfeit some US$26 million annually. Clearly they should be compensated – but by whom? Finally, if conservation interests were to *deny* to them these benefits of development *without* compensation, then they would be condemning the Maasai to a poverty trap on behalf of conservation (Homewood & Rogers 1991).

11.6 A review of policy options

The main conservation problem facing Kenya is the loss of wildlife on land outside the PAs. The challenge is to devise and implement policies which create incentives for landowners and landusers to maintain the wildlife resource and to invest in conservation. To succeed, policy must be effective at both the micro-economic (individual) and macro-economic (whole economy) levels.

At the micro-economic level, three main policy prescriptions are being relied upon to create a partnership between the private and public sectors in achieving national conservation objectives: first, to encourage landowners and landusers to set up and manage their own tourism ventures so they receive a greater proportion of wildlife revenues; second, to permit some consumptive use of wildlife (ranching and culling) given very specific circumstances; third, to distribute social benefits to communities in the form of wildlife development funds.

Equation 11.1 shows that while these should all enhance the net benefits of wildlife and should provide appropriate incentives to landowners, there are two clear policy deficiencies. First, the rights to consumptive use are held at the discretion of the KWS and can be withdrawn at any time for any reason while the most profitable form of consumptive use, sport hunting, is still forbidden: this does not encourage investment. Second, the policy does not address at all the problem of the indirect costs of wildlife on ranch production ($IDirC_W$ in Equations 11.3 and 11.4). If the net benefits from wildlife are less than these indirect costs of production then wildlife will remain a net loss to the landowner and it will be in his best interests to eradicate it.

A subsidy scheme along the lines of the former grazing compensation scheme (Croze *et al*. 1978; FAO 1978) could be reconsidered, for it is practicable to calculate the indirect costs of each wildlife species on ranch production in terms of grazing offtake, veterinary challenge and danger to life and property. Landowners could then be compensated depending upon the numbers and species mix of wildlife on their land and the time they spend there. However, subsidies often tend to become abused, and lead to uneconomic outcomes. For example, landowners might well over-invest in conservation just to get the subsidy, much like farmers over-invest in wheat production.

An economically preferable option would be to allow landowners to maximise the net benefits of wildlife through direct use, but to achieve this the KWS would almost certainly have to be much more radical in its approach. It might well have to relinquish all property rights to wildlife outside the PAs, even to species under protection through international treaty, and remove all restrictions on wildlife utilisation, including sport hunting and trade. It would be left to the landowners to decide how best to use their wildlife resource, including the option to eliminate it. This would open conservation to the full force of the market and lead to maximum economic efficiency.

However, macro-level policy deficiencies undermine this approach. Figure 11.3 and Table 11.3 show how macro-economic forces can compromise efforts to make wildlife profitable to the private sector, for in the face of expanding markets and real gains from production the marginal benefits of development overwhelm those of conservation. Current conservation policy does not address this important issue at all and policies based simply on allowing landowners to utilise wildlife will be continuously undermined by such powerful economic forces.

Policy initiatives are needed to redress this upward trend to the benefits of development. Fortunately the vast array of direct and indirect subsidies to agricultural production are gradually being scrapped as part of contemporary structural adjustment programmes in Kenya (IBRD 1992, 1995; KWS 1995b), so this may favour conservation in the long run by reducing the marginal benefits of production (Mugabe & Wandera 1993). This process would be enhanced by differential land use taxes, specifically taxes designed to reflect the marginal social costs of land development. Similar taxes have proved effective in Germany and Thailand (Panayotou 1994), but require quite sophisticated tenure, legal and enforcement systems.

Conservation policy in Kenya does not address the problem of opportunity costs, namely the benefits from development forgone by a landowner who keeps his land undeveloped to maintain conservation values. Net benefits from wildlife may more than match these opportunity costs on land of low potential, especially if landowners are left alone to maximise such net benefits in any way they want. It will be much more difficult, however, to match these opportunity costs on land of high potential, and here policy options might include lease backs or easements.

A lease-back policy would pay annual economic rents to landowners for not developing their land, the rent reflecting some proportion of the opportunity costs. This is quite similar to the current EU policy of 'set aside' (Adger & Brown 1994) where farmers receive some 75% of net benefits for each hectare taken out of production (MAFF 1993). In contrast, a conservation easement would aim to purchase the development rights to land from the landowner, the price reflecting the net present value of the opportunity costs (Panayotou 1994). Neither policy is particularly easy to implement, and both are expensive as they recognise the true costs of conservation to landowners. However, both can be effective, and undoubtedly both will be needed in Kenya.

These discussions demonstrate the complexity of the interactions between conservation and development. Conservation is indeed a matter of development, and sadly the complexities of the interlinkages are unfamiliar to most conservationists. Environment policy, of which conservation is just one part, needs to be an integral part of the economic development policy of the country, and the policy decisions about Kenya's wildlife and other conservation interests must be taken within the central planning environment. While it is absolutely correct to concentrate first on the fundamental problem of creating incentives for landowners to look after wildlife and other biodiversity, conservation policy needs a much greater flexibility and conceptual depth, and at each level it needs more sophisticated policy initiatives.

Chapter 12: Gorilla Tourism: A Critical Look

Thomas M. Butynski and Jan Kalina

12.1 Introduction

Although nature tourism has sometimes been viewed as a sustainable, important and necessary 'tool' for conserving species, it is now often considered a conservation problem. A growing number of well documented cases link nature tourism to both the loss of species and the degradation of natural habitats (Boo 1990; Butler 1991; Duffus 1993; Ceballos-Lascurain 1996). In this context, the viewing of wild great apes is a highly controversial kind of nature tourism.

The viewing of free-ranging, habituated great apes is a relatively recent and expanding tourist activity. Promoters of great ape tourism state that apes and their ecosystems will benefit if tourism generates significant revenues. Additionally, they reason that people will be assured a long-term supply of the many resources provided by those ecosystems that harbour great apes.

As an emotionally appealing, high-profile activity that can generate substantial revenue, and as one that appears to nicely bridge the gap between conservation and economic and social development objectives, ape tourism has been an 'easy sell' to almost everyone; not only politicians, donors and the public, but also conservationists. Indeed, ape tourism has been widely heralded as an important, if not sustainable, conservation activity (Harcourt 1986; Vedder & Weber 1990; Sholley 1991; Stewart 1991; Weber 1993; McNeilage 1996). Less publicised are the several serious problems inherent to tourism based on habituated, free-ranging, apes.

This chapter provides an overview of tourism for one great ape species, the gorilla *Gorilla gorilla*. We examine the benefits, problems and risks of gorilla tourism as now practised, and assess whether such tourism is likely to be a sustainable conservation activity. What is concluded here for gorilla tourism probably also applies to many other large vertebrates that are the focus of intensive, localised tourism programmes, particularly primates, and especially the other three species of great ape.

12.2 The gorilla

The gorilla is an endangered species (IUCN 1996). The total number of

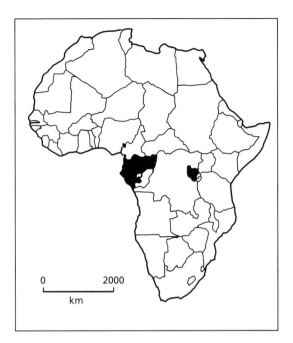

Fig. 12.1 Present distribution of the gorilla *Gorilla gorilla*. (Based on an unpublished map by A. Meder.)

gorillas in Africa is roughly 125 000, about 90% of which live outside of protected areas (Harcourt 1996). Three sub-species of gorilla are currently recognised (Schaller 1963; Groves 1970; Sarmiento *et al.* 1996). Over 80% of gorillas are of the western lowland sub-species *G. g. gorilla*. This sub-species, which is estimated to number more than 100 000 individuals, occurs over an area of approximately 500 000 km^2 (Harcourt 1996) in Angola (Cabinda), Cameroon, Central African Republic, Congo, Equatorial Guinea, Gabon and Nigeria (Fig. 12.1).

Grauer's (or eastern lowland) gorilla *G. g. graueri* numbers roughly 15 000 animals in fragmented populations spread over about 90 000 km^2 (Hall *et al.* 1998) in eastern Democratic Republic of Congo (DRC) (former Zaire), and probably also south-western Uganda (Sarmiento & Butynski 1996). The mountain gorilla *G. g. beringei*, by far the rarest sub-species, is confined to one population of approximately 324 animals (Sholley 1991) in the 447 km^2 Virunga Conservation Area of Rwanda, Uganda and DRC (Fig. 12.2).

The average adult male gorilla weighs roughly 160 kg while the average adult female weighs about 70 kg. Gorilla groups range in size from 2 to 40 animals, but group sizes of 5–12 animals are most typical. Gorillas live in dense primary and secondary forest, lowland swamp, and bamboo at elevations ranging from near sea level to almost 4000 m. The diet is comprised largely of leaves, pith and fruits.

Fig. 12.2 Two juvenile mountain gorillas in the Virunga National Park, DRC. (Photo by Karl Ammann.)

12.3 Threats to gorillas

The two main threats to gorillas are loss of forest habitat and hunting by humans (Lee *et al.* 1988; Harcourt 1996; Butynski, in press). Moist forest is being lost at an estimated average annual rate of 0.5% in Central Africa and 1.9% in West Africa (WRI 1994). At these rates, all of Africa's tropical moist forest will be cleared in 170 years (WRI/IIED 1988). Forest may disappear within 50–70 years in such important gorilla range countries as Cameroon, Equatorial Guinea and DRC.

People hunt gorillas for food over much of the species' range. The bushmeat trade has increased significantly in recent years and is today conducted at an unsustainable level in many places. Hunters have destroyed some gorilla populations and greatly reduced others (Ammann & Pearce 1995; Rose 1996; Hall *et al.* 1998). The commercialisation of the bushmeat trade is probably now a more significant and immediate threat than forest loss for all three species of Africa's great apes.

The human population growth rate in Africa, which exceeds 2.9% per year (WRI 1994), fuels these trends in habitat loss and hunting. At this rate, Africa's population will double to more than 1.5 billion people by the year 2025. As demands for food, clothing, fuel and shelter increase, more gorilla habitat will be lost.

12.4 Gorilla tourism case studies

There are five current gorilla tourism programmes, all located within a small region of the Western Rift Valley in Central Africa (Fig. 12.3). Two new programmes are being developed, one in Odzala National Park, Congo, and one in the Dzanga-Sangha Faunal Reserve, Central African Republic.

The first organised tours to see wild gorillas were initiated in 1955 in Uganda's Mgahinga Game Reserve/Forest Reserve (now the Mgahinga Gorilla National Park). Although based on unhabituated gorillas, this

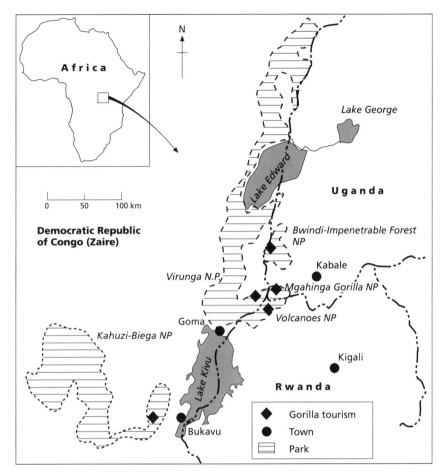

Fig. 12.3 Locations of the five national parks in which tourism on habituated gorillas has been established.

tourism drew much attention from the public and demonstrated that gorilla viewing by tourists was possible, and that financial profits could be obtained.

Kahuzi-Biega National Park (DRC)

Kahuzi-Biega National Park, together with the contiguous Itebero-Kasese region, harbours about 86% of the world's remaining *G. g. graueri* in one large population of roughly 14 550 gorillas (Hall *et al*. 1998). The gorillas visited by tourists are, however, part of a second population of only about 247 individuals (Yamagiwa *et al*. 1993; Vedder 1996), which live in the 600 km² mountain sector of the park. This small population is now isolated from the large population, as a result of habitat destruction by encroachment in large parts of the corridor that connects the mountain and lowland sectors of this park (Hall *et al.*, in press).

The first tourism based on habituated gorillas occurred in DRC's Kahuzi-Biega National Park in 1973 (Table 12.1). This proved popular and by the late 1970s the number of visitors reached several thousand per year. Controls over the guides and tourists, however, were never established. Groups of up to 40 visitors were sometimes taken to view the gorillas. The guides typically cut down vegetation around the gorillas and provoked displays for the benefit of the tourists (Weber 1993; J. Sanderson, pers. comm.).

A major conservation project was initiated in 1985 to, among other activities, train the guides, limit tourist visits to the four groups of habituated gorillas, and generally improve control over the gorilla-viewing programme (von Richter 1991). From 1989 to 1993, about US$210 000 per year were generated through gorilla-viewing fees (Fig. 12.4). The

Table 12.1 Summary of information on gorilla tourism for the five national parks where gorilla tourism has been established (as of 1997).

National park	Mgahinga	Volcanoes[1]	Virunga[1]	Bwindi[1]	Kahuzi[1]	Total
Country	Uganda	Rwanda	DRC	Uganda	DRC	—
Area (km^2)	34	160	240[2]	330	600[3]	1364
Human population density around park (per km^2)	400	400	400	300	300	—
Year gorilla tourism began	1994	1979	1985	1993	1973	—
No. gorillas in area	12[4]	129[4]	181[4]	300	247[5]	869
No. gorilla groups habituated for tourism	1	3–6[6]	4	3–4	4	16–19
No. gorilla groups for research	0	3	0	1	2	6
No. tourists per group	6	8	6	6	8	—
Daily viewing fee for non-resident tourists (US$)	120	126	125	150–180	120	—

[1]Also designated a World Heritage Site.
[2]Portion of the 7800 km^2 Virunga National Park that lies within the Virunga Conservation Area.
[3]Portion of the 6000 km^2 Kahuzi-Biega National Park occupied by the gorilla population that is visited by tourists.
[4]Gorillas in these three parks are all within the Virunga Conservation Area (*c.* 324 gorillas). Most of them move between at least two of the three parks. The number within each of the three parks, therefore, varies considerably and frequently. For example, in 1987, 50 gorillas used the Mgahinga Gorilla National Park but there were usually no more than 20 gorillas in this park at any one time (Butynski *et al.* 1990).
[5]Gorilla tourism in the Kahuzi-Biega National Park is confined to a mountain population of about 247 gorillas (Vedder 1996). There is, however, a lowland population of approximately 14 550 gorillas in this park, and in the contiguous Itebero-Kasese region (Hall *et al.* 1998).
[6]The one gorilla group visited by tourists in Mgahinga Gorilla National Park moves between this park and the Virunga National Park. Therefore, this group is also included among the four gorilla groups visited by tourists in the Virunga National Park.

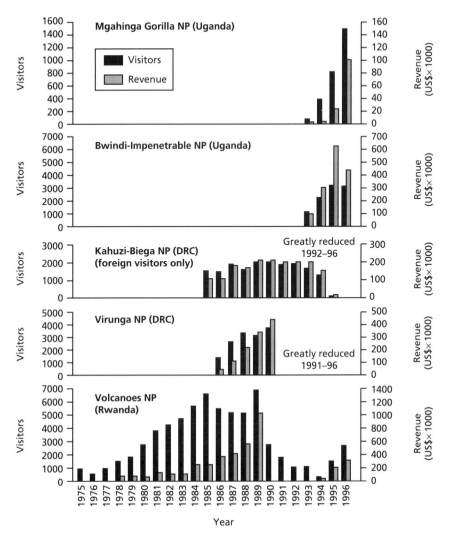

Fig. 12.4 Numbers of visitors, and total revenue from entrance fees and gorilla viewing fees, for each of the five national parks that offer tourism on habituated gorillas. (Data sources: Volcanoes National Park: 1975–84, WWF/IUCN 1985; 1985–89, Weber 1993; 1990–96, Rwanda Office of Tourism and National Parks, unpubl. data. Virunga National Park: Weber 1993. Kahuzi-Biega National Park: B. Steinhauer-Burkart, pers. comm. (Foreign visitors only are shown, and many data are missing.) Bwindi-Impenetrable National Park and Mgahinga Gorilla National Park: Uganda Wildlife Authority, unpubl. data.)

official policy since 1993 is that 40% of park revenues is used for park management, and for development activities in communities near the park (B. Steinhauer-Burkart, pers. comm.).

Due to extreme civil unrest, there has been virtually no tourism in this park since late 1994 (B. Steinhauer-Burkart, pers. comm.). Poaching, agricultural encroachment and insecurity are major problems in Kahuzi-

Biega National Park at this time (August 1997; Basabose & Yamagiwa 1997).

Virunga Conservation Area (Rwanda, Uganda, DRC)

The Virunga Conservation Area (447 km^2) is composed of three national parks, each located in a different country: the Volcanoes National Park (160 km^2) in Rwanda, the Virunga National Park (240 km^2) in DRC, and the Mgahinga Gorilla National Park in Uganda (34 km^2) (Fig. 12.3; Table 12.1). This is where the largest numbers of tourists have visited gorillas over the years.

The Virunga Conservation Area exists as an ecological island completely surrounded by intensive cultivation and a dense human population (*c*. 400 people km^{-2}). In 1978 the average size of family farms in the vicinity of this conservation area was only about 0.5 ha, family income averaged less than US$200 per year, and the annual human population growth rate was 3.7%. Because of these factors, the Virunga Conservation Area is under considerable pressure from agriculturalists, pastoralists, hunters and politicians (Butynski *et al.* 1990; Weber 1993).

The range of the mountain gorilla in the Virunga Volcanoes area has declined from about 500 km^2 in the 1950s to about 375 km^2 today. This decline is largely due to loss of habitat, particularly in Rwanda where the park lost 175 km^2 to agriculture between 1958 and 1969 (Harcourt & Fossey 1981; Vedder & Weber 1990). A market developed for gorilla body parts during the 1970s. Although we do not know how many gorillas were poached, we do know that at least 13 were killed in the three research groups alone (Harcourt & Fossey 1981; McNeilage 1996).

The gorilla population in the Virungas declined from an estimated 400–500 animals in 1960 (Schaller 1963) to fewer than 260 in 1981 (Harcourt 1986; Vedder & Weber 1990). At the time of the last census in 1989 there were about 129 gorillas in Rwanda, 183 in DRC and 12 in Uganda, totalling 324 for the entire Virunga Conservation Area (Sholley 1991). This is a 25% increase over the 1981 census figure. The gorillas in this population move between countries so that, for example, the Uganda portion of Virungas may, at times, hold from zero to 30 or more gorillas.

Volcanoes National Park (Rwanda)

Gorilla tourism began in Rwanda's Volcanoes National Park in 1979 (Vedder & Weber 1990). Some of the more important regulations governing these visits are as follows:
• Every visitor must be at least 15 years of age and in good health.
• Only one group of not more than eight tourists is permitted to visit each gorilla group per day.

• Visits with the gorillas are limited to 1 hour.
• The distance between the visitors and the gorillas must not be less than 5 m.
• All human faeces must be buried at a depth of at least 25 cm.

As of 1989, four groups of gorillas were habituated for tourism and with more than 6900 people visiting them, nearly US$1 million were left in park revenues (Fig. 12.4). This money passed directly to central government. A small portion of these revenues was returned to the park to help cover operating costs (Vedder & Weber 1990). In 1990, foreign tourists coming to Rwanda to visit the gorillas spent an additional US$3–5 million per year elsewhere in the country on accommodation, food, rentals, fuel and other goods and services (Weber 1993).

Political instability in Rwanda since October 1990 (Sholley 1991; McNeilage 1996; Plumptre 1996) has led to the periodic suspension of gorilla viewing. Although only one gorilla is known to have died in the Volcanoes National Park as a direct result of military action, there has been a rise in illegal activities within the park, such as poaching, encroachment by farmers, livestock grazing, and wood and bamboo cutting (Anon. 1994; Williamson *et al*. 1997). A proposal was put forward in March 1996 to give over 2.5 km^2 of the park for agriculture.

Since 1990, the expatriate staff of the Karisoke Research Centre and Mountain Gorilla Veterinary Centre have been evacuated twice, and both facilities, together with the park's headquarters, were looted and destroyed. The park was without surveillance or protection for more than 2 months. Conditions began to stabilise in late 1994 and the research, monitoring, anti-poaching and tourism programmes were re-established to some degree. At this time (August 1997) there is a renewed high level of insecurity in the region, there is no gorilla tourism, the park remains dependent on emergency donor funds, and there is a serious shortage of experienced and trained national park personnel (IGCP 1995; Williamson *et al*. 1997). In recent years, two groups of gorillas habituated for tourism in Rwanda have continued to increase the portion of their range that lies within DRC, and therefore are no longer available to tourists in Rwanda.

Virunga National Park (DRC)

In 1985, tourism began on four groups of gorillas in the Virunga National Park (former Albert National Park) (Aveling & Aveling 1989). In 1990, the 3728 tourists visiting these four groups left more than US$418 000 in park fees (Fig. 12.4).

Since 1991 this park has been affected by the political instability in the region. From 1995 to 1997, at least seven gorillas were shot and one infant taken. As a result, 24-hour surveillance was initiated for all seven habituated gorilla groups in the DRC and Uganda portions of the Virunga

Volcanoes, while all six of the habituated groups in Rwanda were placed under 12-hour surveillance (IGCP 1996; Sikubwabo & Mushenzi 1996). An area of approximately 20–40 m radius around each gorilla group is patrolled by seven trackers and rangers during these surveillance periods.

Gorilla tourism came to a halt in 1996 as the civil war intensified. All of the infrastructure within this park was looted or destroyed. Vehicles and communications equipment were lost and some park staff were killed (Mushenzi 1996). In the absence of park staff, local residents earned money by taking tourists to visit habituated gorillas.

Mgahinga Gorilla National Park (Uganda)

Tourism based on habituated gorillas began in the Mgahinga Gorilla National Park in 1993. One group of gorillas is habituated for tourism but this group moves between Uganda and DRC. Roughly 1100 tourists visited this park in 1996 and paid approximately US$140 000 (Fig. 12.4). Overall, about 20% of the revenue is currently used to manage the park.

Bwindi-Impenetrable National Park (Uganda)

The Bwindi-Impenetrable National Park (former Impenetrable Forest Reserve/Game Reserve) is located about 25 km to the north of the Virunga Conservation Area (Fig. 12.3; Table 12.1). Approximately 300 gorillas live in this park (Butynski & Kalina 1993). Like the nearby Virunga Conservation Area, the Bwindi-Impenetrable Forest is surrounded by a dense human population (Fig. 12.5). A survey of this forest in 1983–84 found much illegal logging, hunting and gold mining, and some agricultural encroachment (Butynski 1984, 1985). Conservation activities were greatly increased in 1986 and by 1989 all illegal activities were reduced to insignificant levels (Butynski & Kalina 1993). An adult male gorilla was killed in this forest in 1979. There were no further incidences of gorilla poaching until 1995 when four gorillas were killed.

Fig. 12.5 The boundary of the Bwindi-Impenetrable National Park. There is intensive cultivation all along the boundary of this park. Most of the original gorilla habitat in the Western Rift Valley has been converted to agriculture. (Photo by Tom Butynski.)

Gorilla tourism began in April 1993. There are presently two groups of gorillas available for tourists and one for research. Two other groups are being habituated for tourism. There are presently 64 gorillas in these five groups, representing about 20% of the population. In 1995 more than 3300 visitors paid about US$600 000 in park fees (Fig. 12.4). One outcome of the first 3 years of this gorilla tourism programme is that the park is able to fund its own recurrent costs, as well as contributing to the operating budgets of other national parks in Uganda.

12.5 Benefits of gorilla tourism

Gorilla tourism has the capacity to promote government and public support for gorilla conservation, increase foreign exchange earnings, generate employment, and attract financial support and investment capital. Revenues from gorilla viewing are also sometimes used directly to support conservation and community development activities in and around areas where gorillas live. By helping to integrate protected areas into the local and national economies, gorilla tourism can provide economic incentives and justification for supporting the conservation of gorillas and their species-rich habitats.

Poaching of gorillas declined throughout the Virunga Conservation Area during 1979–83 and no gorillas are known to have been killed directly by poachers between 1983 and 1994. Most important, the last two censuses (1986 and 1989) show this population to be increasing (Sholley 1991).

Several researchers have stated or implied that tourism has been critical to the conservation of gorillas and ecosystems at sites where such tourism is practised (Vedder & Weber 1990; Stewart 1991; Weber 1993; McNeilage 1996). However, one problem faced by anyone attempting to assess the impact of gorilla tourism on gorilla conservation is that tourism is usually just one of several, often complementary, activities initiated almost simultaneously under new conservation projects. In other words, gorilla tourism has been implemented in ways that make it difficult to evaluate whether it benefits gorilla conservation. An example of this point is presented by the mix of activities initiated or greatly expanded upon during the second half of the 1970s in the Volcanoes National Park (Table 12.2). Taken together, these activities have led to a marked improvement in the conservation of this park. However, the relative contributions of each of the activities to this success remains unknown.

In addition to the mix of activities presented in Table 12.2, the conservation situation in the Virunga Conservation Area was further enhanced (and complicated) by the establishment of many of these same activities in the contiguous Virunga National Park (starting in 1984) and Mgahinga Gorilla National Park (starting in 1986) (Harcourt 1986; Sholley 1991; Butynski & Kalina 1993; Weber 1993).

Table 12.2 Summary of the main conservation activities in the Volcanoes National Park and the year each was initiated or greatly expanded.

Year	Action	Summary
1976	Cattle and herders removed	All of the many cattle and herders that were in the park were removed. They had been causing considerable disturbance up to an elevation of at least 3500 m since at least the 1950s, including serious damage to the vegetation (Schaller 1963).
1978	Donor support enhanced	Donor support for conservation activities in and near the park greatly increased. Although published amounts are not available, the figure from 1978 to 1997 is likely to be well over US$6 million.
1978	Lobbying increased	Lobbying government at various levels for improved support and management of the park expanded substantially as the numbers of conservationists, NGOs, embassies and aid agencies involved in gorilla conservation increased.
1978	Security expanded	Park security activities were greatly expanded. The security force doubled, and the wardens and rangers were trained, well equipped and supervised.
1978	Controlled tourism established	Controlled tourism on habituated gorillas began and generated approximately US$4 million in gorilla-viewing fees over the next 13 years.
1979	Education programme initiated	An environmental education programme began among the people living in the vicinity of the park. This programme reached hundreds of thousands of Rwandans.
1986	Veterinary programme started	The Mountain Gorilla Veterinary Centre (former Volcano Veterinary Centre) was established, primarily to enhance gorilla survival by monitoring the health of habituated gorillas and to administer emergency treatment. This facility has saved many gorillas from death.

12.6 Problems and risks of gorilla tourism

Gorilla tourism shares many of the problems and risks of traditional tourism. While some of these have been mentioned elsewhere (Vedder & Weber 1990; Stewart 1991; Weber 1993; McNeilage 1996), others seem to have been overlooked, or at least not brought to wide attention.

Unstable source of revenue

Gorilla tourism is a highly uncertain source of revenue (Fig. 12.4). This is a drawback to gorilla conservation when managers become overly dependent upon this source of revenue for park protection and management. Visitation can decline rapidly at a particular site as the result of one, or a combination of undesirable events. These include political and economic instability, travel restrictions, disease epidemics, international currency fluctuations, and military, terrorist or criminal activities. Tourists are affected by such events whether they occur in the host country, in the countries from which they originate, or in countries through which they must travel (Boo 1990; Ceballos-Lascurain 1996).

In January 1991, rebels entered the Volcanoes National Park and gorilla tourism was abruptly suspended. A similar situation developed in eastern DRC where civil war, and the killing of several tourists and at least ten gorillas, have all contributed to the cessation of tourism in the Virunga and Kahuzi-Biega National Parks (Fig. 12.4).

High revenue 'leakage' and economic inequity

One of the requirements for the sustainable use of wild species is that there are positive economic incentives for the people living near target populations to conserve them (Ack 1991; Prescott-Allen & Prescott-Allen 1996; SSN 1996). Generally, gorilla-viewing programmes have not succeeded in meeting the economic expectations of local communities.

The World Bank estimates that on average 55% of gross tourism revenues to developing countries 'leak' out of the host countries. For some countries the leakage is 80–90% (Boo 1990). For gorilla tourism, much of the money that visitors spend never reaches the country where the gorilla viewing occurs, and much of the profit generated in the range country is repatriated.

Likewise, in all cases, most of the direct park revenues flow out of the immediate area of the park to central government, and to a few companies and individuals. Based on all available information, we estimate that less than 5% of the direct and indirect income generated from gorilla tourism since 1973 has been spent on gorilla conservation or received by people living near the parks. With a few recent exceptions (see below), there is little re-investment in the park or in the local people to help ensure the long-term viability of the gorillas, the park, or the gorilla tourism programme. This means that the conservation impact of the revenues from gorilla viewing is minimised and that all five programmes continue to depend on outside funding. Further minimising the positive impact of gorilla tourism is its effect on commodity and property prices in the surrounding area. By pushing up prices, gorilla tourism can disrupt local lifestyles and increase the economic burden on already impoverished people.

There is now some effort to ensure that the economic benefits from gorilla tourism are divided more equitably among people living in the vicinity of the Bwindi-Impenetrable and Mgahinga Gorilla National Parks. Revenue-sharing programmes were initiated for these two parks in 1995 and now 12% of the revenues from gorilla-viewing fees are distributed to local communities (Macfie 1995; S. Werikhe, pers. comm.). So far, people around the Bwindi-Impenetrable National Park have received US$100 000 from this programme and those around the Mgahinga Gorilla National Park have received US$15 000 (L. Macfie, pers. comm.).

Changes in gorilla behaviour

The loss of reproductive fitness as a result of utilisation reduces the likelihood of sustainable use. The effects of use must therefore be properly monitored, assessed and minimised (Prescott-Allen & Prescott-Allen 1996; SSN 1996). Of particular concern in the case of gorilla tourism is the loss of fitness as a result of changes in behaviour, and because of increased stress and disease.

Gorillas habituated to people have undergone a behavioural change. For a group of gorillas to allow eight tourists to approach to 5 m requires more-or-less daily contact with people for roughly 1 year (range 3–24 months). Early on in the process the gorillas often flee from the 'habituators' and hide in deep cover for hours. Unhabituated gorillas invariably have diarrhoea as a result of this trauma (T. Butynski, pers. observ.). Threat vocalisations, charges and other intimidation displays are common, and people are sometimes bitten (Fossey 1983; Anon. 1996/97; Williamson et al. 1997; T. Butynski, pers. observ.). The habituation process is obviously a stressful one for the gorillas but the impact of this stress on the fitness of the gorillas has never been studied.

Once tourists begin to visit a group of habituated gorillas, the visits usually occur daily for about 1 hour. After the tourists depart, one or more guides move with the group for several more hours to determine the approximate site where the group will spend the night. Due to insecurity in the Virunga Conservation Area, the 13 groups of habituated gorillas are sometimes accompanied by rangers for 12–24 hours each day (Sikubwabo & Mushenzi 1996). This is roughly 70% of the gorillas in this population – or 70% of those gorillas belonging to the sub-species G. g. beringei (Fig. 12.6).

The close presence of people interrupts and changes activity, social, ranging and other behaviour patterns in free-ranging primates, even in well-habituated individuals (e.g. Tutin & Fernandes 1991; T. Butynski, pers. observ.). Frequent, close human activity around tourist and research gorillas affects the timing and rate of many gorilla behaviours. Little research has been undertaken, however, to measure and evaluate the impact of these human interventions upon the gorillas.

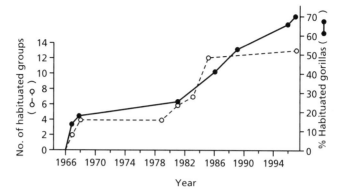

Fig. 12.6 Changes since 1966 in the percentage of gorillas habituated for tourism and research in the Virunga Conservation Area. Also shown are changes in the number of habituated gorilla groups. (Data from Fossey 1983, Sholley 1991, Sikubwabo & Mushenzi 1996, K. Frohardt, pers. comm., K. Stewart, pers. comm., L. Williamson, pers. comm.)

Disease risks

Disease can be a major factor in efforts to conserve endangered species (May 1988; Thorne & Williams 1988). Gorillas are susceptible to numerous human diseases (Benirschke & Adams 1980; Kalter 1980). Many of these can be fatal or cause morbidity with severe consequences for normal behaviour and reproduction. Hence, there is much concern that small populations of gorillas now in frequent, close contact with tourists, guides, rangers, porters and researchers, face severe consequences from introduced diseases (Anon. 1986; Sholley & Hastings 1989; Aveling 1991; Macfie 1991; Sholley 1991; Hudson 1992; Kalina & Butynski 1995; Fig. 12.7).

The mobility of today's human populations, combined with frequent and close contact between humans and wild species, has greatly increased the risks of disease transmission (Holmes 1996). Each year, thousands of tourists from around the world step out of crowded, poorly ventilated airplanes and airports, and within 1 or 2 days are close to (< 6 m), and sometimes touching, gorillas. Another concern is that habituated gorillas are more likely to forage and sleep in the vicinity of people and domestic animals, sometimes leaving the park to do so. These circumstances increase the risk that disease will be transmitted to gorillas.

Disease transmitted into immunologically naive populations can result in an 80–100% mortality (Thorne & Williams 1988; Macdonald 1996). Primates are especially prone to disease epidemics (Young 1994). Small, stressed populations living in fragmented, unstable ecosystems are at particular risk (May 1988; Hudson 1992; Holmes 1996). The stress involved in the habituation process (see above) and frequent visits by

Fig. 12.7 A ranger close to a Grauer's gorilla in the Kahuzi-Biega National Park. Tourist regulations state that a distance of at least 5 m should be kept between people and gorillas. The greatest concern in close encounters such as this one is the transmission of an infectious disease. (Photo by K. Ammann.)

people challenge the well-being of gorillas and, therefore, their ability to respond normally to disease (Hudson 1992; Kalema 1995).

In 1988, in the Volcanoes National Park, six female gorillas in habituated groups died of respiratory illness, and 27 additional cases were treated successfully with injections of penicillin. The illnesses occurred in three of the four tourist groups and in one of the three research groups. This suggests that the disease was new to this population, as 81% of the gorillas in the affected groups became ill. There was serological and pathological evidence of measles or a related morbillivirus in one gorilla. As a result, 65 animals in the seven habituated groups were vaccinated against measles. No further signs of the measles-associated respiratory disease were seen after the initiation of the vaccination campaign, and the disease did not spread to other groups (Hastings *et al.* 1988; Sholley & Hastings 1989). The results of the 1989 census of this population suggest that this disease did not kill gorillas in the unhabituated groups (Sholley 1991).

In 1990, bronchopneumonia affected 26 of the 35 gorillas living in a tourist group in the Volcanoes National Park. Although four of the animals were given antibiotics, two of these died (Macfie 1991).

Disease transmission appears to be the most serious threat that tourism poses for gorillas and, therefore, to the sustainability of gorilla tourism programmes. Indeed, disease may be the most significant factor in the cost–benefit balance for tourism as a gorilla conservation strategy. An appreciable loss due to disease in the small gorilla populations of the Virunga Volcanoes or Bwindi-Impenetrable Forest would make them even more susceptible to extinction from random genetic, demographic and environmental events (May 1988).

Disease transmission from gorillas to humans is also a concern. For example, there is an alpha-herpes virus present among the gorillas of the Virunga Volcanoes that is closely related to the human HSV-2 virus. A substantial portion of this gorilla population is believed to be infected. There is some concern over the potential pathogenicity of this virus for humans, as the alpha-herpes viruses currently present a serious problem from a zoonotic infectious disease standpoint (Eberle 1992).

Need for research and monitoring

Sustainable use of natural resources requires the accumulation and assessment of information on the impact of such use on the target population and ecosystem (Ack 1991; Prescott-Allen & Prescott-Allen 1996; SSN 1996).

Since 1978, millions of dollars have been provided by donors to develop and support tourism based around endangered gorilla populations. Surprisingly little research has been conducted on the effects of tourism on gorilla behaviour, ecology, health and survival. This is especially surprising because all five gorilla tourism programmes are based on small, restricted populations; that is, on populations that are believed to be particularly susceptible to extinction. Moreover, no comprehensive and independent risk assessments, environmental impact studies, or programme evaluations have been undertaken for any gorilla tourism programme. Under these circumstances, it is difficult to advance models for sustainable gorilla tourism, or to be confident that the gorilla tourism programmes are not now, or will not become, detrimental to the gorillas or their ecosystems.

Despite this absence of critical information, most gorilla tourism projects continue to expand. For example, in the Virunga Conservation Area the number of gorilla groups habituated for tourism increased from none in 1978 to 10 in 1997, while the number of tourists visiting some groups has increased from six to eight, and an increase to 10 or more is now being considered. This increase in the size of tourist groups was made despite strong recommendations by scientific advisers to keep group size at six people.

Control of tourists and guides

Sustainable use will not occur unless effective regulatory structures are adopted and enforced, especially by local people and institutions (Ack 1991; Prescott-Allen & Prescott Allen 1996; SSN 1996).

Gorilla-based tourism is exceptionally difficult tourism to control, particularly over the long term. That adequate control over gorilla tourism programmes is often lacking is most clearly demonstrated by the many statements, photographs and videos of tourists and guides close to, or

Fig. 12.8 A juvenile gorilla tugging at the clothes of a guide in the Bwindi-Impenetrable National Park. The gorilla tugged at the clothes of several tourists and guides during this encounter. (Photo by M. Schmitt.)

touching, gorillas (e.g. McBride 1988; Adams & Carwardine 1993; Steele 1994; Wagner 1994; Schmitt 1997; Fig. 12.8).

From an international perspective, park staff concerned with controlling gorilla-based tourism are poorly paid, often earning less than US$1.50 per day. As such, they have problems making ends meet and must be on the look-out for opportunities to increase their income. The majority of tourists, by contrast, have incomes well over 50-fold that of the guides. Two other problems are that some visitors arrive with misconceptions about gorilla tourism and what is permissible, and that the demand to view gorillas often exceeds the number of places available. One result is that some tour operators and tourists pressure and bribe park staff to ignore the rules (Stewart 1992; McNeilage 1996; Macfie 1997). In some cases, tourists are actively encouraged by park staff to break the rules and have more of a 'gorilla experience' than the regulations allow. The benefit to the guide is a larger gratuity at the end of the day.

Infringements of the regulations have been documented in all gorilla-viewing programmes (Aveling 1991; Stewart 1992, 1993; McNeilage 1996; Moulton & Sanderson 1996; Macfie 1997; Schmitt 1997). For example, since 1991, the gorilla-viewing programme in the Virunga National Park has operated without regard for at least some of the rules. The main concerns are physical contact between gorillas and tourists, extended visits with the gorillas (beyond 1.5 hours per visit), large numbers of people in the tourist groups (at least 32 people), twice-daily visits to groups of gorillas, and visits by obviously sick tourists. Another problem is that receipts are not always issued to tourists for the gorilla-viewing fees (Stewart 1992; Wrangham 1992). In such cases, the revenue apparently goes into private pockets.

A similar, but less severe, breakdown in the control of gorilla tourism commenced in the Volcanoes National Park in late 1990 (Stewart 1992; Luebbert 1996; McNeilage 1996; Plumptre 1996). There are cases of groups of 13 or more tourists visiting the gorillas, twice-daily visits to gorilla groups, physical contact between gorillas and tourists, and unauthorised visits to non-tourist gorilla groups.

Money, politics and value diversification

The widespread perception is that gorilla tourism is guided by science and by concern for the survival of the gorilla. A closer look, however, reveals that science frequently has little influence, and that conservation is often relegated to a place behind politics, power struggles and short-term financial gains (Moulton & Sanderson 1996).

For some politicians and tour operators, gorilla viewing is a bonanza from which to reap as much financial profit as possible. Not surprisingly, those calling for more science, for impartial evaluations, and for greater caution and restraint in the development and operation of gorilla tourism programmes have been routinely ignored, and sometimes targeted for attack by those bent on suppressing the problems to make political and monetary gains (McBride 1988; Moulton & Sanderson 1996).

The high demand to see gorillas, and to obtain the money that gorilla tourism brings, are two extremely powerful and destabilising forces that will continue to hamper efforts to make gorilla tourism sustainable. Most seriously affected will be those small gorilla populations whose survival is being tied to tourism revenues (i.e. populations of the Virunga Conservation Area, Bwindi-Impenetrable Forest and Kahuzi-Biega Forest).

Of great concern at the present time is the suggestion to convert some or all of the three research (Karisoke) groups in the Volcanoes National Park to tourism groups and/or to increase further the numbers of tourists visiting each gorilla group (D. Steklis, pers. comm.). If all three research groups become tourist groups, Rwanda would have six groups of gorillas available for tourism. This could increase to eight groups if the two habituated groups that emigrated to DRC expanded their range back into Rwanda. As such, nearly all of the gorillas in the Volcanoes National Park, and about 70% of the world's remaining mountain gorillas (Fig. 11.6), could be visited daily by more than 100 tourists, and by a similar number of guides, porters and rangers. Whatever the risks currently associated with tourism on this small population, these risks would be considerably increased. In addition, the valuable and well-known long-term research on these three groups would be severely restricted and put into jeopardy. Perhaps most importantly, the concept of gorilla tourism as a sustainable activity contributing to the survival of the Virunga gorillas would undoubtedly lose much credibility and support, not only from the international conservation community, but also from those tourists who thought they were benefiting gorilla conservation through their visits.

Also of concern at this time is the current habituation of two new gorilla groups in the Bwindi-Impenetrable forest (Macfie 1997). Both groups live outside the 'tourism zone' as widely agreed upon in the tourism plan (IGCP 1992). In fact, they live within a controlled research area, in which data on unhabituated gorillas were to be collected for monitoring the impact of tourism on gorillas in the tourism zone, although the monitoring programme is yet to begin. This expansion of the gorilla tourism programme, without the benefit of professional independent evaluation, calls into question the sustainability of this tourism programme.

Excessive emphasis on the economic value of gorillas might lead to the belief among decision-makers and local people that gorillas exist primarily to generate financial profits. Indeed the majority of people living near areas with gorilla tourism may already believe that if the gorillas are no longer producing revenue, then neither the park nor the gorillas have any value, and that alternative uses of the land should be considered (e.g. agriculture or livestock production).

One might argue on behalf of gorilla tourism, even if unsustainable, for lack of better solutions for the long-term survival of these small populations. How might highly endangered gorilla populations be protected without subjecting them to the risks presented by tourism? Conservation trust funds may be one less expensive, more manageable, and relatively risk-free approach. The US\$5.5 million Mgahinga and Bwindi-Impenetrable Forests Conservation Trust was established in 1994 and is already generating enough revenue to manage these two parks and attain the needed level of support among local communities. Similar conservation trust funds for the Volcanoes and Virunga National Parks could provide an equally long-term, reliable and ample source of revenue, perhaps replacing less reliable, riskier sources of income such as gorilla tourism.

12.7 Conclusions

Tourism based upon gorilla viewing is not the conservation panacea that many people believe. In particular, there is too much emphasis on generating revenues, while far too little attention is given to either demonstrating or ensuring the long-term sustainability of these programmes. The viewing of free-ranging, habituated gorillas has now been in effect for nearly two decades, yet the recognised cornerstones for ensuring that this activity is sustainable over the long term have not been laid. There continues to be enormous disparity between what needs to be done and what the implementing governments, managers and supporting international conservation bodies are willing or able to accomplish.

Tourism based on small populations of gorillas is likely to be sustainable only:
• Where gorilla conservation is given priority over economic and political concerns.

• Where decisions affecting gorilla tourism are based on sound and objective science.
• Where the scientifically formulated regulations governing this activity are rigorously controlled.
• Where the conservation benefits from gorilla tourism monies are considerably greater than at present.

If these basic prerequisites cannot be met, then tourism on small populations of gorillas should be stopped until they can be met. The stakes are too high to do otherwise.

Given the many problems and the observed management capabilities, we suspect that gorilla tourism, as practised today, is likely to only be sustainable where gorilla populations are large. We suggest that limited tourism on the large lowland population of gorillas (*c.* 14 550 gorillas; Hall *et al.* 1998) of the Kahuzi-Biega National Park and adjacent Itebero-Kasese region of eastern DRC would do little damage. We are concerned that all five of the established gorilla-viewing programmes are based on small populations of gorillas (*c.* 240–340 individuals; Table 12.1). While tourism may contribute to the survival of these small populations, it is at the same time putting them at additional risk. We suggest that further research is needed to assess the arguments, both for and against gorilla tourism, that have been raised.

Advocates of gorilla tourism often claim or imply that this form of tourism is ecologically sound, safe, sustainable and necessary for the continued survival of some gorilla populations. We contend that they have over-stated their case. We hope that this chapter helps to make more people aware of the problems, risks and concerns that surround gorilla tourism, and that doing so gives a beneficial nudge both to gorilla conservation and to gorilla tourism.

Chapter 13: Caribou and Muskox Harvesting in the Northwest Territories

Anne Gunn

13.1 Introduction

Aboriginal people have coexisted with caribou (*Rangifer tarandus*) and muskox (*Ovibos moschatus*) for thousands of years. Many people in the Northwest Territories (NWT) of Canada still rely on hunting and fishing for food and income: over 90% of aboriginal households use meat taken by hunting or fishing [the population of the NWT in 1991 consisted of 35 300 aboriginal and 22 300 non-aboriginal individuals (Bureau of Statistics 1996)]. The estimated value of the annual subsistence harvest is CDN$55–60 million based on its replacement costs in imported food, and about half of this value is from caribou and muskox harvesting. However, these figures only partially convey the importance of hunting. Animals and hunting are an integral part of aboriginal culture and have been for some 8000 years. Non-aboriginals, as well, have their sets of cultural values for hunting, viewing and simply knowing that the caribou and muskox are there.

The context of wildlife harvesting in the 1990s has radically altered in the NWT following land claim settlements for the aboriginal people. Up until the 1990s, northern and aboriginal people had relatively little say in wildlife management, which was centralised and tied to regulations. However, all of this is now undergoing rapid changes with many responsibilities and roles of public government devolving to aboriginal people as land claims are settled. The first land claim settled in the NWT was the Inuvialuit Final Agreement in 1984 with the Gwich'in and Sahtu land claim settlements following in 1992 and 1993 respectively. The settled claims leave the aboriginal groups as the largest private landowners in the NWT. New forms of wildlife management are having to be developed as the political landscape changes. This chapter describes the context of these changes along with the ecology of the caribou and muskox, and the history of their use, and assesses the potential sustainability of wildlife harvesting in the future.

13.2 The system's history

The Northwest Territories covers 3 426 320 km^2 (Law 1995) and is divided by the treeline, running approximately north-west to south-east, where treed taiga becomes treeless tundra. Some barren-ground caribou herds seasonally migrate between the tundra and taiga (Fig. 13.1) while other herds migrate between seasonal ranges on the tundra. Muskox are widely distributed across the tundra as far south as the treeline in the western Arctic.

The prehistoric pattern of nomadic people seasonally hunting caribou and muskox was interrupted by the arrival of European colonisation which began with exploration in the 1700s but gathered momentum with fur trading in the 1800s. By the mid-1900s, fur prices dropped, harvesting equipment costs rose and caribou numbers were low. The government stepped in to help and gathered aboriginal people into communities to receive education, medical services and missionaries. Those communities are small with hundreds to a few thousand people and are supplied by

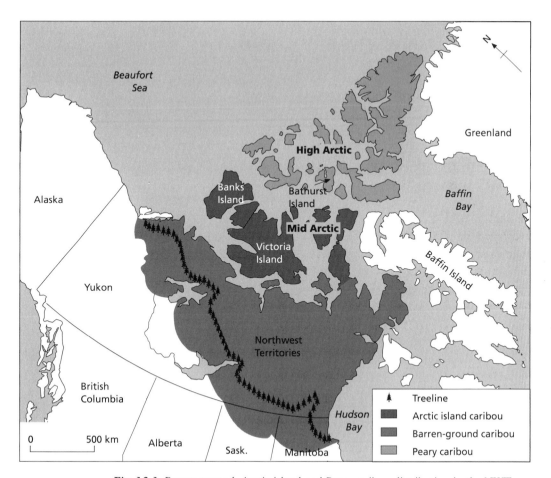

Fig. 13.1 Barren-ground, Arctic island and Peary caribou distribution in the NWT.

aircraft and annual sea-lifts. Only a few communities in the western Arctic are linked by seasonal or year-round roads.

The NWT's political legacy from European colonisation is rapidly changing as the aboriginal groups implement land claims which radically change rights of access to the natural capital (renewable and non-renewable resources). The most conspicuous change will be in 1999 when the NWT divides into a western territory and Nunavut (Our Land) following the settlement of the Inuit land claim. Politically, the NWT's history is typical of many areas in the world with its switch from an aboriginal, dispersed and locally organised society to a centralised society with European-oriented institutions. Most recently through aboriginal land claims, society and government are shifting towards amalgamating previous models.

Socio-economically, the NWT is like many developing countries with a demographically young and rapidly growing population (5% annually), a high dependence on a subsistence lifestyle and geographically localised economic opportunities. Unlike many areas, however, the NWT has not suffered any large-scale habitat modifications as there is almost no agriculture and only small-scale forestry.

Harvesting wildlife in the NWT has gone through booms and busts that brought some wildlife to economic extinction and extinguished local populations. It was unregulated commercial harvesting in the late 1800s and early 1900s that hammered muskox to the point of local extinctions on the NWT mainland (Barr 1991). The cascade of lifestyle changes that accompanied whaling and fur trading along with the introduction of firearms increased caribou harvests from prehistoric levels to beyond those that were sustainable. Fur traders bought and sold thousands of hides as well as dry and fresh meat. By the early decades of this century, the caribou were in decline which then caused problems for the aboriginal people. Without quibbling over the accuracy of population and harvest estimates from the 1940s and 1950s, caribou harvests were high compared to the 1990s, if nothing else because lifestyles were so different. The estimate for annual per-capita caribou needs was 50 in 1950 (Lawrie cited in Kelsall 1968) compared to two or three in the late 1980s (Gunn et al. 1986). In the 1950s, caribou skin was used for clothing and the meat for feeding dog teams; Kelsall (1968) estimated that annual harvests were between 100 000 and 200 000. The government instigated conservation measures (harvesting restrictions and wolf control) but it is unclear how effective they were compared to the lifestyle changes. Caribou numbers in the 1950s did recover, but the measures' legacy was a resentment which was undiminished by administrators', biologists' and aboriginal hunters' difficulty in appreciating each other's viewpoint.

Caribou management from the 1960s onwards depended increasingly on aerial surveys. The surveys emphasised the gap between biologists' and hunters' perceptions of whether the caribou herds were decreasing in the

1970s. Hunters did not believe the surveys. The credibility gap between biologists and hunters began to close in the 1980s, partly when government biologists were living in the communities and biologists and hunters were travelling and working together. Another significant step was the political progress toward settling land claims and the establishment of co-management boards. Caribou and muskox populations were mostly on the increase and both local knowledge and the results of aerial surveys were in agreement.

Most barren-ground caribou herds are relatively large in the 1990s. The four largest herds (Bluenose, Bathurst, Beverly, Qamanirjuaq) total 1.3 million caribou, with a total of about 400 000 caribou in smaller herds. Likewise, most muskox populations have increased since the 1960s and muskox numbers are about 113 000 on the Arctic islands and about 15 000 on the mainland (Ferguson & Gauthier 1992; Government of the Northwest Territories, unpubl. data). On the Arctic islands, the caribou differ ecologically from barren-ground caribou and the most distinctive are the Peary caribou on the High Arctic islands, which are classified as an 'endangered' sub-species – Peary caribou (*R. t. pearyi*). The taxonomy of the Peary-like caribou on the Mid-Arctic islands is uncertain and they are termed 'Arctic island caribou' (Fig. 13.1). Peary caribou numbers on the High Arctic islands have declined from 26 000 in 1961 to 2000–3000 by the mid-1990s (Miller 1991; F.L. Miller, pers. comm. 1996).

This chapter is restricted to migratory barren-ground caribou. Another ecotype (Bergerud 1988) is the 'sedentary' woodland and mountain caribou which are characterised by dispersal at calving, whereas migratory caribou aggregate for calving. Relatively little is known about woodland and mountain caribou, where they occur in the southern and western NWT.

13.3 The system's ecology

The system in this chapter is the Canadian tundra and taiga ranges of caribou and muskox in both the Arctic and sub-Arctic ecosystems. Caribou and muskox are medium- and large-sized herbivores, respectively. A barren-ground caribou cow weighs 90 kg and an adult muskox cow 170 kg in the autumn, when both species reach the annual maximum condition. Their population ecology is characterised by a 10-fold fluctuation in size over decades. The best-known fluctuations are those for the larger herds (Fig. 13.2). The fluctuations are more extreme where the climate restrictions on forage growth are tighter, such as on the Arctic islands, and populations may disappear and then recolonise. For example on Bathurst Island, caribou and muskox populations have twice increased and then crashed since 1961 (Fig. 13.3).

Both caribou and muskox usually calve as 3-year-olds, but with good

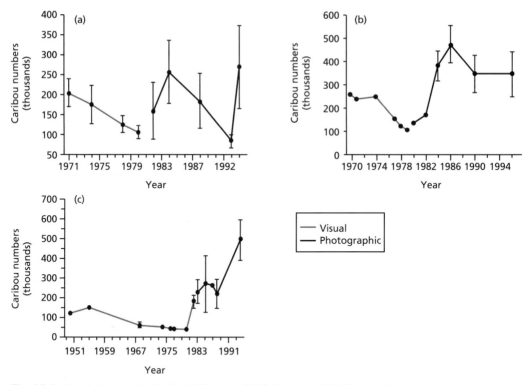

Fig. 13.2 Population trends for the (a) Beverly, (b) Bathurst and (c) Qamanirjuaq herds of barren-ground caribou, NWT, Canada from visual and photographic surveys of calving grounds.

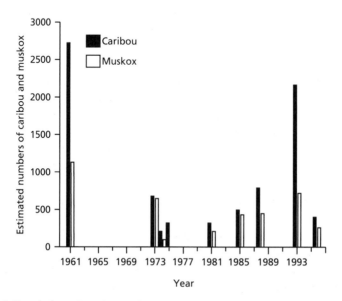

Fig. 13.3 Population of muskox and Peary caribou on Bathurst Island, NWT, Canada. (Data from Miller *et al.* 1977; Miller 1991; F. L. Miller, pers. comm.)

nutrition first calving occurs earlier. Single births are almost invariably the rule and calving frequency is dependent on nutrition (Dauphiné 1976; Thomas 1982; Cameron *et al.* 1993). Under ideal conditions, such as colonising unoccupied ranges, both caribou and muskox can increase at annual rates of 25–30% (Jingfors & Klein 1982; Heard 1990). In stable to declining populations, females younger than 3 years do not breed and females older than 3 years may breed in alternate years or irregularly (Adamczewski 1995; Cameron *et al.* 1993; Valkenburg *et al.* 1996).

Relationships between herbivores and their forage, and the influence of weather on forage, account for variation in fecundity which, together with mortality, drives changes in population size. An elastic tetrahedron is a conceptual model for caribou and muskox ecology with the four apices representing the herbivore numbers, weather, predation and forage. The elasticity serves as a reminder that when relationships are tightly coupled between two apices, a short line will connect those two apices and, correspondingly, the distance between the other apices will be longer (less tightly coupled). When conditions conspire to change the relationships, tightly coupled and loosely coupled can switch.

Weather conspicuously and subtly influences forage and predation and, in the sub-Arctic and Arctic, the weather has two characteristics. Firstly, it varies annually and unpredictably (Maxwell 1980). For example, on Banks Island, the coefficient of variation for snowfall at the end of May is 87% and for the length of the growing season is 47%; serial correlations between years are insignificant (Caughley & Gunn 1993). The significance of variable weather increases in the higher latitudes as the limits of plants and herbivores are reached. The second characteristic of the weather is the short summer. Plant growth is restricted and most nutrients are held in dead material. These factors tie herbivores to 2–3 months to build condition on high quality forage. Nutrient input in tundra ecosystems is low and nutrient cycling through soil organic matter is slow at the characteristic cool temperatures, except for the breakdown of herbivore dung which quickly returns nutrients to the plant. Initially, grazing has a positive feedback effect as the released nutrients stimulate productivity (Jefferies *et al.* 1992). As herbivore densities increase, the positive feedback switches to negative as the removal of forage increases beyond the plant's capability to compensate.

Caribou and muskox respond differently to those two ecosystem characteristics. Muskox are large-bodied grazers with a predominantly conservative lifestyle adapted to buffering much of the variability in weather and forage supplies (Adamczewski 1995). They effectively digest poorer quality forage in the winter and often feed intensively on well-vegetated sites (sedge meadows) where plants are large and packed densely together. Their grazing tends to increase plant productivity through faecal fertilisation and compensatory growth (Henry & Svoboda 1989).

Caribou are probably less strongly coupled by feedback loops to their forage (Jefferies *et al.* 1992). Caribou usually do not feed intensively in one spot but move around, as well as often undertaking long-distance migrations between seasonal ranges. Their gastric tract and dental adaptations indicate that they forage more selectively than muskox and their adaptions and life history indicate that variable weather and forage supplies buffet them more than muskox.

Weather influences forage availability in summer through plant growth (absolute availability) and in winter by snow and ice (relative availability). Absolute forage availability has two components, namely the amount of growth and its timing. Together, the timing and amount of plant growth partially determine the physical condition in summer and calf survival (Eloranta & Nieminen 1986; White *et al.* 1989). A cow's physical condition, pregnancy and calf survival are tightly related (Thomas 1982; Skogland 1985; Eloranta & Nieminen 1986; Cameron *et al.* 1993).

Snow melt triggers green-up, the burst of plant growth which has the highest nutrient value (Chapin *et al.* 1980; Kuropat & Bryant 1983). Caribou calve either just before or during the green-up and they depend on foraging to provide for their calves (Parker *et al.* 1990). The timing of the availability of high quality forage is therefore a key element in calf growth and for the cow in replenishing her body reserves. The timing of green-up also influences forage digestibility (Merrill & Boyce 1991; Langvatn *et al.* 1996).

In contrast to caribou, muskox calve in late April and the cows burn up body reserves to produce milk. Peak milk production is about 2 months after calving, which is later than for caribou whose peak lactation is within 2–3 weeks after calving. Earlier muskox calving means that muskox calves have a 'jump-start' and are able to forage as full ruminants during the summer (Adamczewski 1995).

Summer weather also influences caribou summer condition when temperatures and wind speeds affect mosquito (Culicidae), nasal bot fly (*Cephenemyia trompe*) and warble fly (*Hypoderma tarandi*) activity (White *et al.* 1981; Russell *et al.* 1993). The insects drive caribou to stop feeding in an attempt to escape harassment, either by running from mosquitoes or watching for and evading warble and nasal bot flies. The combined effects can be severe enough to reduce body weight (White *et al.* 1981; Helle & Tarvainen 1984). However, the muskox's dense guard hairs and wool undercoat seem to be adequate protection at least against the mosquitoes.

Conception in August (muskox) or October (caribou) depends on cows reaching a condition threshold. Physical condition during winter, especially in the last third of pregnancy, largely determines fetal birth weight: light caribou calves are less likely to live (Eloranta & Nieminen 1986). Muskox cows are also sensitive to nutrition and body condition after ovulation in early winter and, if subjected to nutritional stress, may terminate their pregnancies (Adamczewski 1995). By contrast, intra-uterine fetal loss is rare in caribou (Dauphiné 1976; Cameron *et al*. 1993).

Relative forage availability depends on the amount of forage (plant biomass) and snow conditions. Caribou are energetically efficient at cratering through the snow to reach forage (Fancy & White 1985). Nevertheless, those costs are some 30% higher than just walking and costs increase with depth, density and whether or not the snow is crusted. Caribou draw on their body reserves during winter when snow reduces forage availability, and movements and digging for forage are energetically costly.

Weather does not push the system too far because the feedback loops (herbivore–forage biomass, forage biomass–forage growth) pull it back toward equilibrium and counteract fluctuations in the weather (Caughley 1987). Caughley and Gunn (1993) suggested that wolves and other predators may accentuate or dampen those swings around equilibrium. The extent of the relationship between caribou and their predators is controversial and the question of predation being a limiting or regulating factor is debatable. Discussions on these subjects usually attempt to correlate changes in caribou numbers with wolf numbers, but Bergerud (1996) has pioneered a description of caribou's evolutionary tactics to reduce predation risks. He has linked caribou migration to the calving grounds as spacing behaviour which reduces their exposure to wolves, whose strategy is to maximise their exposure to caribou by denning along the treeline. Bergerud (1996) aptly sums up the relationship between predation and weather: 'Caribou and wolves are in a predator × prey adaptive race with the extrinsic environment the arena. At times weather favours the prey and at other times the predator in this dynamic competition.' Specifics on how weather influences predation rates are usually lacking. Valkenburg *et al*. (1996) correlated a decline in the Alaskan Delta herd with increased wolf numbers, warm summers and deeper-than-usual snow. The weather reduced caribou physical condition which contributed to increased vulnerability to wolf predation.

Evidence for fluctuations in muskox populations is patchy. In the High Arctic, muskox numbers seemingly fluctuate on an irregular basis. Hubert (1977) describes the dynamics as 'boom or bust' with reproductive failures and adult mortality associated with severe winters (Miller *et al*. 1977; Gray 1987). Older muskox die when ice and snow restrict the availability of forage (Vibe 1967; Miller *et al*. 1977; Gray 1987; Forchhammer & Boertmann 1993); bulls and sub-adult are often the hardest hit (Gunn *et al*. 1989). Miller *et al*. (1977) estimated that the number of muskox on Bathurst Island declined by 70% between late March and late August 1974, with the decline being attributed to deaths following a severe winter. This pattern was repeated in the 1990s (Fig. 13.3).

Implicitly or explicitly, the assumption crept in that it was periodic catastrophic winters that were limiting muskox numbers, but south of the High Arctic whether muskox populations fluctuate or not seems less certain. Prehistorically, muskox were numerous enough to be important in the aboriginal cultures on the Mid-Arctic islands and mainland. Muskox virtually disappeared from Banks and Victoria Islands probably in the late 1800s (Gunn 1990). The decline preceded the changes in hunting patterns that followed the arrival of the whalers and traders but was coincident with possible weather-related catastrophes (Gunn *et al.* 1991b). Muskox rebounded on Banks Island increasing in number from less than 1000 before the 1970s to $53\,000 \pm 2200$ (S.E.) in 1992 (Nagy *et al.* 1996). The muskox increase on the NWT mainland only became apparent in the 1980s and not enough time may have elapsed to know if, and at what amplitude, muskox populations fluctuate. Alternatively, tighter coupling to their forage and more conservative reproductive strategies may imply that muskox are less at the mercy of weather vagaries acting on forage and predation than are caribou.

The relationship between caribou and muskox, weather and their forage is only part of their ecological system. The links between the herbivores and the large carnivores and their suite of scavengers are relatively visible. Wolves and grizzly bears kill, and wolverines, foxes and ravens scavenge. Less conspicuous is the coupling between the large herbivores, their forage and the smaller herbivores, rodents and birds such as ptarmigan, whose feeding on willow and cottongrass flower buds is similar to the caribou and muskox.

The ecological relationship between caribou and their insect harassers and parasites is, curiously, almost unexplored quantitatively. However, mosquitoes may be feeding across the barrens on as few as $100\,000$ to as many as $1\,000\,000$ caribou. What that 10-fold change may imply for mosquito population dynamics is unknown. Mosquitoes at different life-history stages feed fish and birds and, perhaps more significantly, as detritus feeders play a role in the turnover of nutrients in an ecosystem that is very dependent on detritivores to resupply soluble nutrients for plant growth. The links between caribou and other components of the ecosystem are evident, even if unquantified, and attest to a role in the tundra's biodiversity.

13.4 Level and types of utilisation

The NWT Wildlife Act gives the territorial government legislative authority to pass legislation for the 'preservation of game', except that 'restricting or prohibiting Indians or Inuit from hunting on unoccupied Crown lands' is prohibited. However, barren-ground caribou, muskox, polar bear and wood bison are exempt from this prohibition, because in 1961 the federal

Governor-in-Council declared those four species to be in danger of extinction.

The NWT Wildlife Act has not been used to limit aboriginal people when hunting caribou except on Southampton Island. The island was restocked in 1967 after caribou had been hunted to extinction and the harvest was restricted to help caribou numbers increase. Aboriginal people can hunt caribou for subsistence use but commercial use (meat sale) is restricted by quotas. Resident non-aboriginal hunting is limited by a set number of tags per hunter and other tags are allocated for guided sport hunting. The quotas are only allocated if the subsistence (domestic) use is less than the sustainable harvest. The rule of thumb is that if the subsistence harvest does not exceed 3–5% of the most current population estimate, then a commercial quota is possible (Table 13.1). Horns, antlers and skins may be sold or manufactured into goods and small meat plants commercially butcher and process meat from caribou and muskox for local sale.

Annual levels of harvest depend on where the caribou have moved relative to the communities – most hunting is in winter and caribou rotate their winter ranges every few years. The 1990–91 harvest levels are illustrative of the harvest and show a subsistence take of about 55 000 caribou with 4485 commercial tags available (GNWT, unpublished information). The economic value for the subsistence harvest can be estimated using an average of 45 kg of meat per caribou and a replacement cost based on the approximate beef price in smaller communities (CDN$14 per kg).

Table 13.1 Herd size, population trend and harvesting statistics for the large barren-ground caribou herds, NWT (GNWT, unpublished information).

Statistic	Bathurst	Bluenose	Qamanirjuaq	Beverly
Subsistence harvest	16 800	5000	7700	5300
Resident harvest (5 tags per hunter)	1470	202	265	125
Non-resident harvest (tags available)	500 (1170)	30	225	0
Commercial meat harvest (tags available)	(1470)	331 (950)	(285)	(450)
Herd size estimate ±S.E.	355 000 ±95 300	122 300	496 000 ±105 400	276 000 ±106 600
Year	1996	1992	1994	1994
Population trend	Stable	Unknown	Increasing	Stable

The harvests for commercial meat sales are conducted through either organised hunts using portable field processing facilities, or individual hunters harvesting the animals for processing and resale in local stores or local institutions, and for community feasts. The scale and cultural acceptability of commercial harvesting is diverse. The Inuit and Inuvialuit have a greater involvement in commercial harvesting for both caribou and muskox. Dene elders in particular express strong reservations about selling caribou meat and view it as disrespectful use.

Muskox hunting was banned between 1917 and 1967 to allow muskox populations to increase. Muskox harvesting is now regulated through the NWT Wildlife Act, under a quota which applies to aboriginal and non-aboriginal hunters. Quotas are mostly conservative, set at 3–5% of the current population estimate. Quotas are assigned to communities, and it is their decision as to whether the use is subsistence or commercial. Quotas have steadily increased from 2576 tags in 1987 to almost 8000 tags available in 1996. The Wildlife Ordinance was amended in 1979 to allow commercial use either for sport hunting or meat sales. The scale of, and the capital investment in, commercial harvesting varies from frequent small-scale harvests with individual hunters to small portable abattoirs handling up to 100–200 muskox to occasional large-scale harvests taking up to 1800 muskox (Gunn *et al.* 1991a; J. Nagy, pers. comm.). Commercial use remains at low levels and continues to be irregular largely because markets are still undeveloped.

The use of the 3–5% guess for harvest level is the extent of earlier efforts to estimate sustainable harvesting. Harvesting at 3–5% of current popula-tion estimates is low risk, as it is conservative. On the other hand, conservative harvesting strategies carry the risk of jeopardising credibility by missing economic opportunities. Harvesting strategies that trade yield against density were largely ignored, although this type of strategy is recommended for herbivores limited by their forage supplies. Caughley and Sinclair (1994) simplify the strategy so that the population is harvested at its rate of increase, which may require reducing density to accelerate rate of increase.

The current rule of thumb in the NWT is to base quotas and harvest levels on subtracting an assumed natural mortality from the recruitment rate (recruitment being the proportion of 1-year-olds entering the popula-tion). The weak point is the crudeness of the estimate of natural mortality and the fact that we apply an average figure to a range of different situations. Even low harvesting rates can induce over-harvesting if they coincide with increases in deaths from other causes. Caribou on Banks Island may have started to decline during a period when winter snowfall was increasing (Fig. 13.4). Increasing muskox populations may have supported more wolves which also preyed on the caribou, and the proportion harvested effectively increased as the population declined. The

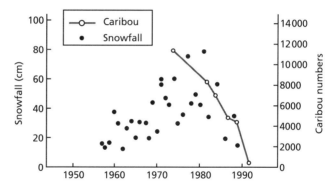

Fig. 13.4 Trends in snowfall (measured as snowfall in September to November at Sachs harbour) and estimated caribou numbers on Banks Island, NWT.

annual harvest during the 1960s and late 1980s was similar at 200–350 caribou but the estimated population was 9000 in 1972 compared to 5000 in 1985, 2600 in 1989 and 1000 in 1992 (Nagy *et al.* 1996).

For most populations, we track size using either visual or photographic counts obtained during aerial systematic linear transect surveys, the precision of which is usually increased by stratifying survey effort according to density (Heard 1985; Graf & Case 1989). Coefficients of variation are used to measure precision and are usually between 10 and 20%. The emphasis is on increasing precision so as to measure rates of population change. Accuracy is usually unknown and varies considerably with survey conditions. When accuracy has been gauged by comparing photographic and visual counts, visual counts of caribou on calving grounds are underestimated by a factor of 1.6 to 2.6 (Heard 1985). Other population parameters (fecundity, adult and juvenile survival) are irregularly obtained and vary in precision and accuracy.

The emphasis on rates of change in population size rather than population size *per se* is based on two concepts. Firstly, that it is not low population size but rather rate of decline that contributes most to extinction risk. Secondly, population trends sum the effects of all population processes so even if we are not monitoring individual factors, we are accounting for their effects.

The management of wildlife and fisheries largely deals with managing harvests but it is evolving to incorporate ecological processes rather than resources. Thinking only of resources can neglect the links between, for example, weather, herbivores and their forage. Management is starting to deal with the natural variability of ecological systems by incorporating risks (events with known probabilities such as years with high insect harassment or severe winters) and uncertainties (events with unknown probability). Inadequate data adds to the uncertainties. One approach to working with uncertainty is Pascual and Hilborn's (1995) use of Bayesian statistics in developing a model for harvesting wildebeest. The model coped with data uncertainties and inadequacies and it incorporated the dynamics of weather influencing forage availability. The conclusion was that harvest rates exceeding 6% ran the risk of population collapse.

Stable harvest levels are unlikely for the reasons prosaically stated by Hilborn and Walters (1992) in their summary for fisheries management: 'Few stocks have the potential for stable yields. Irregularity of production, cyclicity and cussedness are to be expected.' This applies well to managing wildlife. Subsistence harvesters are familiar with the vagaries of wildlife abundance and thus harvesting opportunities, but constancy of yield is sometimes assumed in planning for commercial harvesting.

Caribou and muskox differ in their interactions with their forage and in their ability to buffer environmental variations to the extent that their harvest strategies for sustained yield should differ. More needs to be known about muskox's functional and numerical responses to forage before the implications for sustained yield can be explored, in a model similar to Pascual and Hilborn's (1995) model for harvesting wildebeest. However, if muskox are more tightly coupled to their forage, then they may be more amenable to a conventional yield–density model for sustained yields.

13.5 Politics and institutional factors

Mining and government generate most of the NWT's cash income (Table 13.2). Most mining revenue currently comes from gold, silver, lead and zinc. The scale of non-renewable resource development is still low and has not caused widespread environmental changes, but the consequences of industrial development elsewhere in the world are reaching the NWT. The atmospheric and oceanic transfer of pollutants has reached measurable levels and there are concerns for human and environmental health.

In the 1990s, the dependence on the renewable resource component of the natural capital is high and is dominated by wildlife harvesting and fisheries. Agriculture is essentially lacking and forestry is limited. Forestry was more important when many houses were built out of logs and fuelled by firewood, but by 1995 forestry had shrunk to 225 jobs and a CDN$1.8 million GDP contribution to the economy. Fisheries was worth CDN$5 million in exports from the NWT and CDN$200 000 in local sales.

Table 13.2 Summary of gross domestic product by industry in millions of dollars (CDN$), NWT 1992 (Bureau of Statistics 1996).

Replacement cost of wild meat	Fishing/trapping	Government	Forestry/logging	Mining/oil/gas
60.0	3.3	392.2	1.8	345.7

Settled land claims cover a huge area of the NWT (Table 13.3). They have all selected authorities for resource management as institutions for public government, with authority for wildlife vested in wildlife management boards. The regime is co-operative in that the federal and territorial governments do not have seats on the boards but they nominate appointees to represent the public interest. The boards have authority subject only to overriding conservation, public safety and public health interests and although final authority is invested in the appropriate minister, the boards will make day-to-day decisions on wildlife. Co-management helps to ensure that aboriginal ecological knowledge is not lost and differing cultural views are aired.

The government has agreed with the land claim groups to a settling of priorities with respect to harvesting rights and harvest allocation and the land claim beneficiaries now have harvesting rights, the rights to trade edible products with other aboriginal people and the rights to trade non-edible products, as well as the rights to possess and transport wildlife parts in the NWT and in some cases outside the NWT.

The wildlife management boards are required to calculate a total allowable harvest and the first call on that harvest is the beneficiaries' basic needs level. The basic needs level is determined from harvest studies which rely on personal interviews with hunters to record harvest levels. If the total allowable harvest exceeds the basic needs levels, then the surplus will be allocated to non-beneficiaries, sport and commercial harvesting.

The large barren-ground caribou herds range across jurisdictions – national, territorial, provincial and land claim boundaries (Fig. 13.5). The inter-jurisdictional nature of the caribou herds has been long recognised. The technical and management committees for the Qamanirjuaq herd in the 1960s included provincial, territorial and federal representation. A significant step forward was taken in 1982 when the Beverly–Qamanirjuaq Co-management Board not only included jurisdictional representatives but an equal number of aboriginal users. In 1985, another co-management board was created for the Porcupine Caribou herd. The co-management boards broke new ground in bringing together the users and managers (Peter & Urquhart 1991; Thomas & Schaefer 1991), but the boards are

Table 13.3 Land areas (km^2) of settled land claims in the NWT. Other claims (Treaty 8, Treaty 11 Deh Cho and Treaty 11 Dogribs) are still being negotiated.

Agreement	Total area	Surface only	Surface and subsurface	Number of beneficiaries
Nunavut	1 942 500	315 255	34 395	17 500
Sahtu	280 238	41 400	1813	2000
Inuvialuit	168 350	77 700	12 950	2500
Gwich'in	56 980	16 265	6159	2200

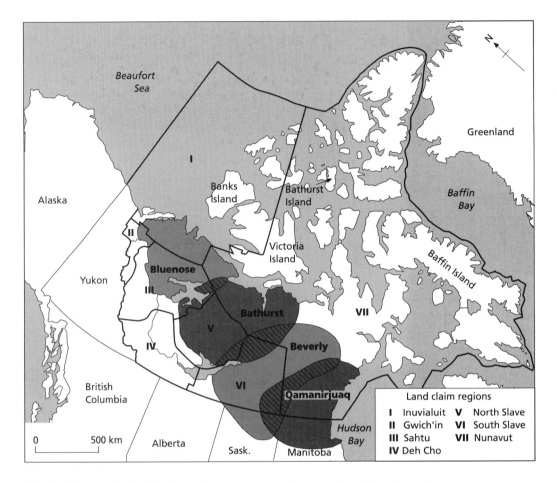

Fig. 13.5 Ranges for the four largest barren-ground caribou herds, and jurisdictional boundaries.

costly. More recently, the co-management for the Bluenose and Bathurst caribou herds is tending away from establishing boards. Instead, workshops and community tours are undertaken to develop management plans that balance aboriginal and scientific knowledge and bring the different users together to negotiate sharing the total allowable harvests for the herds.

13.6 Conservation and harvesting in the NWT

The NWT's socio-economic and political circumstances have changed so rapidly and extensively that neither the past nor the present harvesting patterns are a guide to caribou and muskox's future. The ecological system is itself resilient but a combination of circumstances could destabilise the system and those changes could come from within or outside the NWT or

Canada. For example, a possible combination of events would be global climate change shifting the relationship between herbivores, their forage and, in the case of caribou, insect harassment. The shift could start a caribou decline, while coincidentally, another unrelated influence could also be affecting the caribou. For example, the European wild fur market could have collapsed. Harvesting wolves for their pelts would then dwindle and, consequently, wolf predation would increase, accelerating the caribou decline. At the same time, commercial harvesting could increase to replace the income lost from fur sales. The point is not so much to argue against the anti-fur lobby – the same argument would apply if mineral prices dropped and mining in the NWT slumped – but to emphasise the vulnerability of ecological systems to the cumulative effects of unrelated events.

The co-management boards are beginning their tenure when, fortuitously, most caribou and muskox populations are increasing. The mixture of small-scale commercial harvesting with subsistence harvesting is successful in that it meets community expectations for some cash and a secure meat supply. When the downturn in size begins, as is inevitable with fluctuating populations, and the aboriginal basic needs come close to or even exceed the total allowable harvests, the reality of the change from a common property resource to limited access will hit home. Land claim allocation legislates that non-beneficiary, sport and commercial harvesting would cease before beneficiary hunting was curtailed.

When governments hold the proprietary rights to a resource, the conservation track record is often poor because individuals do not have a proprietary interest. Shifting the proprietary rights for wildlife to the collective rights of the land claim groups increases the incentive for individual harvesters to conserve their resources. However, both guided sport and commercial harvesting generate cash and thus lead to economic expectations and dependence. The temptation will be to wait, perhaps until there are more convincing data or until a decline is glaringly evident. Demonstrating population changes is difficult when population estimates lack precision. The problem then becomes one of timing – to make the politically unpalatable decision to reduce commercial harvests before they accelerate declines and subsistence harvests have to be drastically reduced.

The NWT's economy has slowed since the 1980s resulting in fewer jobs. Meanwhile, we have an increasing population and one with higher expectations of joining the workforce. The government's focus is on developing resources to boost the economy and thus create jobs. Commercial harvesting and guided non-resident hunting will be encouraged beyond the relatively low level at which they now operate.

The history of commercial use in the NWT and elsewhere in the world leaves little doubt that it runs the risk of inducing population declines. The argument that the previous commercial harvesting was unregulated, which is why it caused declines, leaves little room for complacency. Even with regulated quotas, commercial harvesting builds momentum from harvesters' economic expectations, capital amortisation and the discount factor (Hilborn & Walters 1992; Caughley & Gunn 1996). The answer may be that commercial operations have to stay at a small enough level to allow flexibility and prevent capital investment on a scale that then becomes the driving force.

Besides problems stemming from the economic imperative, there is also the problem of how attainable the objective of sustainable harvesting is, either technically or in terms of acceptability to harvesters. Sustainable harvesting does not mean that levels will not vary both proportionately and numerically. The amplitude of the variation will depend on the success of setting harvest levels in step with natural population fluctuations. It is a risk to assert that, given our present understanding, sustainable harvest levels can regulate population fluctuations. It is more realistic to aim to have sustainable harvests track population fluctuations. This is closer to the hunters' presumption, from generations of hunting, that yields vary over time. The need is to build on this experience and scientific understanding of population fluctuations to be able to synchronise harvesting levels and population fluctuations.

Precedents for conservation have already been set by individual hunters and trappers associations even before the establishment of co-management boards. In the High Arctic, the Inuit took it upon themselves to stop hunting Peary caribou when caribou numbers dropped in the 1970s. A recent example was in the 1990s. The Inuvialuit co-management board was involved and supported hunters on Banks Island where they decided to reduce their caribou harvesting when faced with declining numbers (Nagy *et al.* 1996). Compared to the fatalistic attitude that the caribou disappear and return, which weighs against conservation, there is hope that the co-management boards will be able to meld aboriginal and scientific knowledge to balance subsistence use and conservation.

Chapter 14: Hunting of Game Mammals in the Soviet Union

Leonid M. Baskin

14.1 Introduction

This case study aims to reveal the reasons for changes in game mammal abundance in the former Soviet Union over the Soviet period. Most game mammals increased in abundance over this period. Hunting was often a stimulus for mammal conservation and investment in wildlife. By the end of the Russian Empire in 1917, most game mammals were almost extinct due to unchecked exploitation, but in the Soviet period, regulated hunting was helpful in promoting game mammal conservation. The dramatic increase in game mammal numbers was also due to the appearance of refuges where they could breed without disturbance. Nature reserves and areas which humans abandoned, or were expelled from, served as these refuges. In the Soviet Union commercial hunting was under strong state control, and in densely populated regions of European Russia commercial hunting was replaced by recreational hunting. In the Russian forest zone recreational hunting did not significantly reduce game mammal populations, but in Central Asia and the Caucasus, badly regulated recreational hunting and poaching led to the extermination of many species of ungulates and large predators (see Fig. 14.1 for place names). Success in promoting species conservation and re-establishment depends strongly on the ecology of the species concerned; most game mammals are well adapted to the environmental changes caused by human activity.

14.2 The history of game mammal use

The pre-Soviet period

In the past, the hunting grounds of central European Russia were controlled by royalty, aristocrats, monasteries, and peasant communities. The huge hunting grounds of northern Russia, the Urals and Siberia were shared between native families and there was no hunting regulation. For example, on Chukotka, when Chukchi hunters harvested reindeer on river crossings, walls of animal carcasses were observed along river banks (Vdovin 1965). This unchecked exploitation drastically reduced the numbers of most game mammals. The decline of game mammals, starting in

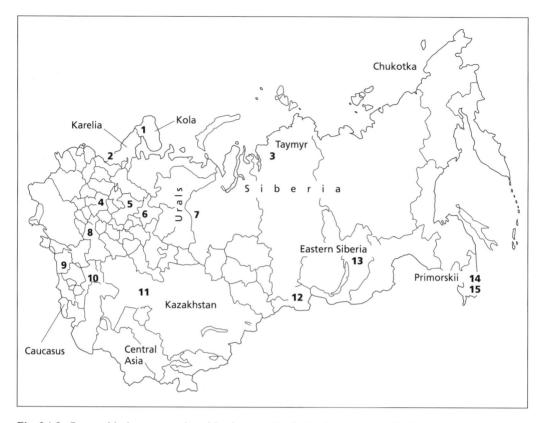

Fig. 14.1 Geographical areas mentioned in the text: (1) the Lapland reserve; (2) St Petersburg; (3) the Pyasina and Dudypta rivers; (4) Moscow; (5) the Kostroma province and the Unzha-Vetluga region; (6) the Kirov province; (7) the Konda-Sos'va river basins and the Kondo-Sos'va reserve; (8) the Voronezh reserve; (9) the Caucasus reserve; (10) Kalmykia; (11) the Torghay lowland; (12) the Sayan reserve; (13) the Barguzin reserve; (14) the Sikhote-Alin reserve; (15) the Lazov reserve.

the 17th century, forced the tsars to restrict hunting. Peter the Great banned moose (*Alces alces*) and roe deer (*Capreolus capreolus*) hunting, and in areas surrounding St Petersburg and Moscow, hunting was permitted only with special licences. Also, Catherine the Great banned hunting from 1 March until 29 June except for brown bear (*Ursus arctos*), wolf (*Canis lupus*) and fox (*Vulpes* sp.).

The most significant and thorough hunting legislation of the Russian Empire was adopted in 1892. It determined, among other things, hunting rights and seasons, and state control. However, like other hunting regulations of the Russian Empire it was only concerned with central European Russia. In most northern European, Ural and Siberian regions hunting restrictions were not established. The legislation of 1892 permitted the hunting of large and small predators, excluding sable (*Martes zibellina*), all year round and using any method except poison (Solov'ev 1926). Game mammal numbers continued to decline, but only the commercially valuable sable was protected.

The Soviet period

In both the Russian Empire and the Soviet Union hunting was practised for subsistence, commercial and recreational purposes, but with the advent of the Soviet Union, the ratio of hunting types changed. In the Soviet Union, subsistence hunting was almost totally replaced by commercial hunting. Recreational hunting involved large numbers of people and had a significant political role; it gave people a pastime, distracting them from political activity. Game mammals as state property, hunting legislation as state business, and the state monopoly in the fur and meat trades all corresponded with communist ideology. The early leaders of the former Soviet Union considered re-establishing game mammals to be part of an ideal image of the future communist paradise. Also, many of the top leaders were enthusiastic hunters, so they reacted favourably to the requests of zoologists and conservationists. Reserves were established in the areas with the most suitable conditions for the species needing protection. In 1924, the Caucasus reserve was established to conserve European bison (*Bison bonasus*), Severtsov goat (*Capra caucasica*), red deer (*Cervus elaphus maral*), and chamois (*Rupicapra rupicapra*). In 1927, the Voronezh reserve was established to conserve beaver (*Castor fiber*) and red deer (*C. elaphus hippelaphus*) populations and the Sikhote-Alin reserve was established in 1935 to conserve tiger (*Panthera tigris*), goral (*Nemorhaedus goral*), red deer (*C. elaphus xanthopygos*), moose (*A. alces cameloides*) and sable. In 1919, when the civil war was at its height, the Soviet government issued a decree banning moose and roe deer hunting. By 1929, the system of hunting management was in place, hunting legislation had been developed, and training centres for hunting managers were established. The law of 1930 laid out the procedure for sharing hunting grounds between the state and hunting co-operatives and societies, the rights and duties of hunters, and the tasks and rights of hunting reserves. Licenses to hunt for ungulates and valuable fur-bearing animals were important for hunting regulation. The fur trade monopoly was given to a special state syndicate which was part of the Ministry of Foreign Affairs. Up until World War II, furs provided very significant export revenues for the country.

After World War II, Stalin conducted a more pragmatic policy. The State Plan of Nature Transformation was announced, included increasing huge forest plantations in steppe areas and building new water reservoirs. At the same time, a significant proportion of game reserves were closed to the public, and from 1946, cropping sable, otter (*Lutra lutra*), marten (*Martes* sp.) and desman (*Desmana moschata*) was licensed. However, Khrushchev gave more preference to logging than nature conservation and the forests of some of the best reserves were cut. Kurushchev also contributed to increasing privileged state hunting management for the top leaders. In the 1960s, the cropping of ungulates (moose, reindeer (*Rangifer tarandus*), saiga (*Saiga tatarica*), wild boar (*Sus scrofa*), roe deer) became more economically significant. Hunts for ungulates and bears, and participation in commercial hunting, became popular among citizens.

In Brezhnev's time, game husbandry, with rigorous protection, intensive management and reintroductions, was organised in all regions. Since 1971, harvesting moose, red deer (*Cervus* sp.), sika deer (*Cervus nippon*), roe deer, reindeer, snow sheep (*Ovis nivicola*), mountain sheep (*Ovis ammon*, *O. vignei*, *O. argali*), goats (*Capra cylindricornis*, *C. sibirica*), saiga, wild boar and musk deer (*Moschus moschiferus*) has been licensed. The law of 1980 reflected most completely the total central control of hunting in the Soviet Union. From Moscow and the capitals of the union republics, authorities determined hunting rules, and confirmed regional proposals on the number of licenses issued to hunt ungulates and fur-bearing animals. The law of 1980 established annual game animal censuses, the single state system of game mammal registers, and annual plans for animal protection.

The totalitarian regime of the Soviet Union was fortunate for game mammal conservation. In particular, the 19 years of Brezhnev's leadership and the 20 years before World War II were especially beneficial for the conservation and sustainable use of game mammals. Particularly in Stalin's time, and despite the protests of hunters, hunting grounds were often withdrawn to establish privileged hunting reserves for top people, military zones, and nature reserves. On these occasions, the disturbance to game mammals lessened and investment in management became plentiful. A total of 150 reserves had been established in the Soviet Union by 1991.

14.3 The ecological features influencing the recovery of game mammal species

The effects of the Soviet period on the abundance of particular game mammal species depended on the peculiarities of the species' ecology. Among species with very narrow ecological niches and small range areas, goral maintained approximately constant numbers and range in spite of protection over about 70 years in the Sikhote-Alin and Lazov reserves (Mislenkov & Voloshina 1989). On the other hand, the racoon dog (*Nyctereutes procyonoides*), which was endemic to the Primorski district, was introduced to European Russia and has unexpectedly spread all over eastern Europe and even become numerous. Omnivorousness and the capacity for winter hibernation gave it advantages compared to fox and lynx. By 1990, there were 191 500 racoon dogs in Russia, and only 26 000 of them were in the Primorski district.

Resilience to hunting was often connected to changes in a species' escape behaviour. Thus, wild boar and brown bear have survived in densely populated areas due to their hiding behaviour. Humans detect animals mainly by vision and hearing, and a hidden animal has a good chance of remaining unnoticed. Very alert mountain sheep run away from their home range when disturbed, cross steep ravines, and become quiet only after escaping from human view. Hunters need to spend a lot of time crossing rugged country, finding the mountain sheep and stalking them again. This behaviour is the main reason for the survival of mountain sheep in many areas of Central Asia, where they suffer from heavy poaching (Baskin 1976). In the Soviet Union, a significant part of the agricultural crop was abandoned on the fields. At first, this happened in the World War II years, and later, since the time of Khrushchev, it has become usual. Crop remains provide a food supply for wild boars. During the 1960s and 1970s, wild boar numbers increased and their range area expanded 800 km to the north.

Being highly intelligent and able to learn quickly, wolves are able to escape poisoned baits and hidden traps. Wolves have been exterminated only in areas where they could not find appropriate shelter. For the wolf, hilly areas with rough relief at the headwaters of big rivers are best for denning and raising cubs. In suitable areas, this species becomes very resistant to extermination and in European Russia some areas are famous as places where the wolf always survives (Pavlov 1990). In open areas, wolves are usually chased by hunters using helicopters and planes. In forested areas a popular method is putting a rope with red cloth flags around a resting group. For unknown reasons wolves are afraid to cross this 'fence' and hunters can chase the wolves towards a line of shooters. However, these methods have never led to the total extermination of the species, only to its numbers becoming small enough to reduce livestock losses.

Some habitats are especially important for game mammal survival. Logging in the Soviet Union increased the area of young forests dramatically. The replacement of mature coniferous forests with young forests influenced the ecology and numbers of forest mammals. At the beginning of the 20th century, squirrel (*Sciurus vulgaris*) was one of the most popular species that was a target of professional hunting. In 1940, the maximum number of squirrel pelts was obtained. Later, cutting of mature coniferous forests where squirrels find cones, their main food, led to a rapid decline in squirrel numbers, as shown by a decline in the number of pelts harvested (Table 14.1).

Martens (*Martes martes*) demonstrate a capacity to adapt to young forests and clearings. An early view was that martens are closely connected to mature coniferous forests, where squirrels are their preferred prey. After

Table 14.1 Data on (a) annual harvests and (b) population sizes of some game mammals in the Russian Federation. Compiled using data from Kaplin (1960); Grakov (1973); Dezhkin (1978); Monakhov & Bakeev (1981); Bannikov *et al.* (1982); Dezhkin *et al.* (1986); Sokolov & Lebedeva (1989); Pavlov (1990); Poletskii (1990); Borisov *et al.* (1992).

(a) Harvest (1000s of individuals)

Year	1910	1930	1940	1950	1960	1970	1980
Squirrel	2027	1809	9120	6985	5234	3823	2934
Beaver	—	—	—	—	—	—	11
Sable	10	11	1	41	153	118	149
Marten	24	33	13	33	57	38	26
Wolf	19*	30*	17	37	9	5	5
Bear	?	5	7	6	3	1	2
Moose	?	?	?	?	22	24	57

*Exported from the Russian Empire or Soviet Union.

(b)	1980 harvest (% of population size)	Population size (1000s)	
		1980	1990
Squirrel	25	11 830	15 324
Beaver	7	154	262
Sable	21	723	710
Marten	15	166	175
Wolf	14	36	23
Bear	2	80	130
Moose	16	351	490

logging, martens have been found to adapt to young forests and clearings where they eat small rodents, hares and berries. Martens find and catch these foods more easily than squirrels and, as a result, their numbers seem to have grown (Table 14.1). A significant correlation (0.7–0.9) has been found between volumes of logged timber and the number of marten pelts harvested. Logging is useful for martens only if 25% of the mature forest is conserved. Old trees are important because martens find shelter in tree hollows (Grakov 1965; Krasovskii 1970).

Since the 1960s, a significant proportion of the forests inhabited by sables have been cut, but sables are able to adapt to life in young forests. However, they need shelters (hollows in old trees, holes under tree roots) because, in spite of the sable's famous fur, it freezes without shelter in the hard Siberian frost. Remnants of mature forests inside large logged areas are important for the welfare of this species (Gusev 1971). For brown bears, large plots of forest (at least 87 km^2) are necessary, as well as quiet, remote places where bears can spend the winter in undisturbed dens (Baskin 1996). However, it seems that brown bears also did not react significantly to the dramatic growth in logging (Fig. 14.2).

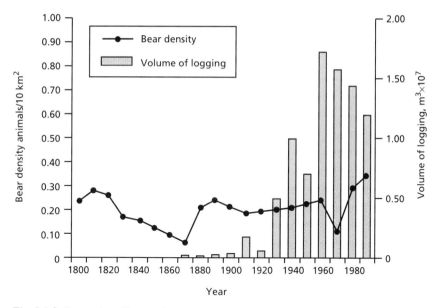

Fig. 14.2 Dynamics of brown bear numbers and logging volumes in Karelia (1800–1990). (Compiled from the data of Silant'ev 1898; Iznar 1909; Bitrakh 1926; Blagoveshchenskii 1956; Pervozvannii 1959; Shcherbakov & Volkov 1985; Danilov & Tumanov 1993.)

14.4 Hunting management

In the Soviet Union a hunter was considered professional if hunting gave him a basic wage. This was possible in areas where fur-bearing animals were numerous, or where the hunter could kill many meat animals or obtain exotic products such as wolf or tiger cubs. For example, commercial hunting of reindeer was productive on river crossings, when hundreds or even thousands of animals could be slaughtered in a day. Professional sable hunters could trap 70–90 animals over a hunting season (October–January), whereas amateurs caught only 5–10 sables. The difference depends on the quality of hunting territories; the best were situated on sable migration routes, and were occupied by the professional hunters. The success of arctic fox (*Alopex lagopus*) hunting depended on the autumn activity of hunters, who needed to distribute pieces of walrus or seal carcasses over their hunting territories, so as to attract arctic foxes to places where traps were situated. A hunter who owned a tractor did this job better and had the chance to trap up to 120 arctic foxes during a hunting season. Obviously, the number of professional hunters could not be too high. Of the 3 million hunters who had hunting licences in the former Soviet Union only 20 000 were professionals, but 80 000–100 000 hunters became professional only for the autumn/early winter period (Poletskii 1990).

Often, commercial hunting used some weak points in a species' ecology. For brown bears, the weak point is the denning period, when the animals are helpless. At the start of the 20th century, brown bears suffered from professional hunters who searched for dens and sold them to rich amateurs (Mel'nitski 1915). Nowadays, there are no professional hunters for brown bears, and fortunately for brown bears, amateurs are not very effctive hunters. For example, in the 1980s amateur hunters killed annually about 1200 brown bears, which comprised 24% of the licences issued. This did not reduce brown bear numbers; conversely, in 1981–90, brown bear numbers in the Soviet Union increased from 74 400 to 130 000 (Sitsko 1983; Borisov *et al.* 1992).

The general approach of the Soviet hunting managers, which corresponded with the state ideology, was 'numbers of populations and volumes of harvest must grow each year'. Obviously, this approach often contradicted reality as there were not adequate conditions for top-class management. In open areas, censuses from planes gave enough reliable data, but in forest areas counting animal tracks was the predominant method and the data were often doubtful (Baskin & Lebedeva 1987). Illegal harvest by poachers was 10–25% of the legal harvest. Because ungulate hunters mainly used rifles, many wounded animals were lost. For these reasons management mistakes were very likely, and managers preferred to set a lower annual harvest than was possible for sustainable use. Often, managers permitted hunting to start too late in populations which had reached high numbers. Also, decisions to stop hunting were often made too late, when populations had dropped below a critical level for survival.

Re-introduction was an important method used by managers to restore the numbers and range of mammals that had become locally extinct. A good example is the re-establishment of the beaver population in the Kostroma district (Fig. 14.3). Another example is the re-establishment of red deer. By 1920, the total number of red deer in the European part of the Soviet Union (excluding the Caucasus population) was about 250. Red deer were re-introduced in several places and captive breeding programmes were instituted. In 1980, the number of red deer in European Russia reached 30 000, and in the Caucasus reserve and neighbouring areas, numbers of red deer increased from 500 in the 1920s to 13 000 in the 1980s. Red deer hunting was always the preserve of important people who hunted a small proportion of the population increment, which was 17% per year on average (Drew *et al.* 1989).

14.5 Examples of management successes and failures

Moose: successful protection

The moose population of the Unzha-Vetluga rivers region is one of the key populations in Central Russia. Forested areas make up more than

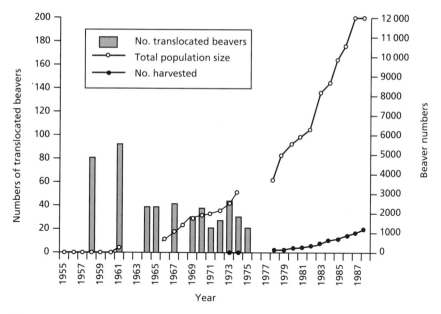

Fig. 14.3 Re-introduction, population growth and harvest of beavers in the Kostroma province.

80% of the territory. In the 1920s, in the period of the great moose depressions of European Russia, the region served as the moose refuge. From here moose settled again over European Russia. Legal cropping of moose started here in 1969; the annual harvest increased slowly from an estimated 3% to 10% of the population size (Baskin 1994), and the population now seems secure.

Reindeer: protection, but with problems

Reindeer can be hunted easily on river crossings (Fig. 14.4). In the past, native people came to river shores to harvest reindeer as they crossed rivers. Often, they obtained thousands of carcasses and used them over the long winter. Unfortunately, this merciless cropping led to reindeer extermination in some places, so in the Soviet Union hunting on river crossings was banned because of the danger that it posed to reindeer populations. It took about 40 years before reindeers became numerous again. The following examples demonstrate what happened to two different populations following this recovery.

By 1961, the Taymyr reindeer population was estimated at 200 000. During the 1960s and 1970s, 96 100 reindeer were harvested in total, but this was not enough to stop rapid population growth, and by 1971 the Taymyr reindeer population exceeded 600 000 animals. Scientists proposed that hunting on crossings should be permitted as the only appropri-

Fig. 14.4 A herd of reindeer (*Rangifer tarandus*). (Photo by L. Baskin.)

ate commercial hunting method (Geller 1967; Pavlov *et al.* 1976). After a special government decision, cropping on rivers started. Reindeer were shot in the water from motor-boats, and towed behind the boats to meat-packing points and refrigerated ships. With this technology, in spite of an annual harvest rate of up to 30%, the population remained stable. It was a good example of sustainable use. However, in 1993 migration routes suddenly changed. Reindeer ceased to cross the Pyasina and Dudypta rivers, where the most appropriate harvesting points were located. Now, herds migrate far from human-populated areas. It is impossible to transport many carcasses by snowmobiles, and using helicopters is unprofitable. By 1993–95, only 10 000–12 000 reindeer were obtained annually, i.e. the harvest decreased to the level observed before 1960–70. Kolpashchikov (1993) estimated that there were 536 000 reindeer in Taymyr. The annual increment is probably 100 000–130 000, and now there is a real danger of a population crash if food resources become exhausted.

The dramatic history of reindeer inhabiting the Lapland reserve on the Kola peninsula and neighbouring areas provides an example of a decision to cease cropping that was delayed too long. The Lapland reserve was established in 1930 with the main task of re-establishing a reindeer population. The protected population grew quickly with an average annual increment of 22%. By 1957, the population numbered 1964 animals; by 1961, there were 4396 animals; and in the late 1960s, more than 12 000. As the pastures had a limited carrying capacity, reindeer hunting on the Kola peninsula was started in 1970. It is a very rough and forested area, where commercial hunting is only possible from planes and helicopters. By 1976, about 10 000 reindeer had been killed (Makarova 1989) and it was then disclosed that almost all the reindeer had been slaughtered. Scientists had believed that the Lapland reserve reindeer would form an emergency reserve for the population on the hunting grounds. The disaster had occurred because most of the reserve reindeer migrated from the reserve into the adjacent areas, where they were killed. By 1977, a total ban on reindeer hunting was imposed. In 1992, we found only 200 reindeer in the Lapland reserve during a plane census.

The saiga antelope: success until the Soviet collapse

In the 19th century, saiga antelopes were numerous, and were the subject of professional hunting. Nebol'sin (1855) states that in the 1840s and 1850s, merchants exported 344 447 saiga horns from the western Asiatic steppes to China. Silant'ev (1898) states that in the late 19th century, some 100 000 saiga antelopes were cropped, but by the 1930s, saiga had become almost extinct. In Kalmykia and Kazakhstan only a few hundred animals survived (Fadeev & Sludskii 1982). Saiga hunting was banned in 1932. In Kazakhstan, the winters of 1927–31 and 1932–33 were extremely severe; millions of domestic animals died, as did saigas (Sludskii 1963). A great starvation of the human population followed and many people died. Most of the nomadic settlements disappeared leaving large areas of the Torghay lowland available for saigas. Furthermore, native Kalmyks were resettled in Central Asia, leaving the pastures of Kalmykia free for saigas.

Saigas recovered quickly and censuses carried out between 1948 and 1950 estimated 100 000 saiga in Kalmykia and 900 000 in Kazakhstan (Sludskii 1955; Bannikov et al. 1961). In 1951, saiga harvesting was permitted in Kalmykia, and in 1954 in Kazakhstan. The populations continued to grow because recreational hunting was not able to stabilise saiga numbers. The breeding rate of the populations was high: in the best years it was 40–50% of winter numbers; in hard years (i.e. no food, diseases) it was 10–20% (Zhirnov 1982). Conflict with agriculture increased and saiga spread north into the agricultural zone. Commercial hunting then began. Teams hunted for saiga at night using trucks with spotlights, as well as net corrals. Hunters chased animals into the corrals using motorcycles and 200 000–250 000 antelopes were slaughtered annually. After censuses, hunting managers and scientists estimated the number of saiga which could be hunted sustainably. According to Zhirnov (1982), in the best years the crop must be 20–25% of winter numbers; in the hard years it needs to be 8–10%. The species has kept its high numbers up to now, in spite of merciless extermination by poachers who are interested in selling saiga horns abroad for eastern medicines.

Tigers: slight improvements

The 1930s and 1940s were critical for tigers. Kaplanov (1948) considered that at that time, there were only 30 tigers left in the Primorski district, and suggested two main reasons for the major decline in tiger numbers. One was the decline of the wild boar population, which was the main prey of tigers, and the other was unchecked professional hunting by Koreans and Chinese hunters for eastern medicine. Two reserves were established where the main tiger habitats were situated. Killing tigers has been banned since 1947, and since 1956 catching tiger cubs alive for zoos has been halted. This was because a female had usually been shot for her cubs to be caught alive. These restrictions had some success: in 1985, Pikunov (1990) estimated about 240–250 tigers inhabited the Primorski district.

Sables and beavers: recovery to sustainable use

Even in the Russian Empire, sable were protected from hunting. Silant'ev (1898) was an initiator of the ban on sable hunting in 1913–16, which included all trade in sable pelts. In 1915–16, the Barguzin and Sayan sable reserves were established. Even so, by the 1930s sable had become almost extinct in many areas, and it had become rare in the others. By the 1950s, numbers had been totally restored, as shown by the rise in the number of pelts obtained by hunters (Table 14.1). The same numbers of pelts were cropped in 1960–80 as in the 18th century, but populations showed no evidence of decline, and some continued growing. For sable to be restored, hunting was banned for some years, and some reserves were established. Also 19 200 sables were translocated to other areas (Monakhov & Bakeev 1981). Sable populations were sustainably used in the former Soviet Union. The ideal cropping rate was 25% of the winter population size although the cropping rate usually fluctuates around 21–30% (Khlebnikov 1977; Monakhov & Bakeev 1981).

Unchecked utilisation led to beavers becoming almost extinct in the Russian Empire. Only four beaver populations with a total number of 800–900 animals were known to have survived. These populations were conserved, and later 12 387 individuals were translocated to other areas (Pavlov et al. 1974). Harvesting restarted in 1963, but hunters had to have a licence to trap beavers. Selling beaver skins other than to the state was banned. The annual harvest was about 5% of total population numbers (Dezhkin et al. 1986). Beaver numbers continued to grow and in 1987, a census of the whole territory of the Soviet Union recorded 194 300 beavers.

14.6 Local people and wildlife

Although many of the changes in game animal status were caused chiefly by the state intervention discussed above, local people were also important to the survival of these species. In Central Asia, for example, high human population density and heavy poaching led to the steady extermination of mountain sheep and goats, as well as large cats. Even reserves did not reliably protect rare species. The fact is, the strip along the Soviet border was their only refuge, because border guards prevented locals from entering and hunting there.

Throughout the Soviet period, social cataclysms and mass human displacements led to the disappearance of human populations over huge territories, leading to enormous refuge areas for wildlife. When Stalin's state policy to root out the nomadic way of life was carried out, many small settlements of nomads disappeared. In Khrushchev's time, masses of rural people abandoned their villages and moved into towns, and in Brezhnev's time, the unreasonable policy of concentrating the rural population into large settlements also made huge areas uninhabited. As a result, the number of hunters, who had been numerous in small remote villages, sharply declined. Some natives were expelled from their territories (Chinese, Koreans, Kalmyks, Finns, Karachaevtsi). Mostly, they lived in remote areas of the Soviet Union in small settlements. Many men of these native tribes were hunters; for the Evenks, Nganasans, Khanti and Mansi, hunting was the national profession. They were turned into reindeer herdsmen and were also forced to live in settlements, where they worked as fishermen and loggers. As a result, the exploitation of hunting grounds around settlements grew but remote areas became totally unused.

In the Soviet period, large predators became very numerous. In 1990, according to official data (Borisov *et al.* 1992), there were 22 470 wolves, 130 000 brown bears, 46 370 lynx (*Felis lynx*) and 29 790 wolverines (*Gulo gulo*) in Russia. Wolves, jackals (*Canis aureus*), wolverines and foxes were considered pests, and a reward was paid for killing them (foxes only in years when rabies occurred). Their population growth depended on their own adaptive capacity. Control of large predator numbers becomes especially vital if a local human population sounds the alarm. In most areas of the former Soviet Union people have an indifferent attitude to neighbouring large predators as long as the predators do not appear in view. For example, it is common on winter nights for wolves to visit Russian villages to catch dogs or sheep, but only predation of domestic animals evokes demands for a reduction in the number of predators.

Also, tigers and polar bears (*Ursus maritimus*) sometimes appear in human settlements. The removal of insolent individuals has been found to help preserve the tolerance of society for the protection of large predators. The biggest problems are in eastern Siberia, where in some years brown bears attack people quite often. There, starvation is a regular event of brown bear ecology, caused by the failure of their basic food crop (Siberian pine cones, berries). Bears cannot go into hibernation, so instead they roam, searching for food. Elimination of roaming, dangerous individuals has become the only method to improve the situation in such years (Ustinov 1972; Zavatskii 1987). In Russian history, there have only been two periods when wolves hunted humans actively: in 1849–51, 260 adult humans and 110 children were killed by wolves in Russia (Koritin 1990); in 1945–51, at least 25 wolf attacks on children up to 15 years old were documented in the Kirov province, an area inhabited by the biggest wolves in Russia (Pavlov 1990). A shortage of hunters during the war was probably a reason for the high wolf density and their abnormal behaviour. In other years, when hunters limited wolf numbers and kept their alertness at the necessary level, attacks against humans were recorded very rarely.

The changes in status of the sable and beaver populations of the Konda and Sos'va river basins provide a good example of the interaction between the influences of local people and the state in the conservation of game animals. Up until the 1960s, the elevated parts of these basins (on the eastern slopes of the Ural mountains) were occupied by excellent mature pine and spruce forests, and the lowlands contained marshes. These were very good conditions for sable and beaver. Before 1929, tribes of Khanti and Mansi native peoples owned these hunting grounds and they protected them against invaders. The tribes, whose population sizes were small, trapped only a limited number of sables on their hunting grounds, aspiring to sustain a constant volume of prey. Beavers were their totem animal, and hunters trapped them only for religious purposes. In 1929, the Kondo-Sos'va reserve was established. Sable and beaver numbers had declined very quickly because the Khanti and Mansi had lost their rights over the hunting grounds. The sable density in 1920 was 2.4 in $100\,km^2$, and in 1929 it was only 0.3–1.2 in $100\,km^2$. Only 200 beavers survived in an area of $8000\,km^2$ at the time the reserve was established. By 1951 (when the reserve was closed by Stalin) sable numbers had increased 40 times and beavers five times. Numbers of both species continued to grow in spite of intensive hunting. Hunters trapped sables preferentially near settlements because other areas were inaccessible (Nasimovich & Shubnikova 1969). In the 1960s mass logging of forests started. Railways and roads were built for timber transportation and many new settlements appeared. Hunters exploited both species without any control. However, although sable disappeared in the $100\,km$ strip along the railway the population kept its high density. Sable found good feeding conditions in young forests and enough shelters, because loggers left many areas of mature forest in places inaccessible for machines (Ponomarev 1971). However, the situation had become worse for beavers. Because of the logging, rivers became shallow and floods continued for long periods. Cutting the watershed forests was especially bad for beavers because rivers which start from watersheds carry out clay alluviums and improve conditions for birch and willow growth, which is important for beaver existence (Skalon 1951).

14.7 Conclusion

Over the 74 years of the Soviet Union, the numbers and ranges of most game mammals achieved maxima. At least, the cropping of most game mammals in the last decades of the Soviet Union was greater than ever before, and seemed sustainable. This contrasts with the over-utilisation which was the most unfortunate factor for game mammals in the Russian Empire. Hunting regulation and the provision of refuges were the main factors leading to the restoration of game mammals in the Soviet Union. The experience of the Soviet Union provides a striking example of the concept of sustainable use as a basis for conservation. Sustainable use was the main stimulus for game mammal protection. Game mammals formed an essential economic resource of the country, that compelled the investment of money in protection and population increase. Besides, nature conservation had an ideological significance as a sign of careful social policies.

Shortages and the bureaucratic inertia of management diminished the results of game mammal utilisation. In republics with high human population densities and a degraded natural environment, as well as in areas with high poaching levels, conservation of game mammals was unsuccessful. Also, the ecological and behavioural features of some species were not conducive to their conservation. As a result, some species stayed endangered and in these cases only a total ban on hunting gave positive results. For some species (tiger, saiga, musk deer) a state monopoly on foreign trade was useful, but now, because the Soviet Union has collapsed and the state monopoly is cancelled, these species are being exterminated for the sake of Far Eastern medicines.

Part 4
Making Conservation Work

Adult male (silverback) mountain gorilla, Gorilla gorilla beringei in the Virunga National Park, Democratic Republic of Congo. (Photo by Karl Ammann.)

Chapter 15: Making Conservation Work

Use, communities and conservation

Conservation and use tend to be seen as opposing forces. Both the body of theory and the examples of practice that we present in this volume illustrate why this is so. When use of a wild population starts, the general trend is a decline in its numbers from the unused state, sometimes leading to extinction. Equilibria can arise, but are rarely maintained, undermined by human population pressure, improving technology or unpredictable ecological processes. This fits with empirical observations of the natural world; there are few wild species that are not affected by human exploitation of the environment, and for those that are used directly, this use can rarely be ignored as an actual or potential threat to a species' long-term survival.

Given that use generally has a negative ecological impact on wild species, it may seem surprising that use, albeit sustainable use, has become an integral part of a large number of conservation initiatives in recent years. The new impetus behind the search for sustainable use reflects both a desperation born of our many and conspicuous conservation failures and a willingness to confront some economic and political realities. We live in a world where the biological status of wild species is not a motivating force for most human action, either at an individual, community or government level. Any conservationists that ignore this fact are either acting in privileged corners of the earth where the conflicting demands of wildlife and humans are low, or they are doomed to meet forces that could totally undermine their efforts. Where people can profitably make use of wild species, they will. Ecological factors are not the only ones that need to be taken into account when considering the conservation of species that are used. We have outlined in this volume many of the political, ethical and practical problems associated with attempts to prevent use or curtail it within the limits of sustainability.

Use and conservation have come to be considered together, as joint parts of the endeavour to achieve sustainable development, in the hope that controlled and sustainable use can be achieved. Wild species and ecosystems are frequently the basis of subsistence for whole communities, who have legitimate rights to their local natural resources. The ideal of sustainable development is that the exploitation of those resources can be managed at a level that has relatively little ecological impact whilst having great social and economic benefit. However, recognition of this ideal is not necessarily an argument for use being a conservation tool. To state that sustainable use is a conservation tool is somewhat tautological – if use is sustainable then conservation is, by definition, being achieved. The real question is how to achieve that conservation. Whether the exploitation of a species can be used to assist its conservation is a question we have explored in Part 2 and the contributing authors of the case studies in Part 3 have all addressed.

On the basis of the evidence presented in this volume, the case for use being a conservation tool is not proven. Use of other species is an inevitable part of human existence. Thus, it is essential that we consider means by which use can be managed sustainably, but the overwhelming conclusion is that this is hard to achieve. The views of the contributing authors are their own, based on their own practical and theoretical work. Their accounts reflect the successes and failures that they see in those systems with which they are familiar. They broadly represent the range of views of professionals working currently in conservation. None of them argue that the systems with which they are involved are free from threat due to human activities, nor that the use to which the species are being put is clearly sustainable. Even in cases that tend to be associated with successful utilisation in the public imagination, such as ecotourism, the practitioners have some misgivings. Gorilla tourism, for example, is frequently feted as a means by which conservation can fund itself to the mutual benefit of people and gorillas; but Thomas Butynski and Jan Kalina highlight how the system entails serious threats to gorilla survival (not the least of which is disease transmission from tourists). Whether the animals are currently under more or less threat than if gorilla tourism had never been developed is not known, but they suggest changes and alternatives that may make their long-term prospects more favourable.

There are some situations where encouraging use is likely to be counter-productive or downright disastrous for conservation. Life history and ecology can make sustainable use through harvesting quite simply impossible. R.E. Gullison describes why this is the case for bigleaf mahogany. The ecology and life history of that species dictates that any sustainable harvesting system based on economic realities is not achievable; the value of the mature tree is exceedingly high, but the time scale over which that value could be realised through any sustainable harvesting scheme is far too slow compared with the other economic options available either in the range states of the species or outside. The maintenance of the species requires a very long adult lifespan, but the trees are very valuable even before maturity. The economic pressure for unsustainable use is therefore overwhelming. Other slow-maturing species, such as other tropical hardwoods, whales, primates, some other large mammals and reptiles, and many other endangered species are other examples of where consumptive sustainable use is likely to be impossible to achieve, for similar reasons. In these cases, the greater the economic value of the species concerned, the stronger will be the threat of unsustainable exploitation.

Even if the life history and value of the species concerned is theoretically compatible with economically viable sustainable use, if use is encouraged without being explicitly linked to conservation measures, then failure is also likely. This realisation has led to efforts to link development and conservation components directly in integrated initiatives. Such efforts will always involve the local community. The case studies in Part 3 reflect the range of degrees of community involvement that can occur. There are now virtually no communities that exist in isolation from national economies and government legislation, so in every case the national government has some stake in any project, even one that is predominantly community-based. Only Anne Gunn describes a case where formal co-management agreements have been established between the national government and the indigenous population of Canada's Northwest Territories. The local population have been granted significant land-rights and major responsibilities in managing the use of their wildlife. These agreements are so recent that it is too early to judge their success as a conservation strategy. She argues that historically low demand for the products of caribou and muskox, combined with the relatively favourable state of the population cycle, are currently producing few conflicts between economic and conservation objectives. However, she notes that the system will not be truly tested until a situation arises when harvesting quotas below the level desired by the community are necessary to preserve the wildlife population. Kathy MacKinnon describes a number of scenarios where development benefits are being linked directly to conservation actions in South-East Asian forests. These include cases where some form of limited use is negotiated as part of a scheme in which other areas are protected.

A practical difficulty with integrated conservation and development projects is that the pressing needs of community development frequently win in the competition for administrators' time and resources, as well as having a higher priority in the eyes of the community, offering more immediately obvious rewards for all concerned than does conservation. Vivienne Solis Rivera and Stephen Edwards describe a project where the sustainable use of two fast-breeding and relatively easily managed species – green and black iguanas – has been used to generate the potential for much-needed local income. They describe how the social objectives of the project have largely been progressing, whereas the conservation objectives and those measures required to monitor them (such as monitoring the wild iguana population, and the status of their habitat) have tended to take a back seat.

It is fairly typical for community-based conservation projects to be able to claim significant benefits in public relations between government officials, resource managers and users in the community, and sometimes successes in rural development, but conservation benefits cannot be clearly demonstrated. Whilst there are many advantages to conservation projects being seen as development projects, this illustrates a potential cost; the criteria of success of funders and beneficiaries may change, so that benefits to wild species may not, in themselves, be sufficient. Achieving rural development may come to be considered a primary goal. Such projects may be looked upon favourably even if conservation benefits are absent. Another risk related to this shift of emphasis is increased politicisation of funding. If governments fall out of favour with foreign donors, for political reasons, development funding is frequently withdrawn. A project whose rationale is the need to conserve wildlife may be less vulnerable to such political action than one that is seen to be funded in order to favour economic development.

There is no doubt that the involvement of the community is a very important part of conservation, particularly when the species concerned are directly threatened by their activities. A range of technical, political and ethical arguments support this view. This is not to say that the involvement of the local community is necessarily a conservation strategy on its own. Conservation can even be easier where local communities are not involved. Two case studies describe situations where either a lack of community involvement or the abuse of community rights have favoured wildlife conservation. Leonid Baskin describes how authoritarian rule in the Communist period generally favoured game conservation in the Soviet Union. Under Stalin, collectivisation of nomads combined with starvation and genocide led to the depopulation of large areas of the Soviet Union, causing wildlife numbers in those areas to rebound dramatically. Border areas, where hunters fear to venture due to aggressive policing, have also proved important wildlife refuges. Sophie des Clers describes how the management of the Falkland Islands fisheries has been greatly simplified by the fact that the local community do not fish; the very rough seas mean that large factory ships from all over the world make up the fishing fleet. Managers can limit the number of fishing licences given out, or close the fishery whenever they feel it necessary, without having to face opposition from the local community arguing that their livelihoods are being threat-ened. One of the significant threats to the sustainability of that fishery is that some boat-owners (many of them subsidised by the European Union to leave overcrowded European waters) are now entering into co-ownership deals with local islanders, in the hope that they can exert the same type of political leverage over the Falkland Islands government that has contributed to the unsustainable harvesting seen in so many fisheries in Europe and elsewhere. This situation is rather unusual. In most cases, the local community comprises an important proportion of the users of local resources, and their rights and contributions will greatly influence the effectiveness of systems of management.

There are numerous situations where local communities have used wildlife sustainably for long periods of their history, but the situation has deteriorated due to the system being put under additional pressure from outside. Joel Freehling and Stuart Marks describe a case from the Luangwa Valley in Zambia, where the local perception is that the reduction in wildlife in the area is due to outsiders – which includes international ivory traders and safari hunters – whilst it is hunting by the indigenous population that has been mostly prohibited by government in recent decades. Other sources of data largely support these local views. Pressures arise from within communities too, typically including increased monetarisation of the economy, population growth and access to improved hunting technology. Whilst turning back the clock may appear to be an attractive conservation strategy, perhaps by enhancing local and lineage authority over resources and protecting areas from outsiders, this is not normally a serious option. Governments are usually hungry for any revenue that more commercial ventures might release, even if many of the benefits do not accrue to locals. In the case of the Maldives, described by Andrew Price, Callum Roberts and Julie Hawkins, the benefits of coral reef tourism have failed to reach many local people, and even for those making use of the tourist trade, social costs can be high, including family separation as young people leave to work in the tourist centres.

Communities cannot be seen as isolated units; individuals within communities have wider horizons and will respond to economic and political changes beyond the local system. Markets are becoming more global – products that were once sold locally may now have acquired great value through rarity, such that hunters are more interested in selling them on to distant urban or even international destinations. Traders from distant countries may be seeking prey in new areas, having exhausted the relevant species' populations nearer to home. The trade in far eastern medicines, many of which are based on exotic wild species, is a classic example of this; this trade comprises the greatest contemporary threat to many mammal species in the former Soviet Union (see Chapter 14) and beyond.

Our conclusions thus far are that use is not a conservation tool; and that even if use is primarily by the local community, who may have a long-term interest in conserving the resource, their interest is not sufficient to ensure its conservation. So what are the prospects for all those wild species used by local or distant communities? The answer, in many cases, is that they are at the mercy of market forces.

Economics and conservation

Economists frequently have faith in the market to reach a socially optimal level of economic activity. This view is applied by environmental economists to the utilisation of wild species. Environmental economists attribute conservation failures (where resources are lost or degraded to the detriment of society in general) to the fact that many of the important values of these resources are outside the economic process. An argument can then be advanced that, if these externalities were valued appropriately, markets would behave in a socially desirable manner. This is rather a big if, worth US$33 trillion ($10^{12}$) per year according to a recent estimate. That is the value that Costanza *et al.* (1997) place on the services of the world's ecosystems, most of which are not captured in commercial markets. Much of this value was calculated by estimating the replacement costs of goods (such as gathered food or fuelwood) or services (such as using rivers for waste disposal) that, at present, are taken from the environment largely at no cost. The value they arrive at is at least double the global gross national product; thus, its inclusion in the balance sheet would indeed revolutionise the manner in which we exploit the environment. Unfortunately this is destined to be largely a paper exercise. Replacement costs are not normally relevant to current users; a woman collecting fuelwood from a local woodland, to cook her family's meals, would have to buy kerosene, possibly at crippling expense, if that woodland were no longer there. However, while the wood is still there, there is no good reason why she should buy and use kerosene, nor are any local traders likely even to provide it if there is no demand. The long-term, or even the short-term, good of society is not a very useful unit of analysis for predicting the likely actions of individuals, or even governments, for reasons that we outline in detail in Part 2. Governments may, in theory, be charged with maximising social welfare, but their political survival is likely to depend on far more short-term considerations, usually weighted according to the power of particular groups within society.

M. Norton-Griffiths uses an economic approach to examine the prospects for wildlife conservation in Kenya outside the country's protected areas. He demonstrates that the conditions under which landowners would prefer to tolerate wildlife, rather than use land for other, conflicting uses, are rather stringent; the position of wildlife in areas with high agricultural (including pastoral) or other development potential is precarious, unless the government, or some other body, is prepared to make very large subsidies that favour wildlife conservation. The magnitude of the required subsidy is so great that it is unrealistic to assume that it will materialise; but he points out how other changes in government policy, both helping wildlife benefits to be realised by landowners and reducing subsidies on agriculture and other conflicting developments, could be helpful for conservation in some areas.

This issue of subsidies is of great importance. There are numerous examples of government subsidies exacerbating situations where market activity already posed some threat to natural resources. Fishing and agriculture are prime examples; forestry, road-building and other building developments are also frequently subsidised. These are all areas where governments are prepared to pay out large amounts of money, distorting the economic optima towards increasing environmentally destructive endeavours. Even if most governments are a long way from subsidising environmentally beneficial activities, many pressing threats to wild species could be greatly reduced if subsidies for over-exploitation and over-intensification were halted. In the case of European agriculture, for example, it is not difficult to imagine means whereby subsidies for intensifying farming could be removed and replaced by subsidies for activities with proven environmental benefits (such as organic farming) at no extra cost to governments, if the political will were there.

Governance

Not only will damaging subsidies have to be removed, but market forces will need to be regulated if environmental degradation is to be checked. Where use threatens species or ecosystems, the establishment of systems of governance limiting that use is necessary for conservation to occur. Regulation needs to apply to use both by local communities and by outsiders. Some such management systems do exist, especially where governments are sufficiently powerful, well-organised and strongly motivated to conserve a particular resource. Frequently governments do not meet these exacting criteria, and even if they do, governance does not depend on the government alone – it may depend on many governments or it may depend on many different non-governmental agents, ranging from the local community to international corporations.

If governance is to be effective, all the parties with direct or indirect interests in the resource should be involved at some level in its management. McGlade (1996) identifies three basic requirements that are essential for successfully establishing workable systems of governance. One is a realistic *evaluation* of the actual state of the resource and of its preferred state. This evaluation does not only include the collection of appropriate data, but also its interpretation. Clearly the chances of co-operation between all the interested parties is enhanced if the interpretation of the data can be agreed upon. Second, appropriate *instruments* for governance need to be developed. Again, communication between parties is important if these are to be successful. It is unlikely that all the structures required will be found within a community, or even within government alone. Formal laws, ecological taxation and voluntary codes of behaviour are examples of regulatory mechanisms that could be combined. The importance of each depends on the ecological, political and historical characteristics of the system concerned. The third requirement is *room for action*. This involves both incentives and constraints on the part of users and managers. It is easy to imagine situations where parties will not co-operate with any system that restricts them from use that they consider essential to their survival, however aware they may be of the unsustainable nature of that use.

The involvement of local communities in resource governance is now accepted as important by most conservationists. The involvement of other elements of the private sector, such as large corporations, is more controversial. Conservationists and private companies are used to fighting each other, and both are wary of the restrictions that the other may impose on their activities, should they enter into formal agreements. However, some agreements between these two traditional enemies have been made, and have the potential to be another important conservation tool. One example is the Forestry Stewardship Council (FSC), established in 1995. Membership of the FSC is voluntary, and enlists members into a labelling scheme designed to inform customers accurately whether the forest products they are purchasing are from a sustainably managed forest. The motivation behind the scheme was the finding that virtually all the impromptu labelling schemes appearing on goods (devised by producers to allay the environmental concerns of their customers) were at best irrelevant and at worst deceitful. Early recruits to the FSC labelling scheme include B&Q and Home Depot (two major retailers of household goods and building materials in the UK and USA respectively). Hence, consumers are given the opportunity, at relatively little cost to themselves, to enhance the economic incentives for suppliers to adopt sustainable management. A similar system is now being devised for marine resources. The Marine Stewardship Council was founded in 1996 by a partnership between WWF and Unilever plc – a major, multi-national supplier of frozen foods (and many other things). Unilever argues that it is in its long-term, commercial interest that the seafood it sells is not over-exploited, and therefore it is willing to use only sustainably harvested supplies. In this way, private companies demonstrate that they can, occasionally, be more forward-looking than governments, with respect to environmental concerns. Clearly there are also public relations benefits to companies with good environmental credentials. Environmental concerns do have the power to harm corporations. Shell – a major petroleum company – has recently suffered two environmental public relations disasters: one concerning the disposal of a disused oil rig that they proposed to dump in the North Sea, the other concerning the execution by the Nigerian government of environmental activists, lead by Ken Sarowewa, whilst Shell was conducting a large amount of business with Nigeria. These events contributed to environmental concerns being raised by a major shareholder, for the first time, at Shell's annual general meeting in 1997.

Political will

At the summit meeting in Rio de Janeiro in 1992, most of the world's heads of state agreed to various actions related to preserving biodiversity, including reducing deforestation, and agreed to tackle the production of greenhouse gases to slow global warming. Five years later, it is clear that very little has been achieved. Most governments have taken no action at all. The few countries that have reduced greenhouse gas emissions have done so largely fortuitously. The world's biodiversity has continued to deteriorate rapidly. The President of the IUCN described her organisation as 'astonished to find that the rate of the world's loss of biodiversity was worse than we had imagined', after their 1996 review of the status of the world's threatened and endangered species. The convening of high profile meetings, and the signing of international agreements and statements of intent are not having a beneficial influence on the environment. This is due to a lack of action; and the lack of real action is indicative of a lack of political will.

However, the situation may not be as bleak as it appears. Changing values in society do influence the political process. Beneficial changes may not have occurred in the environment, but they are apparent at some level in governments and other organisations. The profile of environmental issues has been raised and the time and effort that governments give to environmental issues is increasing rapidly. The reorientation of conservation towards sustainable development has greatly increased its palatability to politicians and to society in general. These changes offer the prospect of a more realistic evaluation of what is needed, and a greater determination by some governments to enable this to occur.

Environmental degradation is the cumulative process of millions of local actions, some with local and some with global effects. An accumulation of millions of local actions will ultimately be the only way that global sustainability can be achieved. Many local and individual initiatives are making small but significant contributions to changing the way we make use of nature. These actions are important, especially where governments are failing to act, but clearly it is governments that have the greatest potential to make the largest contributions. Communities and individuals can change their attitudes towards conservation when appropriate incentives and fair governance are introduced. Protected areas across the world are vital havens of biodiversity; the majority exist due to government action and are reliant on government and community support to survive. Many more wild species are outside protected areas and need some protection from the rush for economic growth. Where resources are being used, new priorities and decisive action at the government level are now needed to establish effective conservation. The state of knowledge and infrastructure that could help make conservation work has been greatly enhanced over the last few years. Where the political will exists, we are in a position to make real steps towards the sustainable use of our environment.

References

Abbot, J. (1996) *Rural subsistence and protected areas: community use of the miombo woodlands of Lake Malawi National Park*. PhD thesis, University of London, UK.

Abbot, J. & Mace, R. (1998) Managing protected woodland: women, woodfuel and law enforcement in Lake Malawi National Park. *Biological Conservation*, (in press).

Ack, B.L. (1991) *Towards Success in Integrated Conservation and Development: Critical Elements in Design, Management, and Implementation*. WWF Wildlands and Human Needs Technical Paper Number 1.

Adamczewski, J. (1995) *Digestion and body composition in the muskox*. PhD thesis, University of Saskatchewan, Saskatoon, Saskatchewan, Canada.

Adams, D. & Carwardine, M. (1993) Meeting a gorilla. In: *The Great Ape Project* (eds P. Cavalieri & P. Singer), pp. 19–23. Biddles Ltd, Guildford, UK.

Adger, W.N. & Brown, K. (1994) *Land Use and the Causes of Global Warming*. John Wiley, Chichester, UK.

Alvard, M. (1993) Testing the 'ecologically noble savage' hypothesis: inter-specific prey choice by Piro hunters of Amazonian Peru. *Human Ecology*, **21**, 355–387.

Alvard, M. (1995) Conservation by native peoples: prey choice in a depleted habitat. *Human Nature*, **5**, 127–154.

Alvard, M. & Kaplan, H. (1991) Procurement technology and prey mortality among indigenous neotropical hunters. In: *Human Predators and Prey Mortality* (ed. M.C. Stiner). Westview, USA.

Alverson, D.A., Freeberg, M.H., Pope, J.G. & Murawski, J.A. (1994) *A global assessment of fisheries by catch and discards*. FAO Technical Paper No. 339, Rome, Italy.

Amilien, C. (1994) International legal issues in tropical timber certification. In: *Timber Certification: Implications for Tropical Forest Management*, pp. 127–134. Yale School of Forestry and Environmental Studies, New Haven, USA.

Ammann, K. & Pearce, J. (1995) *Slaughter of the Apes: How the Tropical Timber Industry is Devouring Africa's Great Apes*. World Society for the Protection of Animals, London, UK.

Anderson, A.B. (1990) Extraction and forest management by rural inhabitants in the Amazon estuary. In: *Alternatives to Deforestation: Steps towards Sustainable Use of the Amazon Rain Forest* (ed. A.B. Anderson), pp. 65–85. Columbia University Press, New York, USA.

Anderson, D.R. (1975) Optimal exploitation strategies for an animal population in a Markovian environment: a theory and an example. *Ecology*, **56**, 1281–1297.

Anderson, R. & Grove, A. (eds) (1987) *Conservation in Africa: People, Policies, and Practice*. Cambridge University Press, Cambridge, UK.

Anderson, R.C. (1992) North–south variations in the distribution of fishes in the Maldives. *RASAIN*, **12**, 210–226.

Anon. (1979) *Atlas of the Oceans. Atlantic and Indian Oceans*. Ministry of Defence, USSR Navy. Pergamon Press, Oxford, UK.

Anon. (1986) Coughs and faeces spread diseases that kill mountain gorillas. *New Scientist*, **109**, 20.

Anon. (1994) Karisoke Research Center, Rwanda, 1993. *Gorilla Conservation News*, **8**, 21–22.

Anon. (1996) Factors influencing sustainability: a report of the activities of the IUCN Sustainable Use Initiative to IUCN members in compliance with Recommendation 19.54. *IUCN–The World Conservation Union, 1st World Conservation Congress, Montreal, Canada.*

Anon. (1996–97) Learning to live with tame humans. *Digit News*, **10**, 3.

Argeloo, M. & Dekker, R.W.R.J. (1996) Exploitation of megapode eggs in Indonesia: the role of traditional methods in the conservation of megapodes. *Oryx*, **30**, 59–64.

Århem, K. (1984) *Pastoral Man in the Garden of Eden: The Maasai of the Ngorongoro Conservation Area.* University of Uppsala, Uppsala, Sweden.

Arnason, R. (1994) On catch discarding in fisheries. *Marine Resource Economics*, **9**, 189–207.

Augustyn, C.J., Lipinski, M.R. & Sauer, W.H.H. (1992) Can the *Loligo* squid fishery be managed effectively? A synthesis of research on *Loligo vulgaris reynaudii. South African Journal of Marine Science*, **12**, 903–918.

Augustyn, C.J., Roel, B.A. & Cochrane, K.L. (1993) Stock assessment in the Chokka squid *Loligo vulgaris reynaudii* fishery off the coast of South Africa. In: *Recent Advances in Fisheries Biology* (eds T. Okutani, R.K. O'Dor & T. Kubodera). Tokai University Press, Tokyo, Japan.

Aveling, C. & Aveling, R. (1989) Gorilla conservation in Zaire. *Oryx*, **23**, 64–70.

Aveling, R. (1991) *'Gorilla tourism': possibilities and pitfalls.* Unpublished report. African Wildlife Foundation, Nairobi, Kenya.

Avio, K.L. & Clark, C.S. (1978) The supply of property offences in Ontario: evidence on the deterrent effect of punishment. *Canadian Journal of Economics*, **10**, 1–19.

Axelrod, R. (1984) *The Evolution of Co-operation.* Basic Books, New York, USA.

Ayres J.M., de Magalhaes Lima, D., de Souza Martins, E. & Barreiros, J.L.K. (1991) On the track of the road: changes in subsistence hunting in a Brazilian Amazonian village. In: *Neotropical Wildlife Use and Conservation* (eds J.G. Robinson & K.R. Redford), pp. 82–92 University of Chicago Press, Chicago, USA.

Banco Central de Bolivia (1994) *Buletin Estadistico*, No. 283. La Paz, Bolivia.

Banco Central de Bolivia (1995) *Buletin Estadistico*, No. 285. La Paz, Bolivia.

Bannikov, A.G. (1954) *The Mammals of the Mongolian People's Republic*, Issue 53. Publishing House of the Academy of Sciences of USSR, Moscow, USSR.

Bannikov, A.G., Zhirnov, L.V., Lebedeva, L.S. & Fandeev, A.A. (1961) *Biology of the Saiga (Biologiya Saigaka).* Sel'khozizdat, Moscow, USSR.

Bannikov, A.G., Pivovarova, E.P. & Fandeev, A.A. (1982) Rational use of some species of ungulates in USSR. *Sbornik Nauchnikh Trudov Moskovskoi Veterinarnoii Akademii Imeni Skrjabina*, **125**, 88–99. Moskovskaya Veterinarnaya Akademiya, Moscow, USSR.

Barnes, R. (1989) The status of elephants in the forests of Central Africa. In: *The Ivory Trade and the Future of the African Elephant.* Ivory Trade Review Group, Queen Elizabeth House, Oxford, UK.

Barr, W. (1991) *Back From the Brink: the Road to Muskox Conservation in the Northwest Territories.* Komatik Series 3. The Arctic Institute of North America, University of Calgary, Canada.

Basabose, A.K. & Yamagiwa, J. (1997) Current situation of conservation and wildlife in Kahuzi-Biega National Park, Eastern Zaire. *Gorilla Conservation News*, **11**, 10–12.

Baskin, L.M. (1976) *Behaviour of Ungulates (Povedenie Kopitnikh Zhivotnikh)*. Nauka, Moscow, USSR.

Baskin, L.M. (1994) Population ecology of the moose in the Russian southern taiga. *Alces*, **30**, 51–55.

Baskin, L. M. (1996) The brown bear in Russia: does it have a future? *Bulleten' Moskovskogo Obshchestva Ispitateley Prirodi, Otdelenie Biologicheskoe*, **101**(2), 18–29.

Baskin, L.M. & Lebedeva, N.L. (1987) Moose management in the USSR. *Proceedings of the Second International Moose Symposium*, Part 2, pp. 619–634. Svenska Jagerforbundet, Uppsala, Sweden.

Bass, S. (1996) Overview presentation on the principles of certification of forest management systems and labeling of forest products. Paper presented at International Conference on Forest Certification. Brisbane, Australia, May, 1996.

Basson, M., Beddington, J.R. & May, R.M. (1991) An assessment of the maximum sustainable yield of ivory from African elephant populations. *Mathematical Biosciences*, **104**, 73–95.

Basson, M., Beddington, J.R., Crombie, J.R., Holden, S.J., Purchase, L.V. & Tingley, G.A. (1996) Assessment and management techniques for migratory annual squid stocks: the *Illex argentinus* fishery in the Southwest Atlantic as an example. *Fisheries Research*, **28**, 3–27.

Becker, G.S. (1968) Crime and punishment: an economic approach. *Journal of Political Economy*, **76**, 168–217.

Beddington, J.R. (1974) Age distribution and the stability of simple discrete time population models. *Journal of Theoretical Biology*, **47**, 65–74.

Beddington, J.R. & Clark, C.W. (1984) Allocation problems between national and foreign fisheries with a fluctuating fish resource. *Marine Resource Economics*, **1**, 137–154.

Beddington, J.R. & Rettig, R.B. (1984) *Approaches to the regulation of fishing effort*. FAO Fisheries Technical Paper 243. Rome, Italy.

Beddington, J.R., Rosenberg, A.A., Crombie, J.A. & Kirkwood, G.P. (1990) Stock assessment and the provision of management advice for the short finned squid fishery in Falkland Islands waters. *Fisheries Research*, **8**, 351–365.

de Beer, J.H. & McDermott, M.J. (1989) *The Economic Value of Non-Timber Forest Products in Southeast Asia*. Netherlands Committee for IUCN, Amsterdam, Holland.

Begg, D., Fischer, S. & Dornbusch, R. (1984) *Economics*. McGraw-Hill, Maidenhead, UK.

Begon, M., Harper, J.L. & Townsend, C.R. (1996a) *Ecology*, 3rd edn. Blackwell Science, Oxford, UK.

Begon, M., Mortimer, M. & Thompson, D.J. (1996b) *Population Ecology: A Unified Study of Animals and Plants*, 3rd edn. Blackwell Science, Oxford, UK.

Bell, P.R.F. Greenfield, P.F., Hawker, D. & Connell D. (1989) The impact of waste discharges on coral reef regions. *Water Science Technology*, **21**, 112–130.

Bell, R.H.V. (1984) Monitoring of public attitudes. In: *Conservation and Wildlife Management in Africa* (eds R.H.V. Bell & E. McShane-Caluzi), pp. 441–450. Office of training for program support, forestry and the natural resource centre. US Peace Corps, Washington DC, USA.

Belovsky, G.E. (1987) Extinction models and mammalian persistence. In: *Viable Populations for Conservation* (ed. M.E. Soulé), pp. 35–38. Cambridge University Press, Cambridge, UK.

Benirschke, K. & Adams, F.D. (1980) Gorilla diseases and causes of death. *Journal of Reproduction and Fertility*, Supplement 28, 139–148.

Bennett, E.L., Nyaoi, A.J. & Sompud, J. (in press) The sustainability of hunting in

Sarawak and Sabah. In: *Hunting for sustainability in Tropical Forests* (eds J.R. Robinson & E.L. Bennett). Columbia University Press, USA.

Berger, D. (1993) *Wildlife Extension: Participatory Conservation by the Maasai of Kenya*. ACTS Environment Policy Series No. 4. ACTS, Nairobi, Kenya.

Bergerud, A.T. (1988) Caribou, wolves and man. *Trends in Ecology and Evolution*, **3**, 68–72.

Bergerud, A.T. (1996) Evolving perspectives on caribou population dynamics. *Rangifer*, Special Issue No. 9, 95–118.

Bergesen, H.O. & Parmann, G. (eds) (1996) *Green Globe Yearbook of International Co-operation on Environment and Development*. Oxford University Press, Oxford, UK.

Binmore, K. (1992) *Fun and Games: A Text on Game Theory*. Heath, Lexington, Mass., USA.

Bitrakh, A.A. (1926) *Fur Hunting in the Northern European Part of USSR. Materiali dlya izucheniya estestvennikh proizvoditel'nikh sil SSSR*, Vol. 61. KEPS, Moscow, USSR.

Blagoveshchenskii, G.D. (1956) *Hunting Wolves and Bears (Okhota na volkov i medvedeii)*. Gosudarstvennoe izdatel'stvo Karel'skoi ASSR, Petrozavodsk, USSR.

Boo, E. (1990) *Ecotourism: The Potentials and Pitfalls*, Vol. 1. World Wildlife Fund, Washington DC, USA.

Borisov, B.P., Gibet, L.A., Gubar, Yu, P. *et al.* (1992) *Resources of Game Animals in Russia (Resursi Okhotnich'ikh Zhivotnikh v Rossii)*. TSNIL Glavokhota, Moscow, Russia.

Bostock, L. & Chandler, S. (1981) *Mathematics: The Core Course for A-level*. Stanley Thornes, Cheltenham, UK.

Boyle, P.R. (ed.) (1983) *Cephalopod Life Cycles*, Vol I. Academic Press, London, UK.

Boyle, P.R. (ed.) (1987) *Cephalopod Life Cycles*, Vol II. Academic Press, London, UK.

Bragdon, S.H. (1990) *Kenya's Legal and Institutional Structure for Environmental Protection and Natural Resource Management: An Analysis and Agenda for the Future*. Economic Development Institute of the World Bank, Robert S. McNamara Fellowship Program, World Bank, Washington, USA.

Brandon, K. (1996) *Ecotourism and Conservation: a Review of Key Issues*. Environment Department Papers No. 33. The World Bank, Washington, USA.

Brodziak, J.K.T. & Rosenberg, A.A. (1993) A method to assess squid fisheries in the north-west Atlantic. *ICES Journal of Marine Science*, **50**, 187–194.

Bromley, D.W. (1991) *Environment and Economy: Property Rights and Public Policy*. Blackwell, Oxford, UK.

Broten, M.D. & Said, M. (1995) Population trends of ungulates in and around Kenya's Maasai Mara Reserve. In: *Serengeti II: Research, Management and Conservation of an Ecosystem* (eds A.R.E. Sinclair & P. Arcese). Chicago University Press, Chicago, USA.

Browder, J. (1992) The limits of extractivism. *Bioscience*, **42**, 174–181.

Browder, J.O., Matricardi, E.A.T. & Abdala, W.S. (1996) Is sustainable tropical timber production financially viable? A comparative analysis of mahogany silviculture among small farmers in the Brazilian Amazon. *Ecological Economics*, **16**, 147–159.

Brown, B.E. & Dunne, R. (1988) The environmental impact of coral mining on coral reefs in the Maldives. *Environmental Conservation*, **152**, 159–166.

Brown, B.E., Dawson Shepherd, A., Weir, I. & Edwards, A. (1990) *Effects of degradation of the environment on local reef fisheries in the Maldives*. Final report to Overseas Development Administration (ODA), London.

Brown, G. (1989) The viewing value of elephants. In: *The Ivory Trade and the Future of the African Elephant*. Ivory Trade Review Group, Queen Elizabeth House, Oxford.

Bucher, E.H. (1992) The causes of extinction of the passenger pigeon. *Current Ornithology*, **9**, 1–33.

Bunce, L., Horwith, B. & Black, D. (in press) Coral reef monitoring and baseline data collection of Antigua's coral reefs. *Caribbean Journal of Science.*

Bureau of Statistics (1996) *Statistics Quarterly*, Vol. 18. Department of Public Works and Services, Government of the Northwest Territories, Yellowknife, Northwest Territories, Canada.

Burghes, D. & Graham, A. (1986) *Introduction to Control Theory Including Optimal Control.* Ellis Horwood, Chichester, UK.

Burkill, I.H. (1935) *A Dictionary of the Economic Products of the Malay Peninsula*, 2 vols. Government Printing Office, Singapore.

Burrill, A. & Douglas Hamilton, I. (1987) *African Elephant Database Project: Final Report.* GRID Case Study series, no. 2. UNEP, Nairobi, Kenya.

Buschbacher, R.J. (1990) Natural forest management in the humid tropics: ecological, social, and economic considerations. *Ambio*, **19**, 253–258.

Butler, R.W. (1991) Tourism, environment, and sustainable development. *Environmental Conservation*, **18**, 201–209.

Butynski, T.M. (1984) *Ecological survey of the Impenetrable (Bwindi) Forest, Uganda, and recommendations for its conservation and management.* Unpublished report. New York Zoological Society, New York, USA.

Butynski, T.M. (1985) Primates and their conservation in the Impenetrable (Bwindi) Forest, Uganda. *Primate Conservation*, **6**, 68–72.

Butynski, T.M. (in press) African primate conservation—the species and the IUCN/SSC Primate Specialist Group Network. *Primate Conservation*, **17**.

Butynski, T.M. & Kalina, J. (1993) Three new mountain national parks for Uganda. *Oryx*, **27**, 214–224.

Butynski, T.M., Werikhe, S.E. & Kalina, J. (1990) Status, distribution and conservation of the mountain gorilla in the Gorilla Game Reserve, Uganda. *Primate Conservation*, **11**, 31–41.

Bythell, J.C. & Sheppard, C.R.C. (1993) Mass mortality of Caribbean shallow corals. *Marine Pollution Bulletin*, **26**, 296–297.

Caddy, J. (ed.) (1983) *Advances in assessment of world cephalopod resources.* FAO Technical Paper 231. FAO, Rome.

Cahyadin, Y., Jepson, P. & Manoppo, B.I. (1994) *The status of* Cacatua goffini *and* Eos reticulata *on the Tanimbar islands.* PHPA/Birdlife International, Bogor, Indonesia.

Caldecott, J. (1988a) *Hunting and wildlife management in Sarawak.* IUCN Tropical Forest Programme, Gland, Switzerland and Cambridge, UK.

Caldecott, J. (1988b) Climbing towards extinction. *New Scientist* 9 June, 62–66.

Cameron, R.D., Smith, W.T., Fancy, S.G., Gerhart, K.L. & White, R.G. (1993) Calving success of female caribou in relation to bodyweight. *Canadian Journal of Zoology*, **71**, 480–486.

Campbell, L.H. & Cooke, A.S. (eds) (1997) *The Indirect Effects of Pesticides on Birds.* Joint Nature Conservation Committee, Peterborough, UK.

Cannon, J.B. (1997) North-east Atlantic review. In: *Review of the State of World Fishery Resources: Marine Fisheries.* FAO Fisheries Circular, No. 920. FAO Marine Resources Service, Fishery Resources Division, Rome. P. 173.

Carvalho, G.R. & Loney, K.H. (1989) Biochemical genetic studies on the Patagonian squid *Loligo gahi* d'Orbigny I. Electrophoretic survey of genetic variability. *Journal of Experimental Marine Biology and Ecology*, **126**, 231–241.

Carvalho, G.R. & Pitcher, T.J. (1989) Biochemical genetic studies on the Patagonian squid *Loligo gahi* d'Orbigny II. Population structure in Falkland waters using isozymes, morphometrics and life history data. *Journal of Experimental Marine Biology and Ecology*, **126**, 243–258.

Caughley, G. (1974) Bias in aerial survey. *Journal of Wildlife Management*, **38**, 921–933.

Caughley, G. (1977) *Analysis of Vertebrate Populations.* John Wiley, London, UK.

Caughley, G. (1987) Ecological relationships. In: *Kangaroos: Their Ecology and Management in the Sheep Rangelands of Australia* (eds G. Caughley, N. Shepherd & J. Short), pp. 159–187. Cambridge University Press, Cambridge, UK.

Caughley, G. & Gunn, A. (1993) Dynamics of large herbivores in deserts: kangaroos and caribou. *Oikos,* **67,** 47–55.

Caughley, G. & Gunn, A. (1996) *Conservation Biology in Theory and Practice.* Blackwell Science, Oxford, UK.

Caughley, G. & Sinclair, A.R.E. (1994) *Wildlife Ecology and Management.* Blackwell Science, Oxford, UK.

Ceballos-Lascurain, H. (1996) *Tourism, Ecotourism and Protected Areas: The State of Nature-based Tourism around the World and Guidelines for its Development.* IUCN, Gland, Switzerland.

Chambers, R. (1983) *Rural Development: Putting the Last First.* Longman, London, UK.

Chapin, F.S. III, Johnson, D.A. & McKendrick, J.D. (1980) Seasonal movement of nutrients in plants of different growth form in an Alaskan tundra ecosystem: implications for herbivory. *Journal of Animal Ecology,* **49,** 189–209.

Child, B. (1996) The practice and principles of community-based wildlife management in Zimbabwe: the CAMPFIRE programme. *Biodiversity and Conservation,* **5,** 369–398.

Child, G. & Grainger, J. (1990) *A System Plan for Protected Areas for Wildlife and Sustainable Rural Development in Saudi Arabia.* National Commission for Wildlife Conservation and Development (NCWCD), Riyadh, Saudi Arabia; IUCN–The World Conservation Union, Gland, Switzerland.

Chin, S.C. (1985) *Agriculture and Resource Utilisation in a Lowland Rainforest Kenyah Community.* Sarawak Museum Journal Special Monograph No. 4.

Chipungu, S. (ed.) (1992) *Guardians in their Time: Experiences of Zambians Under Colonial Rule.* Macmillan, London, UK.

Christensen, V. & Pauly, D. (1995) Fish production, catches and the carrying capacity of the world oceans. *NAGA ICLARM Quarterly,* 18(3), 34–40.

Clark, C.W. (1973) Profit maximisation and the extinction of animal species. *Journal of Political Economy,* **81,** 950–961.

Clark, C.W. (1976, 1990) *Mathematical Bioeconomics: The Optimal Management of Renewable Resources.* Wiley-Interscience, New York, USA.

Clark, C.W. & Kirkwood, G.P. (1986) On uncertain renewable resource stocks: optimal harvest policies and the value of stock surveys. *Journal of Environmental Economics and Management,* **13,** 235–244.

Clark, C.W. & Mangel, M. (1986) The evolutionary advantages of group foraging. *Theoretical Population Biology,* **30,** 45–75.

Clayton, L. & Milner-Gulland, E.J. (1998) Wildlife trade and law enforcement in North Sulawesi, Indonesia. In: *Hunting for Sustainability in Tropical Forests* (eds J.R. Robinson & E.L. Bennett). Columbia University Press, New York, USA (in press).

Clayton, L., Keeling, M. & Milner-Gulland, E.J. (1997) Bringing home the bacon: a spatial model of wild pig harvesting in Sulawesi, Indonesia. *Ecological Applications,* 7, 642–652.

Connell, J.H. (1997). Disturbance and recovery of coral assemblages. *Coral Reefs,* **16** (Suppl.), S101–S113.

Conrad, P. (1995) Harvesting the trees while saving the forest. *Business,* July–August 1995, 32–34.

Cook, P.J. (1977) Punishment and crime. *Law and Contemporary Problems,* **5,** 164–204.

Cook, R.M., Sinclair, A. & Stefansson, G. (1997) Potential collapse of North Sea cod stocks. *Nature*, **385**, 521–522.

Cooke, J.G. (1995) The International Whaling Commission's Revised Management Procedure as an example of a new approach to fishery management. In: *Whales, Seals, Fish and Man* (eds A.S. Blix, L. Walloe & O. Ulltang), pp. 647–657. Elsevier Science, UK.

Copeland, T.E. & Weston, J.F. (1988) *Financial Theory and Corporate Policy*. Addison-Wesley, UK.

Costanza, R. & Patten, B.C. (1995) Defining and predicting sustainability. *Ecological Economics*, **15**, 193–196.

Costanza, R., d'Arge, R., de Groot, R. *et al.* (1997) The value of the world's ecosystem services and natural capital. *Nature*, **387**, 253–260.

Craven, I. & de Fretes, Y. (1987) *Arfak Mountains Nature Conservation Area, Irian Jaya. Management plan, 1988–92*. WWF Indonesia Programme, Bogor, Indonesia.

Craven, I. & Wardoyo, W. (1993) Gardens in the forest. In: *The Law of the Mother* (ed. E. Kemf), pp. 23–28. Sierra Club Books, San Francisco, USA.

Crean, K. & Symes, D. (eds) (1996) *Fisheries Management in Crisis*. Fishing News Books (Blackwell Science), Oxford, UK.

Croze, H.J., Hillman, J.C., Migongo, E. & Sinage, R. (1978) *The Ecological Basis for Calculations of Wildlife-generated Guaranteed Minimum Returns to Landowners in Maasai Mara and Samburu Ecosystems*. Report to the Wildlife Planning Unit of Kenya National Parks. EcoSystems Ltd, Nairobi, Kenya.

Csirke, J. (1987) *The Patagonian Fishery resources and the offshore fisheries in the South-West Atlantic*. FAO Technical Paper 286. Rome, Italy.

Danilov, P.I. & Tumanov, I.L. (1993) North-western European Russia. In: *Medvedi (Bears)* (eds M.A. Vaisfeld & I.E. Chestin), pp. 21–37. Nauka, Moscow, Russia.

Dauphiné, C. (1976) *Biology of the Kaminuriak population of barren-ground caribou*, Part 4. Canadian Wildlife Service Report Series 38.

Davila, P. & Esquivel, F. (1993) *Manejo y aprovechamiento del garrobo* (Ctenosaura similis) *e iguana verde* (Iguana iguana) *en la Cooperativa Omar Baca de la Península de Cosigüina, Dpt. Chinandega*. IRENA, IUCN, UNAN-Leon, Nicaragua.

Dawe, E.G., Shears, J.C., Balch, N.E. & O'Dor, R.K. (1990) Occurrence, size and sexual maturity of long-finned squid (*Loligo pealei*) at Nova Scotia and Newfoundland, Canada. *Canadian Journal of Fisheries and Aquatic Science*, **47**, 1830–1835.

Dawson Shepherd, A.R., Brown, B.E., Warwick, R.M., Clarke, K.R. & Brown, B.E. (1992) An analysis of fish community responses to coral mining in the Maldives. *Environmental Fisheries Biology*, **33**, 367–380.

DeGeorges, P.A. (1992) *ADMADE: an Evaluation Today and the Future Policy Issues and Direction*. Consultant's report prepared for USAID and NPWS, mimeo.

Denevan W. & Padoch, C. (1988) *Swidden Fallow Agroforestry in the Peruvian Amazon*. Advances in Economic Botany 5. New York Botanic Garden, USA.

des Clers, S. (1998) Structural adjustments of the distant fleet of European factory trawlers fishing for *Loligo* squid in Falkland Islands waters. In: *Property Rights and Regulatory Systems for Fisheries* (ed. D. Symes), pp. 142–152. Fishing News Books (Blackwell Science), Oxford, UK.

des Clers, S., Nolan, C., Baranowski, R. & Pompert, J. (1996) Preliminary stock assessment of the Patagonian toothfish longline fishery around the Falkland Islands. *Journal of Fish Biology*, **49** (Suppl. A), 175–181.

de Vaus, D.A. (1996) *Surveys in Social Research*, 4th edn. UCL Press, London, UK.

Dezhkin, V.V. (ed.) (1978) *Hunting Management in the RSFSR (Okhotnich'e Khozyaistvo v RSFSR)*. Lesnaya promishlennost', Moscow, USSR.

Dezhkin, V.V., Dakov, J.V. & Safonov, V.G. (1986) *Beaver (Bobr)*. Agropromizdat, Moscow, USSR.

Dixon, J.A., Fallon Scura, L. & van't Hof, T. (1993) Meeting ecological and economic goals: marine parks in the Caribbean. *Ambio*, **22**, 117–125.

Douglas Hamilton, I. (1988) *Identification Study for the Conservation and Sustainable Use of the Natural Resources in the Kenyan Portion of the Mara-Serengeti Ecosystem.* European Economic Community, Nairobi, Kenya.

Dove, M.R. (1985) *Swidden Agriculture in Indonesia: the Subsistence Strategies of the Kalimantan Kantu.* Mouton Publishers, Berlin, New York, Amsterdam.

Dransfield, J. (1974) *A short guide to rattans.* Regional Center for Tropical Biology (BIOTROP), Bogor, Indonesia.

Dransfield, J. (1988) Prospects for rattan cultivation. *Advances in Economic Botany*, **6**, 190–200.

Dransfield, J. (1992) Rattans in Borneo: botany and utilisation. In: *Forest Biology and Conservation in Borneo* (eds G. Ismail, M. Mohamed & S. Omar), pp. 22–31. Center for Borneo Studies Publication No. 2, Kota Kinabalu, Sabah.

Drew, K.R., Bai, Q. & Fadeev, E.V. (1989) Deer farming in Asia. In: *Wildlife Production Systems* (eds R.J. Hudson, K.R. Drew & L.M. Baskin), pp. 334–346. Cambridge University Press, Cambridge, UK.

Duffus, D. (1993) Tsitika to Baram: the myth of sustainability. *Conservation Biology*, **7**, 440–442.

Duke, J.A. (1992) Tropical botanical extractives. In: *Sustainable Harvest and Marketing of Rain Forest Products* (eds M. Plotkin & L. Famolare), pp. 53–62. Conservation International and Island Press, Washington DC, USA.

Duncan, L. (1995) Closed competition: fish quotas in New Zealand. *The Ecologist*, **25**, 97–104.

Eberle, R. (1992) Evidence for an alpha-herpes virus indigenous to mountain gorillas. *Journal of Medical Primatology*, **21**, 246–251.

Ehrlich, I. (1973) Participation in illegitimate activities: a theoretical and empirical investigation. *Journal of Political Economy*, **81**, 521–565.

Eide, E. (1994) *Economics of Crime: Deterrence and the Rational Offender.* North-Holland, Amsterdam, Holland.

Ellis, J.E. & Swift, D.M. (1988) Stability of African pastoral ecosystems – alternative paradigms and implications for development. *Journal of Rangeland Management*, **41**, 450–459.

Eloranta, E. & Nieminen, M. (1986) Calving of the experimental reindeer herd in Kaamanen during 1970–1985. *Rangifer*, Special Issue No. 1, 115–122.

Engen, S., Lande, R. & Saether, B-E. (1997) Harvesting strategies for fluctuating populations based on uncertain population estimates. *Journal of Theoretical Biology*, **186**, 201–212.

Esquivel, F. (1992) *Estimación poblacional del garrobo negro* (Ctenosaura similis) *en el bosque de 100 manzanas de la Cooperativa Omar Baca.* Informe a UICN. Nicaragua.

Esquivel, F. (1996) *Estimación poblacional del garrobo negro* (Ctenosaura similis) *en el bosque de 100 manzanas de la Cooperativa Omar Baca.* Informe a UICN. Nicaragua.

Fadeev, V.A. & Sludskii, A.A. (1982) *Saigas in Kazakhstan (Saigak v Kazakhstane).* Nauka, Alma-Ata, USSR.

Fancy, S. & White, R.G. (1985) Energy expenditure by caribou while cratering in the snow. *Journal of Wildlife Management*, **49**, 987–993.

FAO (1973) *Zambia Luangwa Valley Conservation and Development Project. Game Management and Habitat Manipulation.* United Nations Development Programme, Rome, Italy. Working Document BP/ZAM/681510.

FAO (1978) *Wildlife Management in Kenya: Plans for Rural Incomes from Wildlife in Kajiado District.* AG/KEN/71/525 Technical Report 1. FAO, Rome, Italy.

FAO (1983) *Report of the Ad hoc Working Group on Fishery Resources of the Patagonian Shelf, Rome, 7–11 Feb 1983. A preparatory meeting for the FAO World Conference on Fisheries Management and Development.* FAO, Inf. Pesca/FAO Fisheries Report 297. FAO, Rome, Italy.

FAO (1984) Fisheries Report 289 Supplement 3. FAO, Rome, Italy.

FAO (1989) *Community forestry: participatory assessment, monitoring and evaluation.* FAO, Rome, Italy.

FAO (1995a) *Review of the state of world fishery resources: marine fisheries.* FAO Fisheries Circular No. 884. FAO, Rome, Italy.

FAO (1995b) FAO Forestry Paper No. 124. FAO, Rome, Italy.

FAO (1997) Technical Guide for Responsible Fisheries, No 4. FAO, Rome, Italy.

Fearnside, P.M. (1989) Forest management in Amazonia: the need for new criteria in evaluating development options. *Forest Ecology and Management*, **27**, 61–79.

Ferguson, M.A.D. & Gauthier, L. (1992) Status and trends of *Rangifer tarandus* and *Ovibos moschatus* populations in Canada. *Rangifer*, **12**, 127–141.

Fernandes, L. (1995) *Integrating Economic, Environmental and Social Issues in an Evaluation of Saba Marine Park, Netherlands Antilles, Caribbean Sea.* East-West Center, Hawaii.

Field, B.C. (1994) *Environmental Economics.* McGraw-Hill, Singapore.

FIGO (1989) *Falkland Islands Interim Conservation and Management Zone.* Fisheries Report, 1987/1988, Fisheries Department, Falkland Islands Government, Falkland Islands Government Office, London, UK.

Fimbel, C. & Fimbel, R. (1997) Rwanda: the role of local participation. *Conservation Biology*, **11**, 309–310.

Firey, W.I. (1960) *Man, Mind, and Land: A Theory of Resource Use.* The Free Press, Glencoe, Illinois, USA.

Forchhammer, M. & Boertmann, D. (1993) The muskoxen *Ovibos moschatus* in north and northeast Greenland: population trends and the influence of abiotic parameters on population dynamics. *Ecogeography*, **16**, 299–308.

Forsythe, J.W. (1993) A working hypothesis of how seasonal temperature change may impact the field growth of young cephalopods. In: *Recent Advances in Fisheries Biology* (eds T. Okutani, R.K. O'Dor & T. Kubodera), pp. 133–143. Tokai University Press, Tokyo, Japan.

Fossey, D. (1983) *Gorillas in the Mist.* Houghton Mifflin, Boston, USA.

Fowler, C.W. (1981) Comparative population dynamics in large mammals. In: *Dynamics of Large Mammal Populations* (eds C.W. Fowler & T.D. Smith), pp. 437–455. John Wiley, New York, USA.

Fowler, C.W. (1984) Density dependence in cetacean populations. *Report of the International Whaling Commission*, Special Issue 6, 373–379.

Framheim, R. (1995) *The Value of Nature Protection. Economic Analysis of the Saba Marine Park.* Saba Marine Park, Saba, Netherlands Antilles.

Galaty, J.G. (1980) The Maasai group ranch: politics and development in an African pastoral society. In: *When Nomads Settle* (ed. P. Salzinan). Praeger, New York, USA.

Galaty, J.G. (1992) Social and economic factors in the privatization, sub-division and sale of Maasai ranches. *Nomadic Peoples*, **30**, 26–40.

Gaston, K. (ed.) (1995) *Biodiversity: A Biology of Numbers and Difference.* Blackwell Science, Oxford, UK.

Gatto, M. (1995) Sustainability: is it a well-defined concept? *Ecological Applications*, **5**, 1181–1183.

Geller, M.X. (1967) The biological basis for regulating numbers of game mammals by cropping in the Extreme North. *Trudi NIISH Krainego Severa*, **15**, 15–22. NIISH Krainego Severa, Norilsk, USSR.

Getz, W.M. (1984) Production functions for non-linear stochastic age-structured fisheries. *Mathematical Biosciences*, **69**, 11–30.

Getz, W.M. & Haight, R.G. (1989) *Population Harvesting*. Princeton Monographs in Population Biology 27. Princeton University Press, Princeton, USA.

Gibson, C.C. & Marks, S.A. (1995) Transforming rural hunters into conservationists: an assessment of community-based wildlife management programs in Africa. *World Development*, **23**, 941–957.

Giesen, W. (1986) *The status of* Scleropages formosus *(Asian arowana) in Indonesia's West Kalimantan province*. World Wildlife Fund, Bogor, Indonesia.

Gilpin, M. & Hanski, I. (1991) *Metapopulation Dynamics: Empirical and Theoretical Investigations*. Academic Press, London, UK.

Ginsberg, J.R. & Milner-Gulland, E.J. (1993) Sex-biased harvesting and population dynamics: implications for conservation and sustainable use. *Conservation Biology*, **8**, 157–166.

Ginsburg, R.N. (ed.) (1994) *Proceedings of the Colloquium on Global Aspects of Coral Reefs: Health, Hazards and History*. Rosenstiel School of Marine and Atmospheric Sciences, University of Miami, Miami, USA.

Gladfelter, W.B. (1982) White band disease in *Acropora palmata*: implications for the structure and growth of shallow water coral reefs. *Bulletin of Marine Science*, **32**, 639–643.

Glantz, M.H. (1986) Man, state and fisheries: an inquiry into some societal constraints that affect fisheries management. *Ocean Development and International Law*, **17**, 191–270.

GOK (1994) *Kenya Population Census 1989*, Vol. 1. Central Bureau of Statistics, Ministry of Planning and National Development, Nairobi, Kenya.

GOK (1995a) *Data Summary Report for the Kenyan Rangelands 1977–1994*. Ministry of Planning and National Development (Department of Resource Surveys and Remote Sensing), Nairobi, Kenya.

GOK (1995b) *National Rangelands Report: Summary of Population Estimates for Wildlife and Livestock*. Ministry of Planning and National Development (Department of Resource Surveys and Remote Sensing), Nairobi, Kenya.

GOK (1995c) *Protected and Adjacent Areas Analysis*. Ministry of Planning and National Development (Department of Resource Surveys and Remote Sensing), Nairobi, Kenya.

GOK (1996) *Statistical Abstract 1995*. Central Bureau of Statistics, Ministry of Planning and National Development, Nairobi, Kenya.

Gomez-Pompa, A. & Kaus, A. (1990) Traditional management of tropical forests in Mexico. In: *Alternatives to Deforestation* (ed. A.B. Anderson), pp. 45–64. Columbia University Press, New York, USA.

Goodwin, H. (1996) In pursuit of ecotourism. *Biodiversity and Conservation*, **5**(3), 277–292.

Gordon, H.S. (1954) The economic theory of a common property resource: the fishery. *Journal of Political Economy*, **62**, 124–142.

Graf, R. & Case, R. (1989) Counting muskoxen in the Northwest Territories. *Canadian Journal of Zoology*, **67**, 1112–1115.

Grakov, N.N. (1965) *Ecology and rational use of forest martens in the northern European part of the USSR. (Ekologiya i ratsional'noe ispol'zovanie zapasov lesnoi kunitsi severa evropeiskoi chasti SSSR)*. PhD Thesis, Kirovskii Sel'skokhozyaistvennii Institut, USSR.

Grakov, N.N. (ed.) (1973) *Hunting Management in the USSR (Okhotnich'e Khozyaistvo SSSR)*. Lesnaya promishlennost', Moscow, USSR.

Gray, D.R. (1987) *The Muskoxen of Polar Bear Pass*. Fitzhenry and Whiteside, Markham, Ontario, Canada.

Grenfell, B.T., Price, O.F., Albon, S.D. & Clutton-Brock, T.H. (1992) Over-compensation and population cycles in an ungulate. *Nature*, **355**, 283–286.

Grigg, R.W. & Dollar, S.J. (1995) Environmental protection misapplied: alleged versus documented impacts of a deep ocean sewage outfall in Hawaii. *Ambio*, **24**, 125–128.

Grist, E.P.M. & des Clers, S. (1997) How seasonal temperature variations may influence the structure of annual squid populations. *Journal of Mathematics Applied in Medicine and Biology*, **14**, 1–22.

Groom, M.J., Podolsky, R.D. & Munn, C.A. (1991) Tourism as a sustained use of wildlife: a case study of Madre de Dios, Southeastern Peru. In: *Neo-tropical Wildlife Use and Conservation* (eds J.G. Robinson & K.H. Redford), pp. 393–412. Chicago University Press, Chicago, USA.

Groves, C.P. (1970) Population systematics of gorilla. *Journal of Zoology, London*, **161**, 287–300.

Gullison, R.E. (1995) *Conservation of tropical forests through the sustainable production of forest products: the case of mahogany* (Swietenia macrophylla *King*) *in the Chimanes Forest, Beni, Bolivia*. PhD thesis, Princeton University, USA.

Gullison, R.E. & Cannon, J.B. (in press) The need for experimental management to determine timber harvest intensities compatible with the objectives of tropical sustainable forestry. *Conservation Biology*.

Gullison, R.E. & Hardner, J.J. (1993) The effects of road design and harvest intensity on forest damage caused by selective logging: empirical results and a simulation model from the Bosque Chimanes, Bolivia. *Forest Ecology and Management*, **59**, 1–14.

Gullison, R.E., Panfil, S.N., Strouse, J.J. & Hubbell, S.P. (1996) Ecology and management of mahogany (*Swietenia macrophylla* King) in the Chimanes Forest, Beni, Bolivia. *Botanical Journal of the Linnean Society*, **122**, 9–34.

Gunn, A. (1990) The decline and recovery of caribou and muskoxen on Victoria Island. In: *Canada's Missing Dimension: Science and History in the Canadian Arctic Islands*, Vol. II (ed. C.R. Harington), pp. 590–607. Canadian Museum of Nature, Ottawa, Ontario, Canada.

Gunn, A., Jingfors, K. & Evalik, P. (1986) The Kitikmeot harvest study as a successful example for the collection of harvest statistics in the Northwest Territories. In: *Native People and Renewable Resource Management* (eds J.E. Green & J.A. Smith). Proceedings of the 1986 Symposium of the Alberta Society of Professional Biologists, Edmonton, Alberta, Canada, pp. 249–259.

Gunn, A., McLean, B. & Miller, F.L. (1989) Evidence for and possible causes of increased mortality of adult male muskoxen during severe winters. *Canadian Journal of Zoology*, **67**, 1106–1111.

Gunn, A., Adamczewski, J. & Elkin, B. (1991a) Commercial harvesting of muskoxen in the Northwest Territories. In: *Wildlife Production: Conservation and Sustainable Development* (eds L.A. Renecker & R.J. Hudson), pp. 197–204. Agricultural and Forestry Experiment Station Miscellaneous Publications 91-6, University of Alaska, Fairbanks, Alaska, USA.

Gunn, A., Shank, C.C. & McLean, B. (1991b) The history, status and management of muskoxen on Banks Island. *Arctic*, **44**, 188–195.

Gusev, O.K (1971) Re-establishing sable in the USSR. *Priroda*, **11**, 68–74.

Gutierrez-Montes, I. (1996) *Aportes de un Proyecto de Manajo de Vida Silvestre a la*

Calidad de Vida de las Poblaciones Rurales – El caso de la Cooperativa Omar Baca. MSc thesis, CATIE-Costa Rica.

Hall, J.S., Saltonstall, K., Inogwabini, B-I. & Omari, I. (1998) Distribution, abundance, and conservation status of Grauer's gorilla (*Gorilla gorilla graueri*). *Oryx*, 32, 122–130.

Hall, P.E. (1910) Notes on the movement of *Glossina morsitans* in the Lundazi District, North-Eastern Rhodesia. *Bulletin of Entomological Research*, 1(3), 183–184.

Hameed, H. (1993) *Implications of tourism for the environment: a Maldives case study*. MPhil thesis, University of East Anglia, UK.

Hames, R. (1987) Game conservation or efficient hunting? In: *The Question of the Commons* (eds B. McCay & J. Acheson). University of Arizona Press, Tucson, USA.

Hamilton, W.D. (1964) The genetical evolution of social behaviour. *Journal of Theoretical Biology*, 7, 1–52.

Hannesson, R. (1996) *Fisheries Mismanagement: The Case of the North Atlantic Cod*. Blackwell Science, London, UK.

Harcourt, A.H. (1986) Gorilla conservation: anatomy of a campaign. In: *Primates: The Road to Self-Sustaining Populations* (ed. K. Benirschke), pp. 31–46. Springer-Verlag, New York, USA.

Harcourt, A.H. (1996) Is the gorilla a threatened species? How should we judge? *Biological Conservation*, 75, 165–176.

Harcourt, A.H. & Fossey, D. (1981) The Virunga gorillas: decline of an 'island' population. *African Journal of Ecology*, 19, 83–97.

Hardin, G. (1968) The tragedy of the commons. *Science*, 162, 1243–1247.

Hardwood Review (1995) *The Hardwood Review Yearbook*. Hardwood Publishing Company, Charlotte, Virginia, USA.

Hart, T. & Hart, J. (1997) Zaire: new models for an emerging state. *Conservation Biology*, 11, 308–309.

Hastings, B.E., Kenny, D., Lowenstine, L.J. & Foster, J.W. (1988) *Mountain gorillas and measles: ontogeny of a wildlife vaccination program*. Unpublished report. Karisoke Research Station, Rwanda.

Hatcher, B.G. (1995) How do marine protected areas benefit fisheries? *Caribbean Park and Protected Area Bulletin*, 5(2), 9–10. Caribbean Natural Resources Institute, US Virgin Islands.

Hatfield, E.M.C. (1991) Post-recruit growth of the Patagonian squid *Loligo gahi* (d'Orbigny). *Bulletin of Marine Science*, 49, 349–361.

Hatfield, E.M.C. (1996) Towards resolving multiple recruitment into loliginid fisheries: *Loligo gahi* in the Falkland Islands fishery. *ICES Journal of Marine Science*, 53, 565–575.

Hatfield, E.M.C. & Rodhouse, P.G. (1991) Biology and fishery of the Patagonian squid *Loligo gahi* (d'Orbigny): a review of current knowledge. *Journal of Cephalopod Biology*, 2, 41–49.

Hatfield, E.M.C. & Rodhouse, P.G. (1994a) Distribution and abundance of juvenile *Loligo gahi* in Falkland Islands waters. *Marine Biology*, 121, 267–272.

Hatfield, E.M.C. & Rodhouse, P.G. (1994b) Migration as a source of bias in the measurement of cephalopod growth. *Antarctic Science*, 6, 179–184.

Hatfield, E.M.C., Rodhouse, P.G. & Porebski, J. (1990) Demography and distribution of the Patagonian squid (*Loligo gahi* d'Orbigny) during the austral winter. *Journal of the Council for the International Exploration of the Sea*, 46, 306–312.

Hawkins, J.P. & Roberts, C.M. (1992a) Effects of recreational SCUBA diving on fore-reef slope communities of coral reefs. *Biological Conservation*, 62, 171–178.

Hawkins, J.P. & Roberts, C.M. (1992b) Effects of recreational SCUBA diving on coral reefs: trampling on reef-flat communities. *Journal of Applied Ecology*, 30, 25–30.

Hawkins, J.P. & Roberts, C.M. (1993) Can Egypt's coral reefs support ambitious plans for diving tourism? *Proceedings of the 7th International Coral Reef Symposium, Guam*, **2**, pp. 1007–1013. University of Guam Press, Guam.

Hawkins, J.P. & Roberts, C.M. (1994) The growth of coastal tourism in the Red Sea: present and future effects on coral reefs. *Ambio*, **23**, 503–508.

Hawkins, J.P. & Roberts, C.M. (1997) Estimating the carrying capacity of coral reefs for scuba diving. *Proceedings of the 8th International Coral Reef Symposium, Panama, June 1996*, **2**, pp. 1923–1926. Smithsonian Tropical Research Institute, Balboa, Panama.

Heard, D.C. (1985) Caribou census methods used in the Northwest Territories. *McGill Subarctic Research Papers*, No. 40, 229–238.

Heard, D.C. (1990) The intrinsic rate of increase of reindeer and caribou populations in arctic environments. *Rangifer*, Special Issue No. 3, 169–173.

Helle, T. & Tarvainen, L. (1984) Effects of insect harassment on weight gain and survival in reindeer calves. *Rangifer*, **4**, 24–27.

Henry, G. & Svoboda, J. (1989) Comparison of grazed and non-grazed high arctic sedge meadows. In: *Proceedings of the Second International Muskox Symposium, Saskatoon* (ed. P.L. Flood), p. A47. National Research Council of Canada, Ottawa, Ontario, Canada.

Hilborn, R. & Walters, C.J. (1992) *Quantitative Fisheries Stock Assessment: Choice, Dynamics and Uncertainty*. Chapman & Hall, New York, USA.

Hill, K. & Hurtado, A.M. (1996) *Ache Life History*. Aldine de Gruyter, New York, USA.

Hodson, T.J., Englander, F. & O'Keefe, H. (1995) Rainforest preservation, markets and medicinal plants: issues of property rights and present value. *Conservation Biology*, **9**, 1319–1321.

Holden, M. (1996) *The Common Fisheries Policy. Origin, Evaluation and Future*. Reissued and updated by D. Garrod. Fishing News Books, Oxford, UK.

Holdridge, L. (1982) *Ecología basada en zonas de vida*. Editorial IICA, Nicaragua.

Holmes, J.C. (1996) Parasites as threats to biodiversity in shrinking ecosystems. *Biodiversity and Conservation*, **5**, 975–983.

Holt, R.D. & Lawton, J.H. (1994) The ecological consequences of shared natural enemies. *Annual Review of Ecology and Systematics*, **25**, 495–520.

Homewood, K.M. & Rogers, W.A. (1991) *Maasailand Ecology. Pastoralist Development and Wildlife Conservation in Ngorongoro, Tanzania*. Cambridge University Press, Cambridge, UK.

House of Representatives (1996) *Three Year Summary of Appropriations: 1996 Budget Explanatory Notes for Committee on Appropriations*. USDA-FS, Washington DC, USA.

Houston, A.I. & McNamara, J.M. (1988) A framework for the functional analysis of behaviour. *Behavioral and Brain Sciences*, **11**, 117–163.

Howard, A.F., Rice, R.E. & Gullison, R.E. (1996) Simulated economic returns and environmental impacts from four alternative silvicultural prescriptions applied in the Neotropics: a case study of the Chimanes Forest, Bolivia. *Forest Ecology and Management*, **89**, 43–57.

HSUS/ISH (1997) *Campfire: A Close Look at the Costs and Consequences*. Humane Society of the USA/International Humane Society, Washington DC, USA.

Hubert, B.A. (1977) Estimated productivity of muskox in Truelove Lowland. In: *Truelove Lowland, Devon Island, Canada – a High Arctic Ecosystem* (ed. L.C. Bliss), pp. 467–491. University of Alberta Press, Edmonton, Canada.

Hudson, E. & Mace, G. (eds) (1996) *Marine Fish and the IUCN Red List of Threatened Animals*. Report of the workshop held in collaboration with WWF and IUCN at the Zoological Society of London 29/04–01/05/96.

Hudson, H.R. (1992) The relationship between stress and disease in orphan gorillas and its significance for gorilla tourism. *Gorilla Conservation News*, 6, 8–10.

Hughes, T.P. (1994) Catastrophes, phase shifts and large-scale degradation of a Caribbean coral reef. *Science*, 265, 1547–1551.

Huppert, D. (1982) Living marine resources. In: *Economics of Ocean Resources. A Research Agenda* (eds G.M. Brown & J.A. Crutchfield), pp. 40–66 University of Washington Press, USA.

Hutton, A. (1985) Butterfly farming in Papua New Guinea. *Oryx*, 19, 158–162.

Hydrographer of the Navy (1993) *South America Pilot, Vol. II: Southern Coasts of South America from Cabos tres Puntas to Cabo Raper, and the Falkland Islands*, 16th edn. Hydrography Department, Ministry of Defence, Taunton, UK.

IBRD (1992) *Re-Investing in Stabilisation and Growth through Public Sector Adjustment.* International Bank for Reconstruction and Development, The World Bank, Washington, USA.

IBRD (1995) *Kenya Poverty Assessment.* Population and Human Resources Division, Eastern Africa Region. International Bank for Reconstruction and Development, The World Bank, Washington, USA.

IFAD (1992) *World Poverty Report.* International Fund for Agricultural Development, Rome, Italy.

IGCP (1992) *Bwindi Impenetrable National Park: a Tourism Development Plan.* Unpublished report. International Gorilla Conservation Programme, Nairobi, Kenya.

IGCP (1995) The International Gorilla Conservation Programme (IGCP) – Rwanda and Zaire, 1994–95. *Gorilla Conservation News*, 9, 13–14.

IGCP (1996) International Gorilla Conservation Programme 1995, annual update. *Gorilla Conservation News*, 10, 17–18.

IIED (1994) *Whose Eden? An overview of community approaches to wildlife management.* International Institute for Environment and Development/Overseas Development Administration, London, UK.

IIED (1995) *Sustainable forest management: an analysis of principles, criteria and standards.* Draft report. IIED, London, UK.

IMF (1981–88) *International Financial Statistics Yearbook.* International Monetary Fund, Washington DC, USA.

IRENA & IUCN (1992) *Estrategia de conservación para el desarrollo sostenible de la Península de Cosigüina.* Managua, Nicaragua.

IUCN (1996) *1996 IUCN Red List of Threatened Animals.* IUCN, Gland, Switzerland.

IUCN Species Survival Commission (1994) *IUCN Red List Categories.* IUCN (World Conservation Union), Gland, Switzerland.

IWC (1994) The Revised Management Procedure (RMP) for baleen whales. *Report of the International Whaling Commission*, 44, Annex H, 145–152.

Iznar, N.N. (1909) Hunting bears in the Olonetskoi region. *Nasha okhota*, 1, 45–74; 2, 33–56; 3, 19–32; 4, 23–32; 5, 21–30; 6, 11–16; 8, 35–40; 11, 21–36; 12, 11–22.

Jachmann, H., Berry, P.S.M. & Imae, H. (1995) Tusklessness in African elephants – a future trend. *African Journal of Ecology*, 33, 230–235.

Jackson, G.D. (1994) Application and future potential of statolith increment analysis in squids and sepiolids. *Canadian Journal of Fisheries and Aquatic Science*, 51, 2612–2625.

Jackson, J.B.C. (1997) Reefs since Columbus. *Coral Reefs*, 16 (Suppl.) S23–S32.

Janzen, D.H. (1988) Management of habitat fragments in a tropical dry forest? *Annals of the Missouri Botanic Garden*, 75, 105–116.

Jaquette, D.L. (1970) A stochastic model for the optimal control of epidemics and pest populations. *Mathematical Biosciences*, 8, 343–354.

Jefferies, R.L., Svoboda, G., Henry, G., Raillard, M. & Ruess, R. (1992) Tundra grazing systems and climatic change. In: *Arctic Ecosystems in a Changing Climate: an Ecophysiological Perspective* (eds F.S. Chapin, R.L. Jefferies, J.F. Reynolds, G.R. Shaver, J. Svoboda & E.W. Chu), pp. 391–412. Academic Press Inc., New York, USA.

Jingfors, K.T. & Klein, D.R. (1982) Productivity in recently established muskox populations in Alaska. *Journal of Wildlife Management*, **46**, 1092–1096.

Johns, A.D. (1988) Effects of 'selective' timber extraction on rainforest structure and composition and some consequences for frugivores and folivores. *Biotropica*, **20**(1), 31–37.

Johns, A.D. (1992) Species conservation in managed tropical forests. In: *Tropical Deforestation and Species Extinction* (eds T.C. Whitmore & J.A. Sayer), pp. 15–53. IUCN, Gland, Switzerland.

Kaitala, V. & Pohjola, M. (1988) Optimal recovery of a shared resource stock: a differential game model with efficient memory equilibria. *Natural Resource Modeling*, **3**, 91–119.

Kalema, G. (1995) Epidemiology of the intestinal parasite burden of mountain gorillas, *Gorilla gorilla beringei*, in Bwindi Impenetrable National Park, south west Uganda. *Zebra Foundation, British Veterinary Zoological Society Newsletter*, Autumn, 19–34.

Kalina, J. & Butynski, T.M. (1995) Close encounters between people and gorillas. *African Primates*, **1**, 20.

Kalter, S.S. (1980) Infectious diseases of the great apes of Africa. *Journal of Reproduction and Fertility*, Supplement **28**, 149–159.

Kaplan, H. & Hill, K. (1992) The evolutionary ecology of food acquisition. In: *Evolutionary Ecology and Human Behaviour* (eds E.A. Smith & B. Winterhalder). Aldine, New York, USA.

Kaplanov, L.G. (1948) *Tiger, Red deer, Moose (Tigr, Isubr', Los')*. Izdatel'stvo Moskovskogo obshchestva ispitatelei prirodi, Moscow, USSR.

Kaplin, A.A. (1960) *Furs of the USSR (Mekha SSSR)*. Vneshtorgizdat, Moscow, USSR.

Kartawinta, K., Soedjito, H., Jessup, T., Vayda, A.P. & Colfer, C.J.P. (1984) The impacts of development on interactions between people and forests in East Kalimantan: a comparison of two areas of Kenya Dayak settlement. *Environmentalist*, **4**, (Suppl. 7), 87–95.

Keith, J.E. & Lyon, K.S. (1985) Valuing wildlife management: a Utah deer herd. *Western Journal of Agricultural Economics*, **10**, 216–222.

Kelsall, J. (1968) *The Migratory Barren-ground Caribou of Canada*. Canadian Wildlife Service Monograph, Ottawa, Canada.

Kemf, E. (ed.) (1993) *The Law of the Mother. Protecting Indigenous Peoples in Protected Areas*. Sierra Club Books, San Francisco, USA.

Kempf, E., Sutton, M. & Wilson, A. (1996) *Marine fishes in the wild*. WWF Species Status Report. World Wide Fund for Nature, Gland, Switzerland.

Kennedy, J.O.S. (1987) A computable game theoretic approach to modelling competitive fishing. *Marine Resource Economics*, **4**, 1–14.

Khlebnikov, A.I. (1977) *Sable Ecology in the Western Sayan (Ekologiia Soboliov v Zapadnikh Sayanakh)*. Nauka, Novosibirsk, USSR.

Kirkwood, G.P. (1996) The Revised Management Procedure of the International Whaling Commission. *Fisheries Management Global Trends Conference, Seattle, USA*.

Kock, R.A. (1995) Wildlife utilisation: use it or lose it – a Kenyan perspective. *Biodiversity and Conservation*, **4**, 241–256.

Kolpashchikov, L.A. (1993) *Status, Spatial Distribution and Details of Wild Reindeer Migrations in the Taymyr Population and Proposals for Cropping the Population Resource*. NIISH Krainego Severa, Norilsk, USSR.

Koritin, S. (1990) Wolves – cannibals. *Okhota i okhotniche khozyaistvo*, **6**, 6–7.

Krasovskii, L.I. (1970) A positive correlation between forest marten numbers and large-scale logging in the European North. *Bulleten' MOIP, Otdelenie biologicheskoe*, **75**(3), 7–15.

Krebs, J.R. & Davies, N.B. (1991) *An Introduction to Behavioural Ecology*, 3rd edn. Blackwell Science, Oxford, UK.

Kula, E. (1992) *Economics of Natural Resources and the Environment*. Chapman & Hall, London, UK.

Kulagin, N.M. (1932) *Moose of the USSR (Losi SSSR)*. Publishing house of the Academy of Sciences, Leningrad, USSR.

Kuropat, P. & Bryant, J.P. (1983) Digestability of caribou summer forage in arctic Alaska in relation to nutrient, fiber, and phenolic constituents. *Acta Zoologica Fennica*, **175**, 51–52.

KWS (1995a) *Wildlife Utilisation Study, Report 5: Policy and Institutional*. African Wildlife Foundation, Nairobi, Kenya.

KWS (1995b) *Wildlife Utilisation Study, Report 2: Economic Analysis*. African Wildlife Foundation, Nairobi, Kenya.

KWS (1995c) *Wildlife–Human Conflicts in Kenya: Report of the Five-Person Review Group*. Kenya Wildlife Service, Nairobi, Kenya.

KWS (1996) *Wildlife Policy 1996*. Kenya Wildlife Service, Nairobi, Kenya.

Lamb, F.B. (1966) *Mahogany of Tropical America: its Ecology and Management*. University of Michigan Press, Ann Arbor, USA.

Langvatn, R., Albon, S.D., Burkey, T. & Clutton-Brock, T.H. (1996) Climate, phenology and variation in age of first reproduction in a temperate herbivore. *Journal of Animal Ecology*, **65**, 653–670.

Larkin, P.A. (1996) Concepts and issues in marine ecosystems. *Reviews in Fish Biology and Fisheries*, **6**, 139–164.

Law, B. (ed.) (1995) *1996 Corpus Almanac and Canadian Sourcebook*. Southam Inc., Don Mills, Ontario, Canada.

Law, R. (1991) Fishing in evolutionary waters. *New Scientist*, 2/3/91, 35–37.

Leach, M. (1994) *Rainforest Relations: Gender and Resource Use among the Mende of Gola, Sierra Leone*. Smithsonian Institution Press, Washington DC, USA.

Leader-Williams, N. & Milner-Gulland, E.J. (1993) Policies for the enforcement of wildlife laws: the balance between detection and penalties in Luangwa Valley, Zambia. *Conservation Biology*, 7, 611–617.

Leader-Williams, N., Albon, S.D. & Berry, P.S.M. (1990) Illegal exploitation of black rhinoceros and elephant populations: patterns of decline, law enforcement and patrol effort in Luangwa Valley, Zambia. *Journal of Applied Ecology*, **27**, 1055–1087.

Lee, P.C., Thornback, J. & Bennett, E.L. (1988) *Threatened Primates of Africa. The IUCN Red Data Book*. IUCN, Gland, Switzerland.

Lessios, H.A. (1988) Mass mortality of *Diadema antillarum* in the Caribbean: what have we learned? *Annual Review of Ecology and Systematics*, **19**, 371–393.

Little, P.D. (1994) The link between local participation and improved conservation: a review of issues and experiences. In: *Natural Connections: Perspectives in Community-based Conservation* (eds D. Western & R. Michael Wright), pp. 347–372. Island Press, Washington DC, USA.

Loftas, T. (1996) Not enough fish in the seas. *Our Planet, UNEP Magazine for Sustainable Development*, 7(6), 29–31.

Losos, E., Hayes, J., Phillips, A., Alkire, C. & Wilcove, D. (1993) *Taxpayers Double Burden, Vol. 3: The Living Landscape*. The Wilderness Society and The Environmental Defense Fund, Washington DC, USA.

Lovo, I. (1994) *Diagnóstico social Cooperativa 'Omar Baca', Cosigüina. Proyecto reproducción de la iguanas y garrobos.* IUCN Regional Office for Mesa America, Costa Rica.

Luce, R.D. & Raiffa, H. (1957) *Games and Decisions.* John Wiley & Sons, Chichester, UK.

Ludwig, D., Hilborn, R. & Walters, C. (1993) Uncertainty, resource exploitation and conservation: lessons from history. *Science*, **260**, 17–36.

Luebbert, C. (1996) The situation in Uganda, Rwanda and Zaire – impressions from a journey. *Gorilla Journal*, **12**, 10–13.

Lushchekina, A.A. (1990) *Ecologo-geograficheskiye osnovi okhrani i ratsionalnogo ispolzovaniya dzerena* (Procapra gutturosa) *v MNR.* [*The eco-geographical basis for the conservation and rational use of the Mongolian gazelle*]. Abstract of PhD Dissertation, Russian Academy of Sciences, Moscow, USSR.

Macdonald, D.W. (1996) Dangerous liaisons and disease. *Nature*, **379**, 400–401.

Mace, R. (1990) Pastoralist herd compositions in unpredictable environments: a comparison of model predictions and data from camel-keeping groups. *Agricultural Systems*, **33**, 1–11.

Mace, R. (1991) Overgrazing overstated. *Nature*, **349**, 280–281.

Mace, R. (1993) Transitions between cultivation and pastoralism in sub-Saharan Africa. *Current Anthropology*, **34**, 363–382.

Mace, R. (1996) When to have another baby: a dynamic model of reproductive decision-making and evidence from Gabbra pastoralists. *Ethology and Sociobiology*, **17**, 263–273.

Mace, R. (1998) The co-evolution of human fertility and wealth inheritance strategies. *Philosophical Transactions of the Royal Society of London, Biological Sciences*, **353**, 389–397.

Mace, R.H. & Houston, A.I. (1989) Pastoralist strategies for survival in unpredictable environments: a model of herd composition that maximises household viability. *Agricultural Systems*, **31**, 185–204.

Macfie, L. (1991) The Volcano Veterinary Center. *Gorilla Conservation News*, **5**, 21.

Macfie, L. (1995) International Gorilla Conservation Programme – Uganda, 1994. *Gorilla Conservation News*, **9**, 17–18.

Macfie, L. (1997) Gorilla tourism in Uganda. *Gorilla Journal*, **15**, 16–17.

MacKenzie, J. (1988) *The Empire of Nature: Hunting, Conservation, and British Imperialism.* Manchester University Press, Manchester, UK.

MacKinnon, J., MacKinnon, K., Child, G. & Thorsell, J. (1986) *Managing Protected Areas in the Tropics.* IUCN, Cambridge, UK.

MacKinnon, K. Hatta, G., Halim, H. & Mangalik, A. (1996) *The Ecology of Kalimantan.* Periplus, Singapore.

MacNab, J. (1983) Wildlife management as scientific experimentation. *The Wildlife Society Bulletin*, **11**, 397–401.

Maddala, G.S. (1989) *Introduction to Econometrics.* Macmillan, New York, USA.

MAFF (1993) *CAP Reform: Arable Area Payments 1993–1994, Explanatory Guide.* Ministry of Agriculture, Fisheries and Food, London, UK.

Magnusson, K.G. (1995) An overview of the multi-species VPA – theory and applications. *Reviews in Fish Biology and Fisheries*, **5**, 195–212.

Makarova, O.A. (1989) Systematic status of wild reindeer on the Kola peninsula. In: *Lesnoi Severni Olen'* (ed. P.I. Danilov), pp. 19–25. Karel'skii filial AN SSSR, Petrozavodsk, USSR.

Mandosir, S. & Stark, M. (1993) Butterfly ranching. In: *The Law of the Mother* (ed. E. Kemf), pp. 114–120. Sierra Club Books, San Francisco, USA.

Mangel, M. (1994) Spatial patterning in resource exploitation and conservation. *Philosophical Transactions of the Royal Society of London B*, **343**, 93–98.

Mangel, M. & Clark, C.W. (1988) *Dynamic Modeling in Behavioral Ecology.* Princeton University Press, Princeton, USA.

Marks, S.A. (1976) *Large Mammals and a Brave People: Subsistence Hunters in Zambia.* University of Washington Press, Seattle, USA.

Marks, S.A. (1979) Profile and process: subsistence hunters in a Zambian community. *Africa*, **41**, 53–67.

Marks, S.A. (1984) *The Imperial Lion: Human Dimensions in Wildlife Management in Central Africa.* Westview Press, Boulder, USA.

Marks, S.A. (1994) Local hunters and wildlife surveys: a design to enhance participation. *African Journal of Ecology*, **32**, 233–254.

Marks, S.A. (1996) Local hunters and wildlife surveys: an assessment and comparison of counts for 1989, 1990 and 1993. *African Journal of Ecology*, **34**, 237–257.

Marsh, C. & Sinun, W. (1992) Pragmatic approaches to habitat conservation within a large timber concession in Sabah, Malaysia. *Paper presented to the 4th World Congress on National Parks, Caracas, Venezuela.*

Martin, E.B. (1983) *Rhino Exploitation.* WWF, Hong Kong.

Martini, A.Z.M., de Arauja Rosa, N. & Uhl, C. (1994) A first attempt to predict Amazonian tree species potentially threatened by logging activities. *Environmental Conservation*, **21**, 152–162.

Maxwell, B. (1980) *The Climate of the Canadian Arctic Islands and Adjacent Waters.* Atmospheric Environment Service, Ontario, Canada.

May, R.M. (1973) *Stability and Complexity in Model Ecosystems.* Princeton Monographs in Population Biology, 6. Princeton University Press, Princeton, USA.

May, R.M. (1975) Biological populations obeying difference equations: stable points, stable cycles and chaos. *Journal of Theoretical Biology*, **49**, 511–524.

May, R.M. (1988) Conservation and disease. *Conservation Biology*, **2**, 28–30.

Maynard-Smith, J. (1982) *Evolution and the Theory of Games.* Cambridge University Press, Cambridge, UK.

Mazany, R.L., Charles, A.T. & Cross, M.L. (1989) Fisheries regulation and the incentives to overfish. Paper at Canadian Economic Association Meeting, June 2–4, 1989. Laval University, Quebec, Canada.

McBride, B. (1988) Gorilla warfare. *The Christian Science Monitor*, 25 May, 16–17.

McCay, B.J. & Acheson, J.M. (1990) *The Question of the Commons.* University of Arizona Press, Tucson, USA.

McGlade, J. (1989) Integrated fisheries management models: understanding the limits to marine resource exploitation. *American Fisheries Society Symposium*, **6**, 139–165.

McGlade, J. (1996) Governance: policy and planning in the coastal zone. In: *Interdisciplinary Scientific Methodologies for the Sustainable Use and Management of Coastal Resource Systems* (eds J. McGlade, D. Pauly & G. Silvestre), pp. 33–57. Final report to the Commission of the European Community, Brussels, Belgium.

McNeilage, A. (1996) Ecotourism and mountain gorillas in the Virunga Volcanoes. In: *The Exploitation of Mammal Populations* (eds V.J. Taylor & N. Dunstone), pp. 334–344. Chapman & Hall, London, UK.

Mead, R. & Curnow, R.N. (1983) *Statistical Methods in Agriculture and Experimental Biology.* Chapman & Hall, London, UK.

Medio, D., Pearson, M. & Ormond, R.F.G. (1997) Effect of briefings on rates of damage to corals by divers. *Biological Conservation*, **79**, 91–95.

Melland, F.H. (1938) *Elephants in Africa.* Country Life Ltd, London, UK.

Mel'nitski, N.A. (1915) *The Bear and Hunting It (Medved' i okhota na nego).* Nasha okhota, Petrograd, Russia.

Menges, F.H. (1974) *Economic Decision-making: Basic Concepts and Models*. Longman, Harlow, UK.

Merrill, E.H. & Boyce, M.S. (1991) Summer range and elk population dynamics in Yellowstone National Park. In: *The Greater Yellowstone Ecosystem: Redefining America's Wildlife Heritage* (eds R.B. Keiter & M.S. Boyce), pp. 263–273. Yale University Press, New Haven, USA.

Mesterton-Gibbons, M. (1993) Game-theoretic resource modelling. *Natural Resource Modeling*, 7, 93–147.

Miller, F.L. (1991) *Peary caribou – a status report*. Canadian Wildlife Service, Ottawa, Canada.

Miller, F.L., Russell, R.H. & Gunn, A. (1977) *Distributions, movements and numbers of Peary caribou and muskoxen on Western Queen Elizabeth Islands, Northwest Territories, 1972–74*. Canadan Wildlife Service Report Series No. 40.

Milner-Gulland, E.J. (1993) An econometric analysis of consumer demand for ivory and rhino horn. *Environmental and Resource Economics*, 3, 73–95.

Milner-Gulland, E.J. (1994) A population model for the management of the saiga antelope. *Journal of Applied Ecology*, 31, 25–39.

Milner-Gulland, E.J. (1997) A stochastic dynamic programming model for the management of the saiga antelope *Ecological Applications*, 7, 130–142.

Milner-Gulland, E.J. & Beddington, J.R. (1993) The exploitation of the elephant for the ivory trade – an historical perspective. *Proceedings of the Royal Society of London B*, 252, 29–37.

Milner-Gulland, E.J. & Leader-Williams, N. (1992) A model of incentives for the illegal exploitation of black rhinos and elephants: poaching pays in Luangwa Valley, Zambia. *Journal of Applied Ecology*, 29, 388–401.

Milner-Gulland, E.J. & Lhagvasuren, B. (1998) Population dynamics of the Mongolian gazelle (*Procapra guttorosa*): an historical analysis. *Journal of Applied Ecology* (in press).

Milner-Gulland, E.J. & Mace, R. (1991) The impact of the ivory trade on the African elephant *Loxodonta africana* population as assessed by data from the trade. *Biological Conservation*, 55, 215–229.

Milner-Gulland, E.J., Beddington, J.R. & Leader-Williams, N. (1992) Dehorning African rhinos: a model of optimal frequency and profitability. *Proceedings of the Royal Society of London B*, 249, 83–87.

Mislenkov, A.I. & Voloshina, I.V. (1989) *Ecology and Behaviour of Goral (Ekologiya i Povedenie Amurskogo Gorala)*. Nauka, Moscow, USSR.

Monakhov, G.I. & Bakeev, N.N. (1981) *Sable (Sobol')*. Lesnaya promishlennost', Moscow, USSR.

Moser, C.A. & Kalton, G. (1971) *Survey Methods in Social Investigation*. Gower, USA.

Moulton, M.P. & Sanderson, J. (1996) *Wildlife Issues in a Changing World*. St Lucie Press, Delray Beach, Florida, USA.

MPE (1992) *Environmental Protocol (Tourism Sector). Section 1–4. Technical Co-operation Programme*. Ministry of Planning and Environment, Republic of Maldives.

MPHRE (1994) *National Development Plan 1994–1996*. Ministry of Planning, Human Resources and Environment, Republic of Maldives.

Mugabe, J. & Wandera, P. (1993) Structural adjustment and wildlife management in Kenya. In: *The Unsteady State: Structural Adjustment and Sustainable Development in Kenya* (ed. C. Juma). African Centre for Technology Studies, Nairobi, Kenya.

Murphy, E.J., Rodhouse, P.G. & Nolan, C.P. (1994) Modelling the selective effects of fishing on reproductive potential and population structure of squid. *ICES Journal of Marine Science*, 51, 299–313.

Murphy, G.I. (1977) Clupeoids. In: *Fish Population Dynamics* (ed. J.A. Gulland). John Wiley & Sons, Chichester, UK.

Mushenzi, L.N. (1996) *Present state of the Virunga National Park in the Central and Southern Sectors*. Unpublished report. Institut Zairois pour la Conservation de la Nature, Goma, Zaire.

Myers, R.A., Barrowman, N.J., Hutchings, S.A. & Rosenberg, A.A. (1995) Population dynamics of exploited fish stocks at low population levels. *Science*, **269**, 1106–1108.

Nagy, J.A., Larter, N.C. & Fraser, V.P. (1996) Population demography of Peary caribou and muskox on Banks Island, NWT, 1982–1992. *Rangifer*, Special Issue No. 9, 213–222.

Nash, R. (1978) *Wilderness and the American Mind*. Yale University Press, New Haven, USA.

Nasimovich, A.A. & Shubnikova, O.N. (1969) Game animal resources in the taiga part of Western Siberia, and their utilization (using the example of the Tyumenskoy region). *Geographicheskii Sbornik*, **3**, 171–208. VINITI, Moscow, USSR.

Nebol'sin, P.I. (1855) Description of trade of Russia and Central Asia. *Zapiski Russkogo Geographicheskogo Obshchestva*, **10**, 45–198.

Nelson, B.W., Kapos, V., Adams, J.B., Oliveira, W.J., Braun, O.P.G. & do Amaral, I.L. (1994) Forest disturbance by large blowdowns in the Brazilian Amazon. *Ecology*, **75**, 853–858.

Nethconsult/Transtec (in association with Borde Failte) (1995) *Maldives Tourism Master Plan: Assessment of Existing Conditions in the Tourism Sector*. EC Report to Ministry of Tourism, Republic of Maldives.

Newton, A.C., Baker, P., Ramnarine, S., Mesen, J.F. & Leakey, R.R.B. (1993) The mahogany shootborer: prospects for control. *Forest Ecology and Management*, **57**, 301–328.

Newton, A.C., Cornelius, J.P., Baker, P., Gillies, A.C.M., Hernandez, M., Ramnarine, S., Mesen, J.F. & Watt, A.D. (1996) Mahogany as a genetic resource. *Botanical Journal of the Linnean Society*, **122**, 61–73.

Norgaard, R.B. & Howarth, R.B. (1991) Sustainability and discounting the future. In: *Ecological Economics* (ed. R. Costanza), pp. 88–101. Columbia University Press, New York, USA.

Norton-Griffiths, M. (1995) Economic incentives to develop the rangelands of the Serengeti: implications for wildlife management. In: *Serengeti II: Research, Management and Conservation of an Ecosystem* (eds A.R.E. Sinclair & P. Arcese). Chicago University Press, Chicago, USA.

Norton-Griffiths, M. (1996) Property rights and the marginal wildebeest. *Biodiversity and Conservation*, **5**, 1557–1577.

Norton-Griffiths, M. & Southey, C. (1993) *The opportunity costs of biodiversity conservation: a case study of Kenya*. CSERGE Working Paper GEC 93-21. University of East Anglia, Norwich, UK.

Norton-Griffiths, M. & Southey, C. (1995) The opportunity costs of biodiversity conservation in Kenya. *Ecological Economics*, **12**, 125–139.

Oates, J.F. (1995) The dangers of conservation by rural development – a case-study from the forests of Nigeria. *Oryx*, **29**, 115–122.

Oelschlaeger, M. (1991) *The Idea of Wilderness: From Prehistory to the Age of Ecology*. Yale University Press, New Haven, USA.

Omondi, P. (1994) *Wildlife–human conflict in Kenya: integrating wildlife conservation with human needs in the Maasai Mara region*. PhD thesis, McGill University, Canada.

Padoch, C. (1992) Traditional management of forests and agroforests in West Kalimantan, Indonesia. Paper presented at the *Proceedings of the Pacific Anthropology Congress, Honolulu, Hawaii*.

Padoch, C. & Peters, C.M. (1993) Managed forest gardens in West Kalimantan, Indonesia. In: *Perspectives on Biodiversity: Case Studies of Genetic Resource Conservation* (eds J.J. Cohen & C.S. Potter), pp. 167–176. American Association for the Advancement of Science, Washington DC, USA.

Pagel, M. & Mace, R. (1991) Keeping the ivory trade banned. *Nature*, **351**, 265–266.

Painter, M. (1995) Anthropology in pursuit of conservation and development. In: *Global Ecosystems: Creating Options Through Anthropological Perspectives* (ed. P.J. Puntenney), National Association for the Practice of Anthropology Bulletin 15, 33–45.

Palmer, J. (1989) Management of natural forest for sustainable timber production: a commentary. In: *No Timber Without Trees: Sustainability in the Tropical Forest* (eds D. Poore, P. Burgess, J. Palmer, S. Rietbergen & T. Synnott), pp. 154–189. Earthscan Publications Ltd, London, UK.

Panayotou, T. (1994) *Economic Instruments for Environmental Management and Sustainable Development*. United Nations Environment Programme, Nairobi, Kenya.

Parker, I.S.C. (1979) *The Ivory Trade*. Department of Fisheries and Wildlife, Washington DC, USA.

Parker, K., White, R.G., Gillingham, M.P. & Holleman, D.F. (1990) Comparison of energy metabolism in relation to daily activity and milk consumption by caribou and muskox neonates. *Canadian Journal of Zoology*, **68**, 104–114.

Pascual, M.A. & Hilborn, R. (1995) Conservation of harvested populations in fluctuating environments: the case of the Serengeti wildebeest. *Journal of Applied Ecology*, **32**(3), 468–480.

Patterson, K.R. (1988) Life history of Patagonian squid *Loligo gahi* and growth parameter estimates using least-squares fits to linear and von Bertalanffy models. *Marine Ecology Progress Series* **47**, 65–74.

Pauly, D. (1995) Anecdotes and the shifting baseline syndrome of fisheries. *Trends in Ecology and Evolution*, **10**, 430.

Pavlov, B.M., Kuksov, B.A. & Savel'ev, V.D. (1976) *Rational Use of Reindeer Resources in Taymyr*. Metodicheskie rekomendatsii, NIISH Krainego Severa. Norilsk, USSR.

Pavlov, M. (1990) *Wolf (Volk)*. Agropromizdat, Moscow, USSR.

Pavlov, M., Korsakova, L. & Lavrov, N. (1974) *Translocation of Game Mammals and Birds of the USSR (Akklimatizatsiya zverei i ptits)*, Vols 1, 2. Kirovskoe Knizhnoe izdatel'stvo, Kirov, USSR.

Pearce, D. & Turner, R.K. (1990) *Economics of Natural Resources and the Environment*. John Hopkins, Baltimore, Maryland, USA.

Pearce, D., Barbier, E. & Markandya, A. (1990) *Sustainable Development: Economics and Environment in the Third World*. Edward Elgar, Cheltenham, UK.

Pearce, F. (1991) *Green Warriors: The People and the Politics Behind the Environmental Movement*. Bodley Head, London, UK.

Pearce, P. (1990) *Introduction to Forestry Economics*. University of British Columbia Press, Vancouver, Canada.

Pearse, P.H. (1980) *Regulation of fishing effort: with special reference to the Mediterranean trawl fisheries*. FAO Technical Paper 197.

Peluso, N.L. (1983) Networking in the commons: a tragedy for rattan? *Indonesia*, **35**, 95–108.

Peluso, N.L. (1986) *Rattan industries in East Kalimantan, Indonesia*. FAO, Jakarta, Indonesia.

Pernetta, J.C. (ed.) (1993) *Marine Protected Areas in the South Asian Seas Region, Vol. 3: Maldives. A Marine Conservation and Development Report*. IUCN – The World Conservation Union, Gland, Switzerland.

Pernetta, J.C. & Elder, D. (1993) *Cross-Sectoral, Integrated Coastal Area Planning: Guidelines and Principles for Coastal Area Development. A Marine Conservation and Development Report.* IUCN – The World Conservation Union, Gland, Switzerland.

Pernetta, J.C. & Sestini, G. (1989) Report of the mission to the Republic of Maldives. *UNEP Regional Seas Reports and Studies*, **104**, 1–84.

Perrings, C. (1987) *Economy and Environment.* Cambridge University Press, Cambridge, UK.

Pervozvannii, I.V. (1959) Description of forest development in Karelia. *Trudi Karel'skogo filiala Akademii Nauk SSSR,* **19**, 5–75. Karel'skii filial AN SSSR, Petrozavodsk, USSR.

Peter, A. & Urquhart, D. (1991) One caribou herd, two native cultures, five political systems: consensus management of the Porcupine Caribou range. *Transactions of North American Wildlife and Natural Resources Conference*, **56**, 321–325.

Peters, C.M. (1996a) *The ecology and management of non-timber forest resources.* World Bank Technical Paper No. 322.

Peters, C.M. (1996b) Illipe nuts (*Shorea* spp.) in West Kalimantan: use, ecology, and management potential of an important forest resource. In: *Borneo in Transition. People, Forests, Conservation and Development* (eds C. Padoch & N.L. Peluso), pp. 230–244. Oxford University Press, Kuala Lumpur, Malaysia.

Peters, C.M., Gentry, A.H. & Mendelsohn, R.O. (1989) Valuation of an Amazonian rain forest. *Nature*, **339**, 655–656.

Pierce, G.J. & Guerra, A. (1994) Stock assessment methods for cephalopod fisheries. *Fisheries Research*, **21**, 225–285.

Pikunov, D.G. (1990) Tiger numbers in the Far East of the USSR. In: *Tezisi 5 c'ezda Vsesoyuznogo Teriologicheskogo Obshchestva*, Vol. 2, pp. 102–103. Nauka, Moscow, USSR.

Pilgram, T. & Western, D. (1986) Inferring the sex and age of African elephants from tusk measurements. *Biological Conservation*, **36**, 39–52.

Plumptre, A. (1996) Gorilla war. *Swara*, **19**, 30–31.

PNUMA (1985) *Manejo de fauna sylvestre y desarrollo rural: información sobre siete especiers de América Latina y Caribe.* Documento téchnico No. 2. FAO, Rome, Italy.

Poletskii, V. (1990) Results of 6 years. *Okhota i Okhotniche Khozyaistvo*, **3**, 6–8.

Polunin, N.V.C. & Roberts, C.M. (eds) (1996) *Reef Fisheries.* Chapman & Hall, London, UK.

Ponomarev, G.V. (1971) Dynamics of game mammal numbers in the Sos'va river basins, under the conditions of many-faceted economic utilization of the taiga area. In: *Dokladi Instituta Geografii Sibiri i Dal'nego Vostoka*, **32**, 26–32. Nauka, Irkutsk, USSR.

Poole, J.H. (1989) The effects of poaching on the age structure and social and reproductive patterns of selected East African elephant populations. In: *The Ivory Trade and the Future of the African Elephant.* Ivory Trade Review Group, Queen Elizabeth House, Oxford, UK.

Poore, D. (1989) Conclusions. In: *No Timber Without Trees: Sustainability in the Tropical Forest* (eds D. Poore, P. Burgess, J. Palmer, S. Rietbergen & T. Synnott), pp. 1–27. Earthscan Publications Ltd, London, UK.

Possingham, H. (1996) Decision theory and biodiversity management: how to manage a metapopulation. In: *Frontiers of Population Ecology* (eds R.B. Floyd, A.W. Sheppard & P.J. de Barro). CSIRO Publishing, Melbourne, Australia.

Prescott-Allen, R. & Prescott-Allen, C. (1996) *Assessing the Sustainability of Uses of Wild Species: Case Studies and Initial Procedures.* IUCN, Cambridge, UK.

Price, A.R.G. & Firaq, I. (1996) The environmental status of reefs on Maldivian resort islands: a preliminary assessment for tourism planning. *Aquatic Conservation: Marine and Freshwater Ecosystems*, **6**(2), 93–106.

Price, A.R.G., Edwards, A., Burbridge, P. & Brown, B.E. (1993) *Coasts*. Environment and Development Briefs No. 6, UNESCO, Paris, France.

Price, C. (1993) *Time, Discounting and Value*. Blackwell, Oxford, UK.

Primack, R.B. (1993) *Essentials of Conservation Biology*. Sinauer Associates, Sunderland, USA.

Quevedo, L. (1986) *Evaluacion del efecto de la tala selectiva sobre la renovacion de un bosque humedo subtropical en Santa Cruz, Bolivia*. MSc thesis, Centro Agronomico Tropical de Investigacion y Ensenanza, Turrialba, Costa Rica.

Rabinovich, J.E., Capurro, A.F. & Pessina, L.L. (1991) Vicuña use and the bioeconomics of an Andean peasant commuity in Catamarca, Argentina. In: *Neo-tropical Wildlife Use and Conservation* (eds J.G. Robinson & K.R. Redford), pp. 337–358 University of Chicago Press, Chicago, USA.

Rainforest Action Network (1996) *Brazil bans mahogany logging in the Amazon*. Press release 29 July 1996.

Ramos, J.M. & del Amo, S. (1992) Enrichment planting in tropical secondary forest in Veracruz, Mexico. *Forest Ecology and Management*, **54**, 289–304.

Rapoport, A. (1960) *Fights, Games and Debates*. University of Michigan Press, Ann Arbor, USA.

Redclift, M. (1987) *Sustainable Development: Exploring the Contradictions*. Methuen, London, UK.

Redford, K. (1991) The ecologically noble savage. *Orion*, **9**, 24–29.

Redford, K.H. (1992) The empty forest. *Bioscience*, **42**, 412–422.

Reed, W.J. (1974) A stochastic model for the economic management of a renewable animal resource. *Mathematical Biosciences*, **22**, 313–337.

Reed, W.J. (1983) Recruitment variability and age structure in harvested animal populations. *Mathematical Biosciences*, **65**, 239–268.

Reining, C. & Heinzman, R. (1992) Nontimber forest products in the Peten, Guatemala: why extractive reserves are critical for both conservation and development. In: *Sustainable Harvest and Marketing of Rain Forest Products* (eds M. Plotkin & L. Famolare), pp. 110–117. Conservation International and Island Press, Washington DC, USA.

Renshaw, E. (1991) *Modelling Biological Populations in Space and Time*. Cambridge Studies in Mathematical Biology, Cambridge University Press, Cambridge, UK.

Reynolds, D., Stokes, D. & Smith, T. (1996) *CAC Global Sea Surface Temperature Annals, Monthly Composite Analyses 1982–1995*. Data Support Section, National Center for Atmospheric Research, USA.

Rice, R.E. (1989) *The Uncounted Costs of Logging: National Forests Policies for the Future*. The Wilderness Society, Washington DC, USA.

Riegl, B. & Velimirov, B. (1991) How many damaged corals in Red Sea reef systems? A quantitative survey. *Hydrobiologia*, **216/217**, 249–256.

Risk, M.J., Dunn, J.J., Allison, W.R. & Horrill, C. (1994) Reef monitoring in Maldives and Zanzibar: low-tech and high-tech science. In: *Proceedings of the Colloqium on Global Aspects of Coral Reefs: Health, Hazards and History* (ed. R.N. Ginsburg), pp. 66–72. Rosenstiel School of Marine and Atmospheric Sciences, University of Miami, Florida, USA.

Roberts, C.M. (1993) Coral reefs: health, hazards and history. *Trends in Ecology and Evolution*, **8**, 425–427.

Roberts, C.M. (1995a) Rapid build-up of fish biomass in a Caribbean marine reserve. *Conservation Biology*, **9**, 815–826.

Roberts, C.M. (1995b) Effects of fishing on the ecosystem structure of coral reefs. *Conservation Biology*, **9**, 988–995.

Roberts, C.M. (1997) Ecological advice for the global fisheries crisis. *Trends in Ecology and Evolution*, **12**, 35–38.

Roberts, C.M. & Hawkins, J.P. (1995) *Status of reef fish and coral communities of the Saba Marine Park, 1995*. Eastern Caribbean Center, University of the Virgin Islands, St. Thomas.

Roberts, C.M., Hawkins, J.P. & Nowlis, J.S. (1995) Economic and social benefits of marine resource management in the Caribbean. *Caribbean Perspectives*, December 1995, pp. 3–8, University of the Virgin Islands, St. Thomas.

Roberts, M.J. & Sauer, W.H.H. (1994) Environment: the key to understanding the south African chokka squid (*Loligo vulgaris reynaudii*) life cycle and fishery. *Antarctic Science*, **6**(2), 249–258.

Robinson, J.G. (1993) The limits to caring: sustainable living and the loss of biodiversity. *Conservation Biology*, **7**, 20–28.

Robinson, J.G. & Bennett, E.L. (eds) (in press) *Hunting for Sustainability in Tropical Forests*. Columbia University Press, New York, USA.

Robinson, J.G. & Redford, K.H. (eds) (1991) *Neo-tropical Wildlife Use and Conservation*. Chicago University Press, Chicago, USA.

Rodan, B.D. & Campbell, F.T. (1996) CITES and the sustainable management of *Swietenia macrophylla* King. *Botanical Journal of the Linnean Society*, **122**, 83–87.

Rodan, B.D., Newton, A.C. & Verissimo, A. (1992) Mahogany conservation: status and policy initiatives. *Environmental Conservation*, **19**, 331–342.

Rogers, A.R. (1994) The evolution of time preference by natural selection. *American Economic Review*, **84**, 460–481.

Rogers, C.S. (1990) Responses of Coral Reefs and Reef Organisms to Sedimentation. *Marine Ecological Progress Series* **62**, 185–202.

Rose, A.L. (1996) The African great ape bushmeat crisis. *Pan Africa News*, **3**, 1–6.

Rosenberg, A.A. & Restrepo, V.R. (1994) Uncertainty and risk evaluation in stock assessment advice for US marine fisheries. *Canadian Journal of Fisheries and Aquatic Science*, **51**, 2715–2720.

Rosenberg, A.A., Kirkwood, G.P., Crombie, J.P. & Beddington, J.R. (1990) The assessment of annual squid species. *Fisheries Research*, **8**, 335–350.

Rosenberg, A.A., Fogarty, M.J., Sissenwine, M.P., Beddington, J.R. & Shepherd, J.G. (1993) Achieving sustainable use of renewable resources. *Science*, **262**, 828–829.

Rouphael, T. & Inglis, G. (1995) *The effects of qualified recreational scuba divers on coral reefs*. CRC Reef Research Centre, Technical Report No. 4, Townsville.

Russell, D.R., Martell, A.M. & Nixon, W. (1993) Range ecology of the Porcupine caribou herd in Canada. *Rangifer*, Special Issue No. 8, 1–167.

Saddler, H., Bennett, J., Reynolds, I. & Smith, B. (1980) *Public choice in Tasmania: Aspects of the Lower Gordon River Hydro-electric Development Proposal*. Centre for Resource and Environmental Studies, Australian National University, Canberra, Australia.

Said, M.Y., Chunge, R.N., Craig, G.C., Thouless, C.R., Barnes, R.F.W. & Dublin, H.T. (1995) *African Elephant Database*. IUCN Species Survival Commission, Gland, Switzerland.

Sarmiento, E.E. & Butynski, T.M. (1996) Present problems in gorilla taxonomy. *Gorilla Journal*, **12**, 5–7.

Sarmiento, E.E., Butynski, T.M. & Kalina, J. (1996) Gorillas of Bwindi-Impenetrable Forest and the Virunga Volcanoes: taxonomic implications of morphological and ecological differences. *American Journal of Primatology*, **40**, 1–21.

Schaller, G.B. (1963) *The Mountain Gorilla: Ecology and Behavior*. University of Chicago Press, Chicago, USA.

Schmitt, T.M. (1997) Close encounter with gorillas at Bwindi. *Gorilla Journal*, 14, 12–13.

Scoones, I. (1994) *Living with Uncertainty: New Directions in Pastoral Development in Africa*. International Technology Publications, London, UK.

Seibert, B. (1988) Agroforestry for the conservation of genetic resources. In: *Agroforestry* (eds A.B. Lahije & B. Seibert), pp. 235–251. UNMUL, Samarinda, Indonesia.

Sharp, R. (1997) The African elephant: conservation and CITES. *Oryx*, 13, 111–119.

Shcherbakov, N.M. & Volkov, A.D. (1985) *Forest Resources of Karelia, their Utilization and Reproduction* (*Lesnie Resursi Karel'skoii ASSR, ikh Ispol'zovanie i Vosproizvodstvo*). Karel'skii filial AN SSSR, Petrozavodsk, USSR.

Shepherd, J.G. & Cushing, D.H. (1990) Regulation in fish populations – myth or mirage. *Philosophical Transactions of the Royal Society of London B*, 330, 151–164.

Sheppard, C.R.C. (1981) The reef and soft substrate coral fauna of Chagos, Indian Ocean. *Journal of Natural History*, 15, 607–621.

Sheppard, C.R.C. (1987) Coral species of the Indian Ocean and adjacent seas. *Atoll Research Bulletin*, 307, 1–32.

Sheppard, C.R.C. (1995) The shifting baseline syndrome. *Marine Pollution Bulletin*, 30(12), 766–767.

Sheppard, C.R.C. & Wells, S.M. (1988) *Coral Reefs of the World, Vol. 2: Indian Ocean, Red Sea and Gulf*. IUCN — The World Conservation Union, Gland, Switzerland; World Conservation Monitoring Centre, Cambridge, UK; United Nations Environment Programme, Nairobi, Kenya.

Shoemaker, C.A. (1982) Optimal integrated control of univoltine pest populations with age structure. *Operations Research*, 30, 40–61.

Sholley, C.R. (1991) Conserving gorillas in the midst of guerillas. *Annual Conference Proceedings, American Association of Zoological Parks and Aquariums*, pp. 30–37.

Sholley, C.R. & Hastings, B. (1989) Outbreak of illness among Rwanda's gorillas. *Gorilla Conservation News*, 3, 7.

Siebert, S. (1988) Development for conservation: rattan cultivation and management in Kerinci-Seblat National Park, Indonesia. Paper presented at the Society of American Foresters Annual Conference, Rochester, New York, USA.

Sikubwabo, K.C. & Mushenzi, L.N. (1996) The mountain gorillas of the Mikeno in Zaire. *Gorilla Journal*, 13, 9–10.

Silant'ev, A.A. (1898) *Review of professional hunts in Russia* (*Obzori promislovikh okhot v Rossii*). Sankt-Petersburg, Russia.

Sitsko, A. (1983) The brown bear: resources, rational harvest. *Okhota i Okhotniche Khozyaistvo*, 11, 6–7.

Skalon, V.N. (1951) River beavers of Northern Asia. *Materiali k Poznaniyu Fauni i Flori SSSR, Novaya Seriya, Otdelenie Zoologicheskoe*, Vol. 25(40), pp. 45–57. MOIP, Moscow, USSR.

Skogland, T. (1985) The effects of density-dependent resource limitation on the demography of wild reindeer. *Journal of Animal Ecology*, 54, 359–374.

Sladek Nowlis, J., Roberts, C.M., Smith, A.H. & Sirila, E. (in press) Human-enhanced impacts of a tropical storm on nearshore coral reefs. *Ambio*.

Sludskii, A.A. (1955) Saiga in Kazakhstan. *Trudi Instituta Zoologii AN Kazakhskoii SSR*, 4, 1–78. Nauka, Alma-Ata, USSR.

Sludskii, A.A. (1963) Dzhuts in Eurasian steppes and deserts. *Trudi Instituta Zoologii AN Kazakhskoii SSR*, 20, 5–88. Nauka, Alma-Ata, USSR.

Snook, L.K. (1991) Opportunities and constraints for sustainable tropical forestry: lessons from the Plan Piloto Forestal, Quintana Roo, Mexico. Presented at Humid Tropical Lowlands Conference on Development Strategies and Natural Resource Management, Panama.

Snook, L.K. (1993) *Stand dynamics of mahogany* (Swietenia macrophylla *King*) *and associated species after fire and hurricane in the tropical forests of the Yucatan Peninsula, Mexico*. PhD thesis, Yale School of Forestry and Environment Studies, USA.

Snook, L.K. (1996) Catastrophic disturbance, logging and the ecology of mahogany (*Swietenia macrophylla* King): grounds for listing a major tropical timber species in CITES. *Botanical Journal of the Linnean Society*, **122**, 35–46.

Sokolov, V.E. & Lebedeva, N.L. (1989) Commercial hunting in the Soviet Union. In: *Wildlife Production Systems* (eds R.J. Hudson, K.R. Drew & L.M. Baskin), pp. 170–176. Cambridge University Press, Cambridge, UK.

Solov'ev, D.K. (1926) *Foundations of Hunting Management* (*Osnovi Okhotovedeniya*). Novaya derevnya, Moscow, USSR.

Soulé, M.E. (1987) *Viable Populations for Conservation*. Cambridge University Press, Cambridge, UK.

Southgate, D. (1996) *Can habitats be protected and local living standards improved by promoting ecotourism, non-timber extraction, sustainable timber harvesting and genetic prospecting?* Report to the Inter-American Development Bank, Washington, USA.

Southgate, D., Coles-Ritchie, M. & Salazar-Canalos, P. (1996) *Can tropical forests be saved by harvesting non-timber products?* CSERGE Working Paper GEC 96-02. University of East Anglia, Norwich, UK.

Southwood, T.R.E. (1995) Ecological processes and sustainability. *International Journal of Sustainable Development and World Ecology*, **2**, 229–239.

SSN (Species Survival Network) (1996) *Criteria for Assessing the Sustainability of Trade in Wild Fauna and Flora*. The Humane Society of the United States/Humane Society International, Washington DC, USA.

Start, A.N. & Marshall, A.G. (1976) Nectarivorous bats as pollinators of trees in West Malaysia. In: *Tropical Trees: Variation, Breeding and Conservation* (eds J. Burley & B.T. Styles), pp. 141–150. Academic Press, London, UK.

Statham, D.C. (1994) The farm scheme of the North York Moors National Park, UK. In: *Natural Connections: Perspectives in Community-based Conservation* (eds D. Western & R. Michael Wright), pp. 282–299. Island Press, Washington DC, USA.

Steele, D. (1994) The Ugandan alternative. *Getaway*, April, 31.

Stewart, K.J. (1991) Editorial. *Gorilla Conservation News*, **5**, 1–2.

Stewart, K.J. (1992) Gorilla tourism: problems of control. *Gorilla Conservation News*, **6**, 15–16.

Stewart, K.J. (1993) Gorilla tourism: a reply to Zaire. *Gorilla Conservation News*, **7**, 12–13.

Stewart, P.J. (1985) The dubious case for state control. *Ceres*, **18**, 14–19.

Stump, K. & Batker, D. (1996) *Sinking Fast: How Factory Trawlers are Destroying US Fisheries and Marine Ecosystems*. Greenpeace Report, Greenpeace, Washington DC, USA.

Styles, B.T. & Khosla, P.K. (1976) Cytology and reproductive biology of Meliaceae. In: *Tropical Trees: Variation, Breeding and Conservation* (eds J. Burley & B.T. Styles), pp. 61–68. Academic Press, London, UK.

Sugihara, G. & May, R.M. (1990) Nonlinear forecasting as a way of distinguishing chaos from measurement error in a time-series. *Nature*, **344**, 734–741.

Sutinen, J.G. & Anderson, P. (1985) The economics of fisheries law enforcement. *Land Economics*, **61**, 387–397.

Sutinen, J.G. & Gauvin, J.R. (1989) An econometric study of regulatory enforcement and compliance in the commercial inshore lobster fishery of Massachusetts. In: *Rights Based Fishing* (eds P.A. Neher, R. Arnason & N. Mollett), pp. 415–431. Kluwer Academic Publishers, Dordrecht, The Netherlands.

Swaine, M.D. & Whitmore, T.C. (1988) On the definition of ecological species groups in tropical forests. *Vegetatio*, 75, 81–86.

Swanson, T.M. (1989) International regulation and national resource management: policy options in elephant conservation. In: *The Ivory Trade and the Future of the African Elephant*. Ivory Trade Review Group, Queen Elizabeth House, Oxford, UK.

Swanson, T.M. (1994) *The International Regulation of Extinction*. Macmillan, Basingstoke, UK.

Symens, P.S. (1996) *Status and conservation of seabirds in the Chagos Archipelago, British Indian Ocean Territory*. MSc dissertation, University of Warwick, UK.

Talbot, L. & Olindo, P. (1992) The Maasai Mara and Amboseli Reserves. In: *Living with Wildlife: Wildlife Resource Management and Local Participation in Africa* (ed. A. Kiss). Technical Paper 130. World Bank, Washington, USA.

Thomas, D.C. (1982) The relationship between fertility and fat reserves of Peary caribou. *Canadian Journal of Zoology*, 60, 597–602.

Thomas, D.C. & Schaefer, J. (1991) Wildlife co-management defined: the Beverly and Kaminuriak Caribou Management Board. *Rangifer*, Special Issue No. 7, 73–89.

Thorne, E.T. & Williams, E.S. (1988) Disease and endangered species: the black-footed ferret as a recent example. *Conservation Biology*, 2, 66–74.

Thouless, C.R. (1993) *Community Wildlife and Game Utilisation in Laikipia and Samburu Districts, Kenya*. KWS/WWF Laikipia Elephant Project Technical Report. African Wildlife Foundation, Nairobi, Kenya.

Tingley, G.A., Purchase, L.V., Basson, M., Rosenberg, A.A., Anderson, J. & Beddington, J.R. (1990) *The Falkland Islands Fishery: 1989 Annual Report*. Renewable Resources Assessment Group, Imperial College, London, UK.

Tingley, G.A., Basson, M., Purchase, L.V., Bravington, M.V., Diaz de Leon, A.J. & Beddington, J.R. (1992) *The Falkland Islands Fishery: 1991 Annual Report*. Renewable Resources Assessment Group, Imperial College, London, UK.

Tingley, G.A., Purchase, L.V., Basson, M., Holden, S.J., Crombie, J.A. & Beddington, J.R. (1994) *The Falkland Islands Fishery: 1993 Annual Report*. Renewable Resources Assessment Group, Imperial College, London, UK.

Trivers, R.L. (1971) The evolution of reciprocal altruism. *Quarterly Review of Biology*, 46, 35–57.

Tuck, G.N. & Possingham, H.P. (1994) Optimal harvesting strategies for a meta-population. *Bulletin of Mathematical Biology*, 56, 107–127.

Tutin, C.E.G. & Fernandes, M. (1991) Responses of wild chimpanzees and gorillas to the arrival of primatologists: behaviour observed during habituation. In: *Primate Responses to Environmental Change* (ed. H.O. Box), pp. 187–197. Chapman & Hall, New York, USA.

UNEP (1995) *Global Biodiversity Assessment*. Cambridge University Press, Cambridge, UK.

UNICEF (1995) *Ellos, ellas y sus derechos: 'Análisis de la situación Nicaragüense'* (versión preliminar), Nicaragua.

Ustinov, S.K. (1972) Cannibalism and human attacks by the Eastern Siberian brown bear. In: *Ekologiya, Morphologiya, Okhrana i Ispol'zovanie Medvedei* (ed. V.E. Sokolov), pp. 85–87. Nauka, Moscow, USSR.

Vail, L. (1977) Ecology and history: the example of eastern Zambia. *Journal of Southern African Studies*, 3, 129–155.

Valdes-Pizzini, M. (1995) La Parguera Marine Fishery Reserve: involving the fishing community in planning a marine protected area. *Caribbean Park and Protected Area Bulletin*, **5**(2), 2–3. Caribbean Natural Resources Institute, US Virgin Islands.

Valkenburg, P., Davis, J.L., Ver Hoef, J.M., Boertje, R.D., McNay, M.E., Eagan, R.M., Reed, D.J., Gardner, C.L. & Tobey, R.W. (1996) Population decline in the Delta caribou herd with reference to other Alaskan herds. *Rangifer*, Special Issue No. 9, 53–62.

Vanclay, J.K. (1991) Research needs for sustainable forest resources. In: *Tropical Rainforest Research in Australia* (eds N. Goudberg & M. Bonell), pp. 133–143. Institute of Tropical Rainforest Studies, James Cook University, Townsville.

Vanclay, J.K. & Preston, R.A. (1989) Sustainable timber harvesting in the rainforests of northern Queensland. In: *Forest Planning for People, Proceedings of 13th biennial conference of the Institute of Foresters of Australia, Leura, NSW, 18–22 September 1989*, pp. 181–191. IFA, Sydney, Australia.

Varangis, P. (1992) *Tropical Timber Prices: Own Trends and Comparisons Among them and other Timber Prices*. World Bank, Washington DC, USA.

Vdovin, I.S. (1965) *Descriptions of the History and Ethnography of the Chukchi (Ocherki istorii i etnografii Chukchey)*. Nauka, Moscow, USSR.

Vedder, A. (1996) *Projet inventaire des gorilles et d'autres grands mammiferes de l'est du Zaire: Secteur original du Parc National de Kahuzi-Biega*. Unpublished report. Institut Zairois pour la Conservation de la Nature/Wildlife Conservation Society, Bukavu, Zaire.

Vedder, A. & Weber, W. (1990) The Mountain Gorilla Project (Volcanoes National Park). In: *Living with Wildlife: Wildlife Resource Management with Local Participation in Africa* (ed. A. Kiss), pp. 83–60. World Bank Technical Paper No. 130. World Bank, Washington DC, USA.

Verboom, J., Lankester, K. & Metz, J.A.J. (1991) Linking local and regional dynamics in stochastic meta-population models. *Biological Journal of the Linnean Society*, **42**, 39–55.

Verissimo, A., Barreto, P., Tarifa, R. & Uhl, C. (1995) Extraction of a high-value resource from Amazonia: the case of mahogany. *Forest Ecology and Management*, **72**, 39–60.

Vibe, C. (1967) Arctic animal in relation to climatic fluctuations. *Meddelelser Om Grønland*, **170**(5), 1–227.

von Richter, W. (1991) Problems and limitations of nature conservation in developing countries: a case study in Zaire. In: *Tropical Ecosystems* (eds W. Erdelen, N. Ishwaran & P. Müller), pp. 185–194. Margraf Scientific Books, Weikersheim, Germany.

Wagner, P. (1994) Gorilla trekking in Zaire. *Getaway*, April, 22–33.

Walters, C.J. (1978) Some dynamic programming applications in fisheries management. In: *Dynamic Programming and its Applications* (ed. M.L. Puterman), pp. 233–246. Academic Press, New York, USA.

Walters, C.J. (1986) *Adaptive Management of Renewable Resources*. Macmillan, New York, USA.

Walters, C.J. & Maguire, J.J. (1996) Lessons for stock assessment from the northern cod collapse. *Reviews in Fish Biology and Fisheries*, **6**, 125–137.

Watling, R. (1983) Sandbox incubator. *Animal Kingdom*, **53**, 28–35.

Watson, F. (1996) A view from the forest floor: the impact of logging on indigenous peoples in Brazil. *Botanical Journal of the Linnean Society*, **122**, 75–87.

Weaver, P.L. (1987) Enrichment plantings in tropical America. In: *Management of the Forests of Tropical America: Prospects and Technologies* (eds J.C. Figueroa, C.F.H. Wadsworth & S. Branham), pp. 259–277. Institute of Tropical Forestry, San Juan.

Weber, W. (1993) Primate conservation and ecotourism in Africa. In: *Perspectives on Biodiversity: Case Studies of Genetic Resource Conservation and Development* (eds C.S. Potter, J.I. Cohen & D. Janczewski), pp. 129–150. AAAS Press, Washington DC, USA.

Weinstock, J.A. (1983) Rattan: ecological balance in a Borneo rainforest swidden. *Economic Botany*, **37**(1), 58–68.

Wells, M. & Brandon, K. (1992) *People and parks. Linking protected area management with local communities.* World Bank, Washington DC, USA.

Wells, S. & Hanna, N. (1992) *The Greenpeace Book of Coral Reefs.* Blandford, London, UK.

Werner, D.I. (1986) Iguana management in Central America. *BOSTID Developments*, **6**(1), 1, 4–6.

Werner, D.I. & Rey, D. (1985) *Green iguana management. Biology of the green iguana.* Fundación pro Iguana Verde, Instituto de Investigaciones Smithsonian, Panamá.

Werner, D.I. *et al.* (1987) Kinship recognition and grouping in hatchling green iguanas. *Behavioral Ecology and Sociobiology*, **21**, 83–89.

Western, D. (1989a) The ecological value of elephants: a keystone role in Africa's ecosystems. In: *The Ivory Trade and the Future of the African Elephant*. Ivory Trade Review Group, Queen Elizabeth House, Oxford, UK.

Western, D. (1989b) The undetected trade in rhino horn. *Pachyderm*, **11**, 26–28.

Western, D. (1994) Ecosystem conservation and rural development: the case of Amboseli. In: *Natural Connections: Perspectives in Community-based Conservation* (eds D. Western & R. Michael Wright), pp. 15–52. Island Press, Washington DC, USA.

White, A., Begon, M. & Bowers, R.G. (1996) Host–pathogen systems in a spatially patchy environment. *Proceedings of the Royal Society of London B*, **263**, 325–332.

White, R.G., Bunnell, F.L., Gaare, E., Skogland, T. & Hubert, B. (1981) Ungulates on arctic ranges. In: *Tundra Ecosystems: a Comparative Analysis* (eds L.C. Bliss, W. Heal & J.J. Moore), pp. 397–483. Cambridge University Press, Cambridge, UK.

White, R.G., Holleman, D.F. & Tiplady, B.A. (1989) Seasonal body weight, body condition and lactational trends in muskoxen. *Canadian Journal of Zoology*, **67**, 1125–1133.

Whitmore, T.C. (1984) *Tropical Rain Forests of the Far East.* Clarendon Press, Oxford, UK.

Williams, H.P. (1993) *Model Solving in Mathematical Programming.* John Wiley, Chichester, UK.

Williamson, L., Warren, Y., Greer, D. & Cogswell, J. (1997) Report from the Karisoke Research Centre, Rwanda, 1996. *Gorilla Conservation News*, **11**, 19–20.

Wilson, J.R. & Lent, R. (1994) Economic perspectives and the evolution of fisheries management: towards subjectivist methodology. *Marine Resource Economics*, **9**, 353–373.

Woodroffe, R. & Ginsberg, J.R. (1998) Extinction risks for large carnivores: ranging behaviour and edge effects. In: *Behaviour and Conservation* (eds M.L. Gosling & W. Sutherland). Cambridge University Press, Cambridge, UK, (in press).

World Bank (1995) *Mainstreaming biodiversity in development. A World Bank assistance strategy for implementing the Convention on Biological Diversity.* World Bank, Washington DC, USA.

World Bank (1996) *Mainstreaming biodiversity in agricultural development. Toward good practice.* World Bank, Washington DC, USA.

WCMC (World Conservation Monitoring Centre) (1992) *Global Biodiversity: Status of the Earth's Living Resources.* Chapman & Hall, London, UK.

WRI (World Resources Institute) (1994) *World Resources 1994–95.* Oxford University Press, New York, USA.

WRI (World Resources Institute)/International Institute for Environment and Development (1988) *World Resources 1988–89.* Basic Books Inc., New York, USA.

Wrangham, R.W. (1992) Letter to I.Z.C.N., Zaire. *Gorilla Conservation News,* **6,** 17–18.

WWF/IUCN (1985) *Parc National des Volcans: Plan de gestion.* Unpublished report. WWF/IUCN, Nairobi, Kenya.

Yamagiwa, J., Mwanza, N., Spangenberg, A., Maruhashi, T., Yumoto, T., Fischer, A. & Steinhauer-Burkart, B. (1993) A census of the eastern lowland gorillas *Gorilla gorilla graueri* in Kahuzi-Biega National Park, with reference to mountain gorillas *G. g. beringei* in the Virunga region, Zaire. *Biological Conservation,* **64,** 83–89.

Young, T.P. (1994) Natural die-offs of large mammals: implications for conservation. *Conservation Biology,* **8,** 410–418.

Zambia (Ministry of Tourism), National Parks and Wildlife Services (1994) *Profile of a Hunting Area: The Mainstay of Zambia's Safari Industry.* Chilanga, Zambia: NPWS headquarters.

Zavatskii, B.P. (1987) Behaviour of brown bears during encounters with humans. In: *Ekologiya Medvedei* (ed. B.S. Judin), pp. 153–158. Nauka, Novosibirsk, USSR.

Zhirnov, L.V. (1982) Modelling saiga population dynamics. *Izvestiya Timiryazevskoi Sel'skokhozyaistvennoi Akademii,* **5,** 157–166. Kolos, Moscow, USSR.

Glossary of Terms

The terms are defined here in the sense in which they are used in the book. Italicised words have their own entries in the glossary. [Section in which the term first appears]

Adaptive management A management strategy whereby the manager deliberately perturbs the system, in order to learn about its *dynamics*. It is a *Bayesian* approach, in which the management strategy evolves over time as knowledge about the system improves. Championed by Walters (1986). [Section 4.1]

Admissible decisions A set of decisions which are all as good as each other, in classical decision analysis. Subjective criteria must be used to choose the best decision among them. [Section 3.1.1]

Bayesian analysis A way of analysing problems in decision-making, in which semi-subjective probabilities are initially used as the probabilities of particular outcomes occurring. Over time, these probabilities are refined in the light of experience. [Section 3.1.1]

Biodiversity The variety of living organisms, usually measured as the number of different species in an area, and the number of individuals of each species. The more species there are, and the more similar in population size they are, the higher the biodiversity. However, biodiversity is actually variability in a broader sense, including the genetic and ecosystem levels. [Section 2.2.3]

Carrying capacity The mean population density of a species that can be supported by its environment indefinitely, in the absence of human interference. [Section 1.2]

Catch per unit effort (CPUE) The amount of biomass that is harvested divided by the amount of *harvesting effort* used to get it. Effort is measured in any appropriate unit (e.g. fishing days). Changes in CPUE are often used as an indirect measure of changes in population size. [Section 1.3.1]

Catchability coefficient A constant that varies from species to species, relating the *harvesting effort* to the population size. It represents the amount of biomass that would be harvested by one unit of effort from a population at *carrying capacity*. [Section 1.3.1]

Chaotic dynamics The *dynamics* of deterministic systems in which future values are extremely sensitive to initial conditions. This means that future values become unpredictable very soon after the present. [Section 2.1.4]

Co-management agreement A legal agreement setting up structures by which a government and local communities govern their natural resources as a partnership. Because the arrangements for control of resource use are legally

389

binding, this should be a relatively stable form of resource *governance*. [Section 4.3.2]

Conservation Preventing the loss of *biodiversity* and biological processes (in the broadest senses of the words). This loss is generally human-induced. Conservation is distinct from preservation because it involves recognising the *dynamic* nature of biological systems, and allowing them to change and evolve. [Title]

Constant escapement A management strategy that controls the amount of harvesting, so that a constant proportion of the population is left after harvesting. This contrasts with most harvesting strategies which involve keeping the number or proportion of the population removed by harvesting constant. [Section 1.3.2]

Consumer surplus The value that a person places on their consumption of a good, over and above the amount that they paid for it. This can be measured by subtracting the maximum amount an individual would be prepared to pay for something from the price that they actually pay. [Section 1.6]

Darwinian fitness The representation achieved by an individual's genes in future generations by direct descent. This is a measure of how successful the individual has been or will be as a result of natural selection. [Section 3.1.1]

Density dependence The way in which the growth rate of a population depends on the size of that population. Generally, the population growth rate slows down as *carrying capacity* is approached (positive density dependence, or compensation). [Section 1.2.1]

Depensation The growth rate of the population increases as the size of the population increases. This is the opposite of normal *density dependence*, and generally only happens at very small population sizes. For example, problems such as being unable to find a mate may worsen the smaller the population gets. Also called the Allee effect. [Section 2.1.1]

Deterministic processes Processes with no randomness involved, so that the exact value of a quantity can be calculated. Deterministic models are often used to describe nature because of their simplicity, although nature is never deterministic. [Section 1.3.2]

Discount rate The rate at which the value of a cost or benefit is reduced depending on how far in the future it is experienced. With a positive discount rate, the further into the future the cost or benefit is experienced, the less it is valued in the present. Value is usually assumed to decline exponentially with time. [Section 1.5]

Dynamics The way in which a system behaves over time. [Section 1.1]

Economic efficiency A strategy is economically efficient if it leads to the socially *optimal* level of resource use. This means that the benefit to human society from the resource is maximised. A strategy can be economically efficient without being equitable, if the benefit to society is maximised but is not divided fairly among people. Without full *economic valuation* of resources, the economically efficient level of resource use cannot be accurately identified. This is likely to be the case for many natural resources. [Section 1.1]

Economic valuation The process of assigning a monetary value to natural resources. Some aspects of natural resources can be easily valued monetarily (e.g. the market price of the timber from a tree). Others are extremely difficult or impossible to value monetarily, or their monetary value may be a meaningless concept (e.g. a tree's value as a habitat for other species). This means that a 'full' economic valuation of a resource is always an unquantifiable underestimate of its true value. [Section 3.2.3]

Elasticity A parameter that describes the responsiveness of one variable to changes in another, defined as the percentage change in one variable with a 1% change in the other. In economics, elasticity is used to describe how the quantity of a good supplied or demanded varies with parameters such as the price of the good, the price of related goods and consumer income. [Section 1.6.1]

Enforcement costs The costs of ensuring that the law is upheld. In resource management, this involves costs such as monitoring harvesters, and taking them to court if they violate the regulations. Enforcement costs are important and often overlooked costs of resource management. [Section 3.1.4]

Equi-marginal principle When there are several producers with different production costs (different boat sizes, for example), the overall profit-maximising level of production is when the *marginal* costs of production are equal. This means that the cost of producing one extra unit is the same for each producer. The equi-marginal level of production is equivalent to the 'ideal free distribution' of predators in ecology. [Section 4.3.2]

Externalities Costs and benefits that are not represented in the supply and demand curves. These costs or benefits are experienced by people other than those deciding the level of production or consumption of a good, so they are not taken into account in the market. External costs occur if the supply curve does not reflect all the costs of production; external benefits occur if the demand curve does not reflect all the good's benefits. Natural resource-based markets are particularly prone to externalities. [Section 3.2.1]

Governance The way in which a system is governed. It includes any actions of national governments, international agreements, communities, businesses and individuals, which have an impact on the functioning of the system. Thus, governance is broader than resource management, and works on several levels. [Part 4]

Harvesting effort The amount of input that harvesters put into resource harvesting. Effort can be measured in many ways, such as the number of days spent hunting, the number of fishing boats working in the area, or the number of each type of gun or snare used. [Section 1.3.1]

Inter-generational equity The issue of whether future generations are treated fairly by current generations. This is particularly relevant in discussions of how costs and benefits are *discounted* over time, because discounting leads to costs being delayed and benefits being taken early. [Section 3.2.3]

Intrinsic rate of population increase The maximum rate at which a population can grow (when it is very small and growing exponentially). Mortality from factors other than resource constraints (such as predators or climate) is ignored, although in reality these factors might be very important. [Section 1.2]

Keystone species Keystone species have a disproportionate effect on their ecosystem, due to their size or their activities. Any changes in their population size have correspondingly large effects on the ecosystem. For example, beavers change water flow substantially by building dams. [Section 2.2.3]

Logistic growth Population growth that follows the logistic equation. The population grows exponentially at the *intrinsic rate of population increase* when it is very small. Growth slows at a constant rate as the population gets larger, and stops when *carrying capacity* is reached. [Section 1.2]

Marginal analysis Analysis of the effect of increases of one unit; equivalent to differentiation if one unit is a very small amount. For example, the marginal cost of producing a good is the cost of producing one more unit of the good, and depends on how much is already being produced. [Section 1.5]

Market failure This occurs if there are *externalities*; the supply curve doesn't represent all costs involved in producing the good, or the demand curve doesn't represent all the benefits from the good. If this happens, the market equilibrium (the point at which the supply and demand curves cross) is not the *economically efficient* production level. [Section 3.2.1]

Maximin A pessimistic strategy in game theory, in which player A assumes that because player B's incentives are directly contrary to theirs (the game is *zero-sum*), player B will try to minimise the benefit that player A receives from the game. Thus, player A choses a strategy that maximises the benefit they receive from the game, given that player B is trying to minimise this benefit. The symmetrically opposite strategy, played by player B, is called minimax. [Section 3.1.3]

Maximum sustainable yield (MSY) The largest harvest that can be taken from a population indefinitely, without driving the population to extinction. In the logistic model, the maximum sustainable yield is the harvest that keeps the population at half the *carrying capacity*. [Section 1.3]

Meta-population A population that is divided among several patches of habitat. The sub-populations inhabiting these patches go extinct with a relatively high frequency, but patches are also repopulated at high frequency. The balance between extinction and recolonisation determines the total population size. [Section 1.2.1]

Monopoly A market in which there is only one producer of goods or services. The producer can choose how much of the good to produce, and at what price to sell it. [Section 1.4]

Non-market value Value that is not expressed in economic markets; that part of value which people do not put a price on. Goods like natural resources often have a lot of non-market value. Some non-market values can be estimated indirectly; others are intractable to economic analysis. [Section 3.2.1]

Oligopoly A market in which a small number of firms produce most of the good, and the price of the good depends both on an individual firm's output, and the output of the other firms. This kind of market leads to tension, involving competition between firms to maximise market share and collusion to obtain *monopoly* profits. [Section 1.4]

Open access A resource which anyone can harvest. This is equivalent to having a *perfectly competitive* market structure. Open access is not the same as common property; the latter term is ambiguous. If access is open only to members of a group (e.g. members of a village), the resource is communally controlled rather than being open access. [Section 1.4]

Opportunity cost The cost of not doing whatever you could have done instead of what you decide to do. For example, the opportunity cost of going hunting for a day might be the money that could have been earnt that day by working as a casual labourer. [Section 1.4]

Optimal The optimal (best) strategy is the option which maximises the value of the variable that is used to represent the decision-maker's objectives. An optimum may be subject to constraints. For example, the optimal management strategy for a natural resource may be to harvest at the rate that maximises the manager's profits, or it may be to manage it so as to minimise the probability of extinction. A typical constraint in this case would be that the population should not fall below a particular size. [Section 1.1]

Participatory approach An approach to *conservation* and/or rural development in which the local community is involved in decision-making and resource management. This enables local communities to identify their needs, beliefs and opinions to outsiders attempting to develop and/or conserve their area. This hopefully leads to more appropriate, and thus more *sustainable*, resource management. [Section 4.2.5]

Perfect competition A market in which individual buyers and sellers have no influence on the market price for the good. Individuals can assume that the price of the good is given, however much of it they personally buy or sell. This happens when there are a large number of individual buyers and sellers, each selling an identical product, with perfect information available about the market, and no barriers to entry or exit. [Section 1.4]

Present value The current worth of a resource. Generally used when the resource will produce benefits and/or costs in the future, which are *discounted* and added together to give the present value. The present value is then used as the value of the resource for decision-making purposes. [Section 1.5]

Prisoner's dilemma A famous non-*zero-sum* game without communication or bargaining. The game has two decisions – co-operate or defect. If both players defect, they both get a negative payout. If they both co-operate, they both get a positive payout. If one co-operates and the other defects, then the co-operator gets a very low payout, and the defector a very high payout. The game has been used to explain why natural resources can be over-exploited, and is linked to the *Tragedy of the Commons*. [Section 3.1.3]

Regulatory instrument A method for controlling resource use, for example, *transferable permits* or a trade ban. An economic instrument is one that uses the market to control resource use, such as a tax on harvesting, so it changes users' incentives rather than preventing them from doing something. [Section 4.3]

Renewable resource A resource that is able to reproduce itself, and thus can theoretically be used in a *sustainable* way. Generally, this means the resource is a population of living organisms. If the rate of use is consistently higher than the rate of renewal, the resource will eventually be exhausted. [Section 1.1]

Risk In decision thory, risk is when the probability of any particular outcome of a decision is known, so that the expected value of the decision can be calculated by multiplying each possible outcome by the probability that it will occur and summing over all possible outcomes. People's behaviour can be categorised by their attitude to risk – a risk-neutral person will take a gamble if its expected value is at least zero, a risk-prone person will gamble even if the expected value is negative, and a risk-averse person will only gamble if the expected value is positive (the more risk-averse they are, the higher the expected value must be for them to take the risk). [Section 3.1.1]

Stochasticity Randomness. A stochastic system is one in which one or more of the variables are described by probability distributions rather than *deterministic* functions. [Section 1.2.1]

Sustainable use Use that can be continued indefinitely. Because ecosystems and human societies are *dynamic*, it is not possible to claim that use will be sustainable in the future, although it is possible to say that it has been sustainable in the past, and appears sustainable at present. Three facets of sustainable use are usually identified: ecological (the ecosystem can continue to function, *biodiversity* is not reduced unacceptably), economic (the use is profitable), and social (the use is culturally acceptable). All need to be fulfilled for sustainability to be likely. Sustainability is not equivalent to stability. In fact, a use strategy that can adapt to change is likely to be more successful than a static one. [Part 1]

Time horizon The distance into the future a person looks. The time horizon may be a genuine period over which a decision has effects (e.g. farmers planting crops each spring have a time horizon of one year because they can change their decision next year), or it may be the point at which the long-term returns from an action become negligible due to *discounting*. [Section 3.1.4]

Time preference rate A measure of how a person's (or animal's) valuation of benefits or costs depends on the time from the present at which they experience them. Generally, it is assumed that people prefer benefits as close to the present as possible, and costs as far into the future (or past) as possible. Also, that people value costs or benefits less the further from the present they are experienced. This is one theoretical justification for *discounting* the future. [Section 1.5]

Tragedy of the Commons The problem of reciprocal *externalities* in *open-access* resources, as stated by Hardin (1968). It has links to the *prisoner's dilemma*. An example of the problem concerns herders keeping animals on common grazing land, who individually decide whether or not to increase their herd sizes. Rational decision-making by individuals leads to resource over-exploitation. [Section 3.2.2]

Transactions costs The cost of reaching and enforcing agreements. If a lot of people are involved, or if complicated legislation is needed, transactions costs may be high enough to prevent effective resource management. [Section 4.3.2]

Uncertainty In decision theory, uncertainty is when the set of possible outcomes of a decision is known, but the probability of any one outcome is unknown. It is realistic to assume uncertainty, but analysis is difficult. One possible approach is *Bayesian* analysis, in which semi-subjective probabilities are as-signed to the outcomes, so the problem can be analysed as if it were a *risk* problem. [Section 3.1.1]

Utility A vague concept in economics, being the unit of measurement for human happiness. A common proxy for utility is money, so that, for a decision about harvesting, the objective becomes maximising the monetary yield from the resource. Another measure for utility is *Darwinian fitness*; an organism acts to maximise its genetic representation in future generations. [Section 3.1.1]

Zero-sum game A game in which any gain to one player is a direct loss to another. Thus, the total amount of benefit available from the game is zero. In this case, the interests of the players are directly opposed, and co-operation is not possible. In a non-zero-sum game, there is some benefit available that does not come directly from other players. In this case, co-operation between players, so that this benefit can be obtained, may be possible. [Section 3.1.3]

Index